William H. Besant

Solutions of Examples in Conic Sections

Treated Geometrically. Third Edition

William H. Besant

Solutions of Examples in Conic Sections
Treated Geometrically. Third Edition

ISBN/EAN: 9783337251673

Printed in Europe, USA, Canada, Australia, Japan

Cover: Foto ©berggeist007 / pixelio.de

More available books at **www.hansebooks.com**

SOLUTIONS OF EXAMPLES

IN

CONIC SECTIONS,

TREATED GEOMETRICALLY

BY

W. H. BESANT, D. Sc., F.R.S.

FELLOW OF ST JOHN'S COLLEGE, CAMBRIDGE.

THIRD EDITION, REVISED.

CAMBRIDGE:

DEIGHTON, BELL AND CO.

LONDON: GEORGE BELL AND SONS.

1890

𝕮𝖆𝖒𝖇𝖗𝖎𝖉𝖌𝖊:

PRINTED BY C. J. CLAY, M.A. & SONS,

AT THE UNIVERSITY PRESS.

PREFACE TO SOLUTIONS.

I HAVE frequently received requests for a book of Solutions of the Examples in my treatise on Conic Sections, but have never been able to find time to prepare them.

Mr Archer Green, B.A., Scholar of Christ's College, volunteered to undertake the task, with the aid of my notes and his own, and, with the exception of a few at the end, wrote out the solutions entirely.

Mr Green was however prevented by illness from completing the revision of the proofs, and I am much indebted to Mr J. Greaves, Fellow of Christ's College, who kindly undertook to examine the rest of the sheets.

The book will, I hope, prove useful both to students and teachers, as a companion volume to the treatise on Conic Sections.

W. H. BESANT.

Sept. 1881.

PREFACE TO THE THIRD EDITION.

THE solutions have been revised, and many additions have been made to them. They will now be found to be in complete accordance with the sixth edition of the Geometrical Conics.

W. H. BESANT.

Jan. 1890.

CONIC SECTIONS.

SOLUTIONS OF EXAMPLES.

——◆——

CHAPTER I.

1. IF the tangent at P meet the directrix in Z, and S be the focus, PSZ is a right angle;

∴ S lies on the circle of which PZ is diameter.

2. Let PN and QM be the ordinates at P and Q.

Then $PN : QM :: SP : SQ :: XN : XM$;

∴ the triangles PXN and QXM are similar and PX, QX equally inclined to XS.

3. By Art. 8, FS is the external bisector of the angle PSQ.

4. $SP : PK :: SA : AX :: SE : EK$;

∴ EP bisects the angle SPK.

5. Since F, S, P and K lie on a circle,

the angle KSF = the angle FPK = the angle FTS.

6. $PN : P'N' :: SP : SP'$;

∴ $XK : XN :: XK' : XN'$;

∴ the angle LNN' = the angle $K'N'X$ = the angle $LN'N$.

7. Let Q be the point where the tangent at R meets NP.

Then $NQ : NX :: SR : SX :: SA : AX :: SP : NX$;
$$\therefore SP = QN.$$

8. Let SY be perpendicular to the tangent at P and GL perpendicular to SP.

Then, since the triangles PSY, GPL are similar,
$$PG : PL :: SP : SY,$$
or $\qquad\qquad PG : SR :: SP : SY.$

9. If the tangent meet the directrix in Z, and SP be drawn such that ZSP is a right angle meeting the tangent in P,

then P will be the point of contact of the tangent ZP.

10. If P, Q be the extremities of the chord, and PK, QL be perpendicular to the directrix,
$$SP : PK :: SA : AX :: SQ : QL;$$
$$\therefore SP + SQ :: PK + QL :: SA : AX.$$

Now the distance of the middle point of PQ from the directrix is equal to half $PK + QL$, and is therefore least when $SP + SQ$ is least, that is, when PQ goes through the focus.

11. If TP, TP' be the fixed tangents, and the tangent at Q meet them in E, E',

the angle $PSE =$ the angle ESQ, and the angle $QSE' =$ the angle $E'SP'$;
$$\therefore \text{ the angle } ESE' = \text{half the angle } PSP'.$$

12. If perpendiculars from the given points PK, QL be drawn to the directrix and S be the focus,
$$SP : SQ :: PK : QL, \text{ a constant ratio};$$
$$\therefore \text{ the locus of } S \text{ is a circle.}$$

13. Let the normal at P meet the axis in G.

Taking O as the fixed point in the axis, it is obvious that the triangles OSR, GSP are similar;

$$\therefore SO : SR :: SG : SP :: SA : AX;$$

$\therefore SR$ is constant, and R lies on a circle of which S is the centre.

14. $AT : AX :: SR : SX :: SA : AX;$

$$\therefore AT = AS.$$

15. ST bisects the angle between SP and SQ, Art. 12, and SR bisects the angle between QS, and SP produced, Prop. II., Art. 5;

$$\therefore RST \text{ is a right angle.}$$

16. The triangles EAT, ERS are similar ;

$$\therefore AT : SR :: EA : ER :: AX : SX;$$

$$\therefore AT : AX :: SR : SX :: SA : AX;$$

$$\therefore AT = AS.$$

17. If TL be perpendicular to the directrix,

$$SR : TL :: SA : AX :: SM : TL ;$$
$$\therefore SM = SR.$$

18. FS is the external bisector of the angle QSP, and $F'S$ of QSP' ;

$\therefore$ the angle $FSF' = $ half the angle PSP'.

19. Since the triangles SPN, SGL are similar,

$$\therefore GL : PN :: SG : SP :: SA : AX.$$

20. If the normals PG, $P'G'$ meet in Q, and QV be drawn parallel to the axis to meet the chord in V,

$$VQ : VP :: SG : SP :: SA : AX :: SG' : SP' :: VQ : VP' ;$$
$$\therefore VP = VP', \text{ or } V \text{ bisects } PP'.$$

1—2

21. DS is the external bisector of the angle PSQ, and ES of pSQ;

$$\therefore DSE \text{ is a right angle.}$$

22.   The semi-latus rectum is an harmonic mean between SP and SP';

$$\therefore 2SP . SP' = SR . PP'.$$

23. $PE : PL :: PQ : PG :: PV : PS :: PP' : 2SP$, see Ex. 20 ;

$$\therefore PE : SR :: SP' : SR;$$

$$\therefore PE = SP'.$$

Similarly $P'E = SP.$

24.   The right-angled triangles DSQ, DSE have a common hypotenuse.

Also $SE = SR = SQ$;

$$\therefore \text{ the angle } QSE = \text{the angle } ESP.$$

25.   Let S be the focus and P and Q the given points. Through P draw a straight line PK so that SP may bear to PK the given ratio of the eccentricity.

Through Q draw a straight line QL so that $SQ : QL$ in the same ratio.

With centres P, Q and radii PK, QL respectively describe circles.

The perpendicular from S on a common tangent to these circles will be axis.

26.   Let the tangents at P and Q intersect in T.

Draw TN perpendicular to directrix and TM perpendicular to SP.

Then $SM : TN :: SA : AX.$

But ST bears a constant ratio to SM, since angle $TSM =$ half PSQ ;

$$\therefore ST \text{ bears a constant ratio to } TN.$$

27. Let T be the intersection of the tangents at P and p.

Draw TK perpendicular to Pp.

Then $TK : PL :: TP : PG$ and $TK : pl :: Tp : pg$.

Again, draw GM, gm perpendicular to SP, Sp respectively, and TN, Tn perpendicular to SP, Sp respectively.

Then $TP : PG :: TN : MP :: Tn : mp :: Tp : pg$;

$$\therefore TK : PL :: TK : pl;$$

$$\therefore PL = pl.$$

CHAPTER II.

THE PARABOLA.

1. THE distance of the centre of the circle from the fixed point is equal to its distance from the fixed straight line, and therefore its locus is a parabola of which the fixed point is focus and the fixed straight line directrix.

2. Through the vertex draw a straight line making the given angle with the axis; the tangent at the point where the diameter bisecting this chord meets the curve will be the tangent required.

Or, draw a radius vector from the focus, making twice the given angle with the axis.

3. Since $TA = AN$, $PN = 2AY$; $\therefore AY^2 = AS \cdot AN$.

4. Let SY' be drawn perpendicular to the line through G parallel to the tangent.

Then in the right-angled triangles YST, $Y'SG$, $ST = SG$, and the angles YST, $Y'SG$ are equal;

$$\therefore SY = SY'.$$

5. Draw SY perpendicular to the tangent and YA perpendicular to the axis.

Produce SA to X, making AX equal to SA.

Then the straight line through X perpendicular to SX is the directrix.

6. Let the circle touch the fixed circle in Q, and the straight line in R; let P be its centre, and S the centre of the fixed circle.

Produce PR to M, making RM equal to SQ, then the

straight line MX drawn through M parallel to the given line is a fixed straight line.

Then, since SP is equal to PM, the locus of P is a parabola of which S is focus and MX directrix.

7. Draw SY, SY' perpendiculars on the two tangents.

Then, if SA be perpendicular to YY', A is the vertex.

Produce SA to X, making AX equal to AS; X is the foot of the directrix.

8. If the tangent at the end of the latus rectum meet PN in Q,
$$QN = XN = SP.$$

9. Since SYP and PNS are right angles, P, N, S, Y lie on a circle ;
$$\therefore TY . TP = TS . TN.$$

10. SE is half TP,

and $\qquad PT^2 = PN^2 + TN^2 = 4AS . AN + 4AN^2$;
$$\therefore SE^2 = AN . XN = AN . SP.$$

11. If SY be drawn perpendicular to the tangent and A be vertex, SAY is a right angle ;
$$\therefore A \text{ lies on the circle of which } SY \text{ is diameter.}$$

12. Draw SY perpendicular to the tangent, then if the circle described with centre S and radius equal to a quarter of the latus rectum meet the circle described on SY as diameter in A, A is the vertex.

Produce SA to X, making AX equal to SA, then X is the foot of the directrix.

13. $\qquad SN : SP :: SN' : SP'$,

or $\quad AN - AS : AN + AS :: AS - AN' : AS + AN'$;
$$\therefore AN : AS :: AS : AN' ;$$
$$\therefore AN . AN' = AS^2.$$

Again, $4AS \cdot AN : 4AS^2 :: 4AS^2 : 4AS \cdot AN'$;

$$\therefore PN^2 : SR^2 :: SR^2 : P'N'^2,$$

or $PN : SR :: SR : P'N'$;

$\therefore$ the latus rectum is a mean proportional between the double ordinates.

14. Let V be the middle point of the focal chord PSP', and let the diameter through V meet the curve in Q ; then, if QT, QM be the tangent and ordinate at Q, and VL be ordinate of V,

$$VL = QM \text{ and } TM = SL \text{ ; }$$

$\therefore VL^2 = QM^2 = 4AS \cdot AM = 2AS \cdot TM = 2AS \cdot SL.$

Hence the locus of V is a parabola of which S is vertex and SL axis.

15. If P, P' be the given points, PK, $P'K'$ perpendiculars on the directrix, the focus is the point of intersection of a circle centre P, radius PK with a circle of which P' is centre and $P'K'$ radius.

In general two circles intersect in two points, therefore two parabolas can be drawn satisfying the given conditions.

16. If PG be normal at P, the triangles PNG, pPR are similar;

$$\therefore Pp : PN :: RP : NG \text{ ; }$$

$$\therefore RP = 2NG = \text{latus rectum ; }$$

$\therefore$ the locus of R is an equal parabola having its vertex A' on the opposite side of X, such that AA' is equal to the latus rectum.

17. Let P, P' be the given points, S the given focus.

A common tangent to the circles described with centres P, P' and radii PS, $P'S$ respectively will be the directrix.

18. If SP be the focal distance and SY perpendicular to the tangent at P, Y lies on the circle of which SP is diameter.

Also the angle $AYS=$ the angle SPY;

$\therefore AY$ touches the circle.

19. The tangents at the ends of the focal chord PSP' meet in F on the directrix at right angles : also the straight line through F at right angles to the directrix bisects PP' in V;

$$\therefore FV = VP = VP';$$

$\therefore$ the directrix touches the circle described on PP' as diameter.

20. Draw the farther tangent to the circle parallel to the given diameter, then the locus of the point is a parabola of which the centre is focus, and the tangent thus drawn directrix.

21. Draw a straight line parallel to the given straight line, on the farther side of it, and at a distance from it equal to the radius of the circle, then the locus of the point is a parabola of which the centre of the circle is focus, and the straight line thus drawn directrix.

22. Let Q be the centre of the circle touching the sector in R and AC in M.

Through C draw CB at right angles to AC, and on the same side of it as Q, and draw QN perpendicular to the tangent at B.

Then $$NQ + QM = BC = CQ + QR;$$
$$\therefore CQ = QN;$$

$\therefore Q$ lies on a parabola of which C is focus and BN directrix.

23. Y is the middle point of TP and Z of PG; therefore YZ is parallel to the axis.

24. If SQ be perpendicular to the normal PG,
$$PQ = QG,$$
and if QM be the ordinate, $NM = MG$;
$$\therefore SM = AN \text{ and } PN = 2QM;$$
$$\therefore QM^2 = AS . AN = AS . SM;$$

$\therefore Q$ lies on a parabola of which S is vertex and SG axis.

25. The triangle PSG is isosceles; therefore GL is equal to PN.

26. If the circle described with centre S and radius equal to the perpendicular from S on the tangent at P meet the circle of which SP is diameter in Y, and the angle SYA be made equal to the angle SPY, then the foot of the perpendicular SA on YA will be the vertex.

27. Since SQ is double SA, ASQ (and likewise QSP) is equal to the angle of an equilateral triangle;
therefore SP and SQ are equally inclined to the latus rectum.

28. $\quad QX^2 = SX^2 + SQ^2 + 2SX \cdot SQ$
$$= 4AS^2 + QG^2 + 2SQ \cdot NG$$
$$= 4AS^2 + QN^2 + 2QN \cdot NG + NG^2 + 2SQ \cdot NG$$
$$= 4AS^2 + QN^2 + NG^2 + 2NG \cdot SN$$
$$= 4AS^2 + NQ^2 + 2AN \cdot NG$$
$$= 4AS^2 + QN^2 + PN^2 = 4AS^2 + QP^2.$$

29. The angle
$$SPF = SPG - FPG = SGP - GPH = SHP;$$
therefore the triangles SPF, SHP are similar;
$$\therefore\ SF \cdot SH = SP^2 = SG^2.$$

30. A, B, C, S lie on a circle; therefore, if D be the end of the diameter drawn through S, DA, DB, DC are perpendicular to SA, SB, SC respectively.

31. Since PQ and PR are equally inclined to the axis, the circle through P, Q, R touches the parabola at P; therefore PQ is a diameter of this circle.

Therefore PRQ, the angle in a semicircle, is a right angle.

32. Let MR and AQ meet in V.
Draw the ordinates VW, RZ.

Then $MW : MZ :: WV : RZ :: AW : AN$;

$\therefore MW : AW :: MZ : AN$;

$\therefore AN : AW :: AZ : AN.$

Again, $VW^2 : QN^2 :: AW^2 : AN^2$;

$\therefore VW^2 : RZ^2 :: AW : AZ$;

$\therefore V$ lies on the curve.

33. Let P, Q be the given points. Bisect PQ in V, and draw VT parallel to the axis meeting the given tangent P in T.

Draw PS, QS such that TP, TQ may be equally inclined to the axis and to SP, SQ respectively. PS, QS meet in the focus.

Through P draw a straight line PK parallel to the axis, making PK equal to SP, then the straight line through K at right angles to PK will be the directrix.

34. Let P be the vertex and QVQ' be corresponding ordinate.

Take M in VP produced such that

$$QV^2 = 4MP . PV.$$

Make angle TPS equal to the angle MPT, PT being parallel to QQ', and make PS equal to PM.

Then S is the focus, and the straight line drawn through M at right angles to PM is the directrix.

35. $PM^2 : QN^2 :: AM : AN$, QN being the ordinate of Q;

$\therefore AM = 4AN$ and $3AM = 4NM$;

$\therefore 3AT = 4QN = 2PM.$

36. Draw PN perpendicular to AB.

Then $AN : NP :: CQ : AC :: NP : AC$;

$\therefore PN^2 = AC . AN.$

Therefore the locus of P is a parabola of which A is vertex and AB axis.

37. The triangles LKP, PSK, KSA and TKA are similar ;

$$\therefore KL^2 : SP^2 :: KP^2 : KS^2 :: KA^2 : AS^2 :: TA : AS$$
$$:: SP - AS : AS.$$

38. With centre S and radius one-fourth of the chord describe a circle meeting the parabola in P. The chord through S parallel to the tangent at P will be the chord required.

39. $PN^2 = 4AS.AN = 4AS.AN' + 4AS^2$
$$= P'N'^2 + N'G'^2 = P'G'^2.$$

40. If Pp, $P'p'$ be two parallel chords, and V, V' their middle points, VV' is a diameter. Let VV' meet the curve in Q.

Draw QT parallel to Pp, then QT is the tangent at Q.

Produce VV' to a point M such that $PV^2 = 4QM.QV$, then the straight line drawn through M at right angles to MV is the directrix.

Make the angle TQS equal to the angle TQM.

Then if QS be made equal to QM, S is the focus and the straight line through S perpendicular to the directrix is the axis.

41. Let the tangents at P and P' intersect in T.

Then $\qquad 4SP.PV : 4SP'.P'V'$
$$:: P'V^2 : PV'^2 :: TP^2 : TP'^2 :: SP : SP';$$
$$\therefore PV = P'V'.$$

42. If in the preceding Example $P'T$ meets PV in Z and the sides of the triangle ABC are parallel to ZP, PT and TZ respectively,

$$AB : AC :: TZ : TP :: TP' : TP.$$

43. If U, V be the vertices of the diameters bisecting Pp, Qq,

$$PS.Sp : QS.Sq :: SU : SV :: Pp : Qq.$$

44. Draw RW, LZ parallel to QQ'.

Then $\qquad PL^3 : PR^2 :: LZ^3 : RW^3$

$$:: QV^2 : RW^2 :: PV : PW :: PN : PR ;$$

$$\therefore PL^2 = PR . PN.$$

45. This question is solved in Conics, Art. 212, p. 217.

46. $\qquad PN^2 : AN^2 :: AM^3 : QM^3 ;$

$$\therefore 4AS : AN :: AM : 4AS.$$

47. Let AP, Ap meet the latus rectum in L and l respectively.

Then $PN^3 : SL^2 :: AN^3 : AS^3 :: AN : An$, (Example 13)

$$:: PN^2 : pn^3 ;$$

$$\therefore SL = pn.$$

In like manner $\qquad Sl = PN.$

48. If PK, QL be perpendicular to the directrix, and QL' to PK produced, the angle $SPQ =$ the angle QPL' ;

$$\therefore PL' = SP = PK ;$$

$$\therefore SQ = QL = KL' = 2PK = 2SP.$$

49. Is equivalent to Example 32.

50. Let Q be the point of intersection, and let QK be the ordinate of Q.

Then $\qquad AK : QK :: PN : NT ;$

$$\therefore 2AK . AN = QK . PN = PN^3 = 4AS . AN ;$$

$$\therefore AK = 2AS,$$

or Q lies on a fixed straight line parallel to the directrix.

51. Let Tp, Tq be the fixed tangents, and let PQ touch the curve in R.

Then $\qquad SP^2 = Sp . SR = Sq . SR = SQ^2 ;$

$$\therefore SP = SQ.$$

52. Let TM be the ordinate of T, and TW perpendicular to SP.

$$TM^2 = ST^2 - SM^2 = TW^2 + SW^2 - SM^2$$
$$= TW^2 + XM^2 - SM^2 = TW^2 + XS^2 + 2XS \cdot SM ;$$

∴ the locus of T is a parabola of which XS is axis.

If $TW = 2AS$, $TM^2 = 4AS \cdot XM$, or X is the vertex.

53. Let the chord PQ meet the axis in O, and the tangent at A in V.

Then by Art. 48, $VO^2 = VP \cdot VQ$;

∴ V is a fixed point, and the locus of A is the circle of which OV is diameter.

54. Let the diameter TV meet the curve in R.

Then the tangent at R, being parallel to PQ, meets TP at right angles in Z on the directrix.

Also $\qquad TZ : ZP :: TR : RV ;$

$$\therefore TZ = ZP.$$

Therefore T and P arc equidistant from the directrix.

55. Let PT meet the axis in t.

Then $\qquad PQ : PT :: 2PV : PT :: 2PG : Pt,$
$$:: 2PN : Nt :: PN : AN.$$

56. If the tangents TP, TQ are equal, T lies on the axis.

Let the tangent at R meet them in p and q.

Then, since T, p, q and S lie on a circle, the triangles SqT, SpP are similar ;

$$\therefore Tq : pP :: TS : SP ;$$
$$\therefore Tq = pP.$$

So $\qquad\qquad Tp = qQ.$

57. $$AN . NL = PN^2 = 4AS . AN;$$
$$\therefore NL = 4AS.$$

But $$NG = 2AS;$$
$$\therefore LG = \text{half the latus rectum.}$$

58. By Art. 5, $P'S$, $Q'S$ are the external bisectors of the angles PSA, QSA;

therefore $P'SQ'$ is a right angle.

59. The angles TCS, DRS are equal, being supplements of equal angles SCP. SRC, Art. 35.

And the angle $CTS = TQS = RDS$;

$\therefore$ the triangles TCS, DRS are similar;

$$\therefore DR : TC :: RS : SC :: RC : CP;$$
$$\therefore PC : CT :: CR : RD.$$

Similarly $$TD : DQ :: CR : RD.$$

60. Let AD and XP intersect in Q, and let QM be the ordinate.

Then $QM : DS :: AM : AS$ and $QM : PN :: XM : XN$;

$$\therefore AM : XM :: AS : XN;$$
$$\therefore AM : AS :: AS : AN;$$
$$\therefore QM^2 : PN^2 :: AM^2 : AS^2 :: AM : AN,$$

or Q is on the parabola.

61. By Example 18, YY', the tangent at the vertex, is a common tangent.

$$SY^2 = AS . SP, \qquad SY'^2 = AS . Sp;$$
$$\therefore YY'^2 = SY^2 + SY'^2 = AS . Pp.$$

62. If PV be the diameter bisecting AQ,

$$AM = 4AN.$$

Also $\qquad AM.MR = QM^2 = 4AS.AM$;

$$\therefore MR = 4AS.$$

Now focal chord parallel to AQ

$$= 4SP = 4XN = 4AS + AM = AR.$$

63. Let AR, CP meet in p.

Draw pN, pD perpendicular to CA, CR, and let Dp meet the tangent at A in M.

$$Cp : CP :: CD : CR :: Np : CR :: AN : AC;$$

$$\therefore Cp = AN = pM.$$

Therefore the locus of p is a parabola of which C is focus and AM directrix.

64. If QMQ' be the common chord,

$$9AS^2 = 4AQ^2 = 4AM^2 + 4QM^2 = 4AM^2 + 16AM.AS;$$

$$\therefore AM \text{ is half } AS.$$

65. Let the fixed straight line BK meet the tangent at P in K.

Draw KY' at right angles, and SY' parallel to KP.

Draw $Y'A'$ perpendicular to the axis, and KL parallel to BA.

Then, since $KY = SY'$, $SA' = KL = BA$

therefore A' is a fixed point.

Therefore KY' touches the parabola of which S is focus and A' vertex.

66. $QD.DR = QM^2 - DM^2 = QM^2 - PN^2$

$$= 4AS.AM - 4AS.AN = 4AS.PD.$$

67. Draw the double ordinate QMq; then, if the diameter through Q' meet Qq in D',

$$QD'.D'q = 4AS.Q'D'.$$

Now $\quad NT : PN :: Q'D' : QD' :: D'q : 4AS$;

$\therefore D'q : 4AS :: 2AN : PN :: PN : 2AS$;

$\therefore 2PN = D'q = D'M + Mq = QM + Q'M'$.

Therefore the line through P bisecting QQ' is parallel to the axis.

Hence the locus of the middle points of a series of parallel chords is a straight line parallel to the axis.

68[1]. Take CP, CQ two tangents such that PCQ is two-thirds of a right angle; join SC cutting the curve in R, and draw the tangent ARB. Then, Art. 38,

$$CSP = CSQ = 120^0, \text{ and } CAR = \tfrac{1}{2}CSQ = 60^0;$$

$\therefore CAB$ is equilateral.

69. Draw AZ, AN perpendicular to the tangent and SY respectively, and draw SM perpendicular to ZA.

Then $\quad SM^2 = AN^2 = YN . NS = ZA . AM$.

Therefore the locus of S is a parabola of which A is vertex and ZM axis.

70. If GZ be drawn parallel to PY and SZ to PG, then SY, SZ are equal.

Therefore, if ZB be perpendicular to the axis, $BS = AS$.

Hence GZ touches an equal parabola of which B is vertex and S focus.

71. If pqr be the triangle formed by the given straight lines, describe a parabola passing through p, q and r having its axis parallel to AS. (Ex. 45.)

If s be the focus of this parabola, draw SP parallel to sp, PQ to pq, and PR to pr.

Then $\quad PQ : pq :: SA : sa :: PR : pr$,

and the angles QPR, qpr are equal.

---

[1] If a parabola touch the sides of an equilateral triangle, the focal distance of any vertex of the triangle passes through the point of contact of the opposite side.

72. Let RW be the ordinate of R.

Then

$$AN^2 : AW^2 :: PN^2 : RW^2 :: PN^2 : QM^2 :: AN : AM ;$$
$$\therefore AN : AW :: AW : AM,$$

or $\qquad WN : AN :: MW : AW ;$

$$\therefore RL : QR :: AN : AW :: PN : NL.$$

73. Let PV be the ordinate to the diameter RQM.

Then $\qquad PM : RM :: PN : TN$

$$:: 2PN.AS : 4AS.AN :: 2AS : PN ;$$
$$\therefore PM.PN = 2AS.RM.$$

But $\qquad PM^2 = 4AS.QV = 4AS.RQ ;$

$$\therefore RM : RQ :: 2PN : PM :: PP' : PM ;$$
$$\therefore QM : QR :: P'M : PM.$$

74. Let PP' be the chord, TWV its diameter, RQM the line parallel to the axis.

Then $\quad PM : RM :: PV : TV :: PV : 2WV$

$$:: 2VP.SW : 4SW.WV :: 2SW : PV ;$$
$$\therefore PM.PV = 2SW.RM.$$

But $\qquad PM.MP' = 4SW.QM,$

$$\therefore RM : QM :: 2PV : MP',$$

or $\qquad RQ : QM :: PM : MP'.$

75. SR, Sr are the exterior bisectors of the angles PSQ, pSQ respectively.

Therefore RSr is a right angle.

Therefore SD, which is half the latus rectum, is a mean proportional between DR and Dr.

76. Let PVP' be parallel to the given straight line, $QV'Q'$ the chord joining the two other points of intersection of the parabola and circle.

Let the diameters through V and V' meet the curve in p and p'.

Then pp' is a double ordinate ; draw $V'H$ parallel to pp' to meet pV.

VV' is perpendicular to QQ', and therefore parallel to the normal at p' ;

$$\therefore VV' : p'g :: V'H : p'n ;$$
$$\therefore VV' = 2p'g.$$

77. The arcs QU and RV are equal, since QV and UR are parallel.

Therefore QR and UV are equally inclined to QV, that is to the axis.

But QR and the tangent at P are equally inclined to the axis ;
therefore UV is parallel to the tangent at P.

78. $$VR : VR' :: VR : V'Q' :: PV : PV''$$
$$:: QV^2 : Q'V'^2 :: QV^2 : VR'^2 ;$$
$$\therefore VR . VR' = QV^2.$$

79. If FR, QE meet the tangent at P in V and T,

$$TE : EQ :: VR : RF :: PF : FQ. \quad (\text{Ex. 74.})$$

Therefore EF is parallel to TP.

80. If Q be the vertex of the diameter bisecting the chord Rr which meets the diameter PW in W,

$$RW . Wr = 4SQ . PW.$$

Therefore the rectangle under the segments varies as the distance of the point of intersection W from P.

81. QS, $Q'S$ are equally inclined to SP, and therefore to the axis.

Therefore $Q'S$ meets the curve at the end of the double ordinate QMq, and, since $AM . AM = AS^2$, the semi-latus rectum is a mean proportional between QM and $Q'M'$.

Also, since the diameter through P bisects QQ', PS is an arithmetic mean between QM and $Q'M'$.

82. BB' will bisect $C'A'$ in V.

Let V' be the middle point of $B'B''$.

VV' is parallel to the axis.

And BB'' is parallel to VV', and therefore to the axis.

Similarly AA'' and CC''' are parallel to the axis.

83.   Let C be the centre of the circle.

The angle between tangents to circle $= PCP' = 2PSP'$
$= 4$ times angle between the tangents to the parabola.

84.   The tangents at the ends of the focal chord PSP'
will meet in T on the directrix.

If the normals at P and P' meet in Q, TQ will be parallel
to the axis.

Let TQ meet the curve in p and PP' in V. Let QM be
the ordinate of Q.

Then $XM = TQ = 2TV = 4Sp = 4Xn.$

Therefore, if we take B in XM such that $XB = 4AS$,

$BM = 4An,\ \ QM^2 = pn^2 = 4AS.An = AS.BM.$

Hence the locus of Q is a parabola of which B is vertex and
BM axis.

85.   Produce PA to P', making AP' equal to AP.

On AP' as diameter describe a circle meeting the tangent
at P in T.

Join TA and produce to N, making AN equal to AT.

In AN take a point S such that $PN^2 = 4AS.AN$, then S
is focus.

86.   If G be the intersection of the normals and Q vertex
of the diameter bisecting PSp,

$$PS.Sp = AS.Pp = AS.TG.$$

87.   If pq be a tangent parallel to PQ, $Tp = pP$, and
T, p, q, S and O lie on a circle.

Therefore the angles TSO, TpO are equal, and TpO is
a right angle.

88. $SM^2 : AN^2 :: QM^2 : PN^2 :: AM : AN$;
$$\therefore SM^2 = AM . AN.$$
So $SM'^2 = AM' . AN$;
$$\therefore MM' . AN = MM' . (SM - SM') ;$$
$$\therefore SM - SM' = AN.$$
$$MM' = SQ - SQ' ;$$
$$\therefore MM' : SM - SM' :: AP : AN ;$$
$$\therefore MM' = AP.$$

89. If P, Q, P', Q' be the points of intersection, PQ, $P'Q'$ are equally inclined to the axis.

Hence the middle points of PQ and $P'Q'$ are equidistant from the axis.

Therefore, if P, Q be on one side of the axis and $P'Q'$ on the other, the sum of the ordinates of P and Q is equal to the sum of the ordinates of P' and Q'.

If P' be on the same side of the axis as P and Q, the ordinate of Q' is equal to the sum of the ordinates of P, Q, and P'.

90. Let the diameter through T meet the curve in W, PQ in V, and PN in t.

Let WZ be the ordinate of W; draw Qq parallel to the axis to meet PN.

$$QM . PN = PN . qN = tN^2 - Pt^2 = WZ^2 - 4AS . WV$$
$$= 4AS . AZ - 4AS . LZ = 4AS . AL.$$

91. $pX : XA :: PN : AN :: 4AS : PN$
$$:: 4AS . QM : 4AS . AL. \quad (\text{Ex. } 90.)$$
So $qX : XA :: PN : AL$;
$$\therefore pX + qX : XA :: PN + QM : AL ;$$
$$\therefore pX + qX : PN + QM :: XA : AL :: tX : TL.$$
But $NP + QM = 2TL$;
$$\therefore pX + qX = 2tX,$$
or $pt = tq.$

92. Let TF, TD be drawn parallel to PE, QE normals at P and Q.

The angle $TFQ = PEQ =$ supplement of $PTQ = TSQ$;

$$\therefore \ Q, S, F, T \text{ lie on a circle.}$$

Therefore TSF is a right angle.

So TSD is a right angle, and DF goes through S.

93. If pq be a tangent parallel to PQ, $Tq = qQ$.

Also, T, p, q and S lie on a circle ;

therefore the angles Tpq, TSq are equal.

Therefore TSq is a right angle.

94. Let RO be the diameter through the given point O.

Take T in OR produced such that $TR : RO$ in the given ratio.

If TP be a tangent, the chord POQ will be divided as required. (Ex. 74.)

95. If QN be the ordinate,

$$BP + PQ = QN + BX - NX = BX + QN - SQ,$$

which is greatest when $QN = SQ$, that is when Q is on the latus rectum.

96. If SZ and PG meet in Q and QT be ordinate,

$$TA \ : \ AS \ :: \ QZ \ : \ ZS \ :: \ QP \ : \ PG \ :: \ TN \ : \ NG;$$
$$\therefore \ TN = 2TA.$$

97. If QV be the ordinate of the point of contact,

$$TP = PV.$$

Therefore the distance of V from TQ is twice the distance of P, or the locus of V is a straight line parallel to TQ.

98. If $TPSQ$ be the parallelogram, the angles TSP, TSQ are equal;

therefore $TPSQ$ is a rhombus and T lies on the axis.

Therefore TSP is the angle of an equilateral triangle.

99. If SZ, SZ' be the perpendiculars on the second tangents, TQ, TQ' and PP' be the common tangent, SY perpendicular to it,

then angle $\qquad A'SY = ASY = YSP$;

$\qquad \therefore A'$ lies in SP, and A in SP' ;

$\qquad \therefore SP = SP'$.

Now $\qquad SQ \cdot SP = ST^2 = SQ' \cdot SP'$;

$\qquad \therefore SQ = SQ'$;

$\qquad \therefore SZ = SZ'$.

100. If the tangent meet AY in Y and the other parabola in Q,

$$QM^2 = \tfrac{1}{2}AS \cdot AM, \quad AY^2 = AS \cdot AT,$$

$$QM : AY = MT : AT ;$$

$$\therefore 2TM^2 = AT \cdot AM.$$

This can be constructed by taking $AM = MT$, or by taking $AM = 2AT$, the two solutions corresponding to the two points in which the parabola is cut by the tangent.

CHAPTER III.

THE ELLIPSE.

1. $$SD^2 = BC^2 = CS \cdot SX.$$

Therefore CDX is a right angle.

2. ST, SP are equally inclined to PT, since pST is parallel to $S'P$.

Therefore $$ST = SP.$$

3. $$PN : PG' :: SY : SP :: BC : CD$$
$$:: PF : AC :: AC : PG'.$$

Therefore $$PN = AC.$$

4. T lies on a circle of which QQ' is a diameter and V centre;

therefore $$VT = VQ.$$

Now $$QV^2 : CP^2 - CV^2 :: CD^2 : CP^2,$$

or $$VT^2 : CV \cdot CT - CV^2 :: CD^2 : CP^2.$$

Therefore $$TV : VC :: CD^2 : CP^2.$$

Therefore $$CT : CV :: CD^2 + CP^2 : CP^2,$$

or $$CT^2 : CV \cdot CT :: AC^2 + BC^2 : CP^2.$$

Therefore $$CT^2 = AC^2 + BC^2.$$

5. Through T draw a straight line at right angles to AA' meeting AP, $A'P$ in E, E'.

Then $$ET : PN :: AT : AN :: CT - CA : CA - CN.$$

Now $$CT : CA :: CA : AN;$$
$$\therefore CT + CA : CT - CA :: CA + AN : CA - CN;$$
$$\therefore ET : PN :: A'T : A'N :: E'T : PN.$$

Hence PT bisects any straight line parallel to ET terminated by $A'P$, AP.

6. Draw CD parallel to the given line, and CP parallel to the tangent at D.

The tangent at P will be parallel to CD and the given line.

7. $$SR : XE :: SA : AX :: SR : SX.$$
Therefore $$XE = SX,$$
and $$AT = AS.$$

8. Draw GL perpendicular to SP.
Then $$PL = SR,$$
and $$SY : SP :: PL : PG :: SR : PG.$$

The angle SPS' is greatest when SPY is least, that is when $SY : SP$ or $BC : CD$ is least.

Hence SPS' is greatest when CD is greatest, that is when
$$CD = AC.$$
Hence SPS' is greatest when P is on the minor axis.

9. $$CE'^2 = CP^2 + PE'^2 + 2PF . PE' = CD^2 + CP^2$$
$$+ 2CD . PF = AC^2 + BC^2 + 2AC . BC;$$
$$\therefore CE' = AC + BC.$$
So $$CE = AC - BC.$$
$$(CP + CD)^2 = AC^2 + CB^2 + 2CP . CD,$$
which is greater than $(AC + BC)^2$, since $CP . CD$ is greater than $PF . CD$ or $AC . BC$.

Similarly $CP \sim CD$ is less than $AC - BC$.

10. Let $S'Q$ drawn parallel to SP meet the normal in K, and SY in Q.

Then $S'K = S'P$ and $KQ = SP$;

therefore $S'Q = AA'$.

11. $SY : S'Y' :: YP : PY' :: TP - TY : TY' - TP$
$$:: PG - SY : S'Y' - PG.$$

12. $PS'Q$ is the supplement of $QPS' + PQS''$,
and is therefore equal to the excess of twice $QPT + PQT$
over two right angles,
that is, is the supplement of twice PTQ.

13.   Since CZ and SP are parallel, the angle
$$CZP = SPY = SNY;$$
therefore Y, Z, C, N lie on a circle.

14.   Let AQ and SP meet in R.

Then $SA : SR :: SG : SP :: SA : AX$.

Therefore R lies on a circle of which S is centre.

15.   Since KPt is a right angle, t lies on a circle which
passes through S, P, S', K,

therefore $GK : SK :: S'G : SP :: SA : AX$,

and $St : tK :: SY : SP :: BC : CD$.

16.   If SP meet $S'Y'$ in Z, then since $S'Y' = Y'Z$,
SY' will bisect PG.

17.   Let the circle whose centre is P touch the circles
whose centres are S, H in Q, R.

Then $SP + PH = SQ + QP + PH = SQ + HR$.

Hence the locus of P is an ellipse of which S and H are
foci.

18. $TN : TC :: PN : Ct$.

Therefore $TN.NG : CT.NG :: PN^2 : Ct.PN$.

But $PN^2 = TN.NG$.

Therefore $CT.NG = Ct.PN = CB^2$.

19. $TP : TQ :: CD : AC :: BC : PF :: PG : BC.$

20. $PN^2 : AF.A'F' :: TN^2 : TA.TA'$
$:: TN^2 : CT^2 - CA^2 :: TN : CT$
$:: CT - CN : CT :: CA^2 - CN^2 : CA^2 ;$
$\therefore AF.A'F' = BC^2.$

21. The perpendiculars from T on SP, SQ, HP, HQ are all equal.
Hence a circle can be described with centre T to touch
$$SP, SQ, HP, HQ.$$

22. If P, Q be two points of intersection,
PC bisects the angle ACa and QC bisects $A'Ca.$
Therefore PCQ is a right angle.

23. If SP, HQ meet in $R,$
$$PSQ + PHQ = 2PRQ - SQH - SPH,$$
and $SQH + SPH + 2RQT + 2RPT = 4$ right angles,
$\therefore PSQ + PHQ =$ twice the supplement of $QTP.$

24. Since t, P, S, g lie on a circle,
the angle $\qquad PSt = Pgt = STP.$

25. $Q'M : PM :: BC : AC :: PN : QN.$
Therefore $\qquad Q'M : CN :: CM : QN.$
Therefore QQ' passes through $C.$

26. $\qquad SY : SP :: BC : CD.$
Therefore $\qquad SY.CD = SP.BC.$

27. If T be intersection of tangents at A and $B,$
then, since TC bisects $AB,$ it is a diameter of the conic.
Therefore the tangent at C is parallel to $AB.$

28. The angles SPT, HPt are equal.
Also $\qquad TP.Pt = CD^2 = SP.PH,$
or $\qquad TP : SP :: HP : Pt.$
Therefore SPT, HPt are similar.

29. $$PE = PE' = AC.$$

Therefore $SE = HE'$, and the angles SCE, HCE' are equal. Therefore the circles circumscribing SCE, HCE' are equal.

30. The angles KPG, GPL are equal ; therefore KL is a double ordinate of the circle of which PG is diameter.

31. If Q be the centre, QN the ordinate, and T, T' the points where the tangent at P meets the tangents at the vertices,

$$QN^2 : SN \cdot NH :: AT \cdot A'T' : AH \cdot A'S :: BC^2 : A'S^2.$$
$$(\text{Ex. 20.})$$

32. Since the tangents are equally inclined to SP, $S'P$ respectively, the bisector of the angle between them bisects SPS', and therefore passes through the point where the axis minor meets the circle.

33. If $PQRS$ be the quadrilateral, p, q, r, s points of contact, H the focus,

the angle $\qquad pHP = PHs, \; pHQ = QHq,$
$$SHr = SHs, \; rHR = RHq.$$

Therefore $PHQ + SHR = PHS + QHR =$ two right angles.

34. $\qquad SG : SC :: SP :: SY$ (see Ex. 15)
$$:: CD : BC :: PV : VA.$$

35. The normals at P and Q will meet on the minor axis in K.

Then angle between the tangents $= PKQ = PSQ$.

36. The auxiliary circle lies entirely without the ellipse except at A and A' ; therefore AA' is the greatest diameter.

The circle described on BB' as diameter lies wholly within the ellipse ; therefore BB' is the least diameter.

37. Let any circle passing through N and T meet the auxiliary circle in Q.

Then $$CN \cdot CT = CA^2 = CQ^2.$$

Hence CQ touches the circle at Q, and the circle cuts the auxiliary circle orthogonally.

38. The angle $PNY = PSY = PS'Y' = PNY'$.

Therefore $$PY : PY' :: NY : NY'.$$

39. $$PQ^2 : TQ^2 :: SY \cdot S'Y' : TY \cdot TY'.$$

But $$TQ^2 = TY \cdot TY'.$$

Therefore $$PQ^2 = SY \cdot S'Y' = BC^2.$$

Therefore $$PQ = BC.$$

40. If QN and PM be the perpendiculars on the given lines passing through C, R their point of intersection,

$$RN : QN :: CP : CQ ;$$

therefore the locus of R is an ellipse of which the outer circle is the auxiliary circle.

41. $$SP : S'P :: SY : S'Y' :: SY^2 : BC^2,$$

and $$S'Q : SQ :: S'Z' : SZ :: BC^2 : SZ^2.$$

Therefore $$SP \cdot S'Q : S'P \cdot SQ :: SY^2 : SZ^2.$$

42. Let Ca, Cb be the conjugate diameters, and Pm, Pn ordinates of P.

Then $$Cm \cdot CM = Ca^2,$$

and $$Cn \cdot CN = Cb^2 ;$$

$$\therefore CM \cdot Pm : Ca^2 :: Cb^2 : Pn \cdot CN ;$$

$\therefore$ the triangle CPM varies inversely as the triangle CPN.

43.

$$CAV : CPT :: CA^2 : CT^2 :: CN : CT :: CPN : CPT ;$$

therefore the triangles CAV, CPN are equal.

44. Let TPQ, Tpq be the tangents intersecting the auxiliary circle in P, Q, p, q.

Let E, e be their middle points.

$$PE^2 + pe^2 = ET^2 + Te^2 - TP . TQ - Tp . Tq$$
$$= CT^2 - 2TP . TQ = CT^2 + 2CA^2 - 2CT^2 = SC^2.$$

45. Let QQ' be a diameter equally inclined to the axis with the conjugate to $P'P$.

Then the circles described through P, P', Q and P, P', Q' will touch the ellipse at Q and Q'.

Hence Q, Q' are the points at which PP' subtends the greatest and least angles respectively.

46. Draw the tangent Qr.

Then, since the angles PSQ, QSr are equal, Q always lies on the tangent at the end of the focal chord RSr.

47. The triangle YCY' will be the greatest possible when YCY' is a right angle : P will then lie on the circle of which SS' is diameter.

This intersects the ellipse in four points, provided SS' is greater than BB'.

48. The points where the lines joining the foci of the two ellipses meet the common auxiliary circle are points through which the common tangents pass.

49. The circle passing through the feet of the perpendiculars is the auxiliary circle of the ellipse.

50. Draw QN perpendicular to AB.

Then $\qquad QN : NA :: BP : AP :: CA : AT$,

and $\qquad QN : BN :: AT : AB$.

Therefore $\qquad QN^2 : AN . NB :: CA : AB$.

Therefore the locus of Q is an ellipse of which AB is major axis.

51. PG, GN, NP are at right angles to CD, DR, RC respectively.

Therefore the triangles CDR, PGN are similar.

Therefore $\quad PG : CD :: PN : CR :: BC : AC.$

52. Let PS, QS meet the ellipse and circle again in p, q.

And let $P'Cp'$ be the diameter parallel to SP.

Then, since pq is an ordinate,

$$SQ : SP :: Sq : Sp :: Qq : pP :: AA' : P'p'.$$

Again, $\qquad PS.Sp : AS.SA' :: CP'^2 : CA^2.$

Therefore $\qquad SR.Pp : 2BC^2 :: P'p'^2 : AA'^2,$

or $\qquad Pp : AA' :: P'p'^2 : AA'^2.$

Therefore $\qquad SQ : SP :: AA' : Qq,$

and $\qquad\qquad Qq = P'p'.$

53. If SP meet $S'Y'$ in L', $SL' = AA'$;

therefore $\qquad\qquad SR = AC.$

54. Since the directions before and after impact are equally inclined to the tangent at the point of impact, the lines in which the ball moves will touch a confocal ellipse or hyperbola.

55. Let the tangent at P meet the tangents at A and A' in F and F'.

Then, since the angles PSF, FSA and $PSF', F'SA'$ are respectively equal, S (and similarly S') lies on the circle of which FF' is diameter.

56. $P', D',$ the two angular points, will lie in PN, DM respectively.

Therefore $\quad P'N : NC :: DM : NC :: BC : AC.$

Therefore P' lies on a fixed straight line through C.

Similarly Q' lies on the other equi-conjugate diameter.

57. The angles SPS', STS' are equal by Ex. 15.

$\therefore\quad SPS' : STS' :: SP.S'P : ST.S'T :: CD^2 : ST^2.$

58. If T be the centre, then, since the angles TSP, TSA are equal, T lies on the tangent at A.

59. Let QL be the ordinate of Q.

Then $\qquad QL : LS :: CN : NP,$

and $\qquad QL : LS' :: CM : MD.$

$$QL^2 : SL.SL' :: CM.CN : PN.DM :: AC^2 : BC^2,$$

or Q lies on an ellipse of which SS' is minor axis.

60. If P is the corresponding point on the ellipse, and SZ the perpendicular on the tangent to the circle,

$$SZ : AC :: CT - CS : CT :: AC^2 - CS.CN : AC^2$$
$$:: SP : AC;$$
$$\therefore SZ = SP.$$

61. The tangent at Q is parallel to the normal at P; $\therefore$ the tangent at P is parallel to the normal at Q.

62. If $PQ\ P'Q'$ be the parallelogram, the angle DHE is the supplement of $HPQ' + HQP'$;

that is, of $SPQ + SQP$; that is, of DSE.

Hence S, H, D, E lie on a circle.

63. Let the line through C parallel to the tangent meet the directrices in Z, Z'.

Since the auxiliary circle is fixed, $SY, S'Y'$ are fixed straight lines meeting ZZ' in fixed points $y\ y'$.

And $\qquad Cy.CZ = CX.CS = CA^2.$

Therefore Z and Z' are fixed points.

64. The angle $S'TZ = STY = SZY =$ complement of YZT.

Therefore YZ and $S'T$ are at right angles.

65. If G be the centre of the circle, GL bisects SP at right angles.

Therefore SP is equal to the latus rectum.

66. If CZ be perpendicular to YY', the perimeter of the quadrilateral is equal to SS' together with twice $CZ + ZY$, which is greatest when $CZ = ZY$, that is when SPS' is a right angle.

67. Draw SZ perpendicular to $S'Z$ the straight line on which S' lies.

Let $PS'P'$ be the chord parallel to SZ.

Produce PP' both ways to M and M', so that
$$S'M = S'M' = AA'.$$
Then the lines drawn through M and M' perpendicular to SZ are fixed, and $\quad SP = PM, \quad SP' = P'M'.$

Hence the ellipse will touch two parabolas having S for focus.

68. Let TQ, TQ' be tangents, V the middle point of QQ'.

Then $\quad QV . VQ' = CP^2 - CV^2 = CV . VT.$

Hence Q, Q', C, T lie on a circle.

69. Draw QM perpendicular to the minor axis.

Then $\qquad QC^2 : AC^2 - CN^2 :: BC^2 : AC^2,$

or $\qquad QC^2 : BC^2 :: AC^2 - CN^2 : AC^2.$

Therefore $\quad BC^2 - CN^2 - QN^2 : CN^2 :: BC^2 : AC^2,$

or $\qquad BC^2 - MC^2 : QM^2 :: AC^2 + BC^2 : AC^2.$

Therefore Q lies on an ellipse of which BB' is minor axis.

70. If QM be the ordinate of Q,

$$AM^2 : CN^2 :: QM^2 : PN^2 :: AM . MA' . AC^2 - CN^2.$$
Therefore $\quad AM . AA' : AM^2 :: AC^2 : CN^2,$

or $\qquad\qquad 2CN^2 = AC . AM.$

But $\qquad AQ . AO : CP^2 :: AM . AC : CN^2;$

therefore $\qquad\qquad AQ . AO = 2CP^2.$

71. $SP : SN :: SC : CQ :: SC : AC-QR.$

Therefore $SP.AC=SP.QR+SN.SC.$

But $SP : XS+SN :: SC : CA$

or $SP.AC=XS.SC+SN.SC.$

Therefore $SP.QR=XS.SC=BC^2.$

72. If the tangent meet the tangent at A in T, and $S'Y$, $S'Z$ be perpendicular to TS, and the tangent, T, A, Y, S', Z lie on the circle of which $S'T$ is diameter.

The angles YTZ, ATS' are equal since $ATS, S'TZ$ are equal. Art. 68.

Therefore the chords YZ, AS' on which these angles stand are equal.

73. If P, Q, P', Q' be the parallelogram, p, q, p', q' the points of contact, $pq, p'q'$ are parallel focal chords bisected by $PCP'.$

But QCQ' bisects pp', qq' and is therefore conjugate to PCP' and parallel to $pq, p'q'.$

Therefore $CQ=CQ'=CA$

74. If T be the point from which the tangents are drawn,

$ST, S'T$ are perpendicular to TP', TP respectively.

Therefore $SP, S'P'$ are both parallel to $CT.$

75. $CS^2 : CA^2 :: CG : CN :: CG.CT : CN.CT.$

Therefore $CG.CT=CS^2.$

76. If PG be the normal at the point of contact,

$$CG.CT=CS^2.$$

Therefore G is a fixed point and P lies on the circle of which GT is diameter.

77. Let the given straight line pq meet the axis $t.$

Let the tangents at p and q meet in $Q.$

Let CQ meet pq in V and the curve in P

Through P, Q draw PG, QG' perpendicular to pq, meeting the transverse axis in G and $G'.$

Then $\quad CG' : CG :: CQ : CP :: CP : CV :: CT : Ct$;

$$\therefore CG' . Ct = CG . CT = CS^2,$$

or G' is a fixed point.

78. Draw SY perpendicular to the tangent; produce SY to L making $LY = YS$.

The point of intersection of the circles described with centres L and P', and radii AA' and $AA' - SP$ respectively will be the second focus.

79. Draw SY, SY' perpendicular to the given tangents.

The point of intersection of circles described with centres Y and Y', and radius equal to CA will be the centre.

80. If OS be drawn perpendicular to PQ, S will be one focus.

If SP, PS' be equally inclined to OP and SQ, QS' to OQ, S' will be the other.

Bisect SS' in C, and take CA in SS'' such that

$$2CA = SP + PS'.$$

If X be the foot of the directrix,

$$CX . CS = CA^2.$$

81. $\qquad Qq : Aq :: PN : AN,$

and $\qquad Rr : rA' :: PN : NA'.$

Therefore $\quad Qq . Rr : Aq . A'r :: PN^2 : AN . NA'$
$$:: BC^2 : AC^2 :: SL : AC.$$

Now $\qquad Aq : qA' :: Aq^2 : Qq^2,$

and $\qquad A'r : rA :: A'r^2 : Rr^2.$

Therefore $\quad Aq . A'r : Ar . A'q :: AC^2 : SL^2.$

82. By Ex. 75. $\quad CT : CS :: CS : CG$;

therefore $\qquad TS : CS :: SG : CG.$

But $\qquad TY : PY :: TS : SG$;

$$\therefore TY^2 : PY^2 :: CS^2 : CG^2 :: CT : CG :: TZ : PZ.$$

83. If O be the intersection of the lines
$$OC^2 = AC^2 + BC^2.$$

84. $$TP \,:\, TQ \,::\, CP' \,:\, CQ',$$
and the angles PTQ, $P'CQ'$ are equal.
Therefore PQ is parallel to $P'Q'$.

85. S, P, t, S' lie on a circle, and the triangles SCt, PYS are similar.

Therefore $St \,:\, Ct \,::\, SP \,:\, SY \,::\, CD \,:\, BC$,

or $St\,.\,PN \,:\, Ct\,.\,PN \,::\, CD\,.\,BC \,:\, BC^2$.

Therefore $St\,.\,PN = CD\,.\,BC$.

86. If the tangent at Q meet the minor axis in t', the angle $SQS' = St'S'$, or t' is on the circle.

Now $QM\,.\,Ct' = BC^2 = PN\,.\,Ct$.

Therefore $QM \,:\, PN \,::\, Ct \,:\, Ct' \,::\, Ct \,:\, Ct + St$
                   $::\, BC \,:\, BC + CD$ by Ex. 85.

87. If $S'Z$ be perpendicular to TY,
the angle $STY = $ complement of TZY', (Ex. 64)

             $= $ half supplement of YCY' the angle at the centre,

             $= CYY'$; and $STY = S'TY'$.

88. The tangents at L and L' are perpendicular to the tangent at P, and therefore D and D' where they meet the tangent at P are on the director circle.

Now $DL \,:\, DP \,::\, D'L' \,:\, D'P$;
therefore PQ bisects the angle LPL'.

Therefore $LP + PL' = $ diameter of director circle.

89. AB, AE are equally inclined to BC,
and $AB^2 = AD\,.\,AE$.
Therefore AB is a tangent.

90. If the tangent at P meet the tangent at A in T, TS, TS' bisect the angles PSA, $PS'A$.

91. If the chords of intersection NO, PQ meet in T and CD, CE, $C'D'$, $C'E'$ are parallel radii,

$$CD^2 : CE^2 :: TN.TO : TP.TQ :: C'D'^2 : C'E'^2.$$

92. If PN meet CD in K,

$$PK : PQ :: SG : SP :: SA : AX,$$

and $PN.PK = PG.PF = BC^2.$

Therefore PQ varies inversely as PN.

93. Draw perpendiculars SY, CE, $S'Y'$, SZ, CF, $S'Z'$ on tangents TP, TQ,

then $CT^2 = CF^2 + TF^2 = CZ^2 + TZ.TZ' = CA'^2 + SY.S'Y'$
$$= CA'^2 + CB^2.$$

94. If the circle meet the minor axis in K and L, the tangents at P and Q meet either in K or L, see Ex. 15.

95. This problem is equivalent to Ex. 45.

96. Let CV bisecting the chord QSQ' meet the curve in P and directrix in T.

Let DCD' be the parallel diameter.

Then $SR : SC :: CS - CR : CS :: CT - CV : CT$

$$:: CS^2 - CG^2 : SC^2 :: SG.GS' : CS^2$$

$$:: SP.PS' : CA^2 :: CD^2 : CA^2 :: QQ'^2 : DD'^2$$

by Art. 76.

97. Let the tangent at Q meet PN in P' and the axis in U.

Then $CT.CN = CA^2 = CM.CU$;

therefore $CT : CM :: CU : CN,$

or $TM : CM :: NU : CN.$

But $PN.Q'M : Q'M^2 :: TN : TM,$

or $PN.Q'M : CM.MU :: CN.NT : CM.NU.$

Hence $PN.Q'M : CN.NT :: CM.QM : CM.P'N$

$$:: QM^2 : P'N.QM.$$

Now $PN^2 : CN.NT :: BC^2 : AC^2 :: QM^2 : Q'M^2.$

Hence $PN.Q'M : PN^2 :: Q'M^2 : P'N.QM.$

Therefore $P'N : PN :: AC : BC,$

or P' is on the auxiliary circle.

98. The diameter bisecting PQ is fixed, hence V the centre of the circle, is a fixed point.

VM bisecting RS at right angles, is a fixed straight line ;

PQ and RS are equally inclined to the axis ;

$\therefore$ CM and CV are equally inclined to the axis.

Therefore M is a fixed point, and RS a fixed straight line.

99. The angle BSC
$$= BAC + SBA + SCA = BAC + HBC + HCB$$
$$= BAC + \text{supplement of } BHC.$$

Hence if BHC is constant, BSC will be constant.

100. The angles SPT, HPt are each equal to SQH ;

also $STP = tQS - PtH$
$$= HPt - PtH = PHt.$$

Therefore $TP : SP :: HP : Pt,$

or $TP . Pt = SP . HP = CD^2.$

Therefore CT, Ct are conjugate.

101. CT bisects PQ and is parallel to SP ;

$\therefore$ T is the foot of the perpendicular from S' on PT. See Cor. (3), Art. (66).

102. $SC : CY$ is a given ratio and SY is fixed.

103. Take p a point near and let the focal chord $p'Sq'$ meet pq in O;
$$PQ : p'q' :: pO . Oq : p'O . Oq' :: ST . Oq : Sp' . Oq'$$
$$:: ST . pq : Sp . p'q' \text{ ultimately.}$$

104. Dropping perpendiculars from the focus on the sides, their feet are the middle points, and, as they lie on a circle, form a rectangle; the diagonals, intersecting in H, are therefore at right angles, and SAD can be proved equal to HAB.

105. If CD, CP meet the directrix in E and G, ES is perpendicular to the chord of contact of tangents from E, which is parallel to CP.

106. If CP, DC meet the tangent at A in Q and R, prove that $AQ \cdot AR = BC^2 = AS \cdot AS'$.

Then $QSA = ARS'$, and $QRA = AQS'$.

CHAPTER IV.

THE HYPERBOLA.

1. If the circle whose centre is P touch the circles whose centres are S and H in Q and R,

$$SP \sim HP = SQ \sim HR.$$

Therefore P lies on an hyperbola of which S and H are foci.

2. $\quad SD^2 = BC^2 = CS^2 - CA^2 = CS^2 - CX \cdot CS = CS \cdot SX.$

Therefore the triangles SCD, SDX are similar.

3. If the straight line meet the curve in P and the directrix in F,

$$SF : SX :: CS : CA :: SA : AX :: SR : SX.$$

Therefore $\qquad SF = SR.$

Draw PK perpendicular to the directrix.

Then $\qquad PF : PK :: SC : CA :: SP : PK.$

Therefore $\qquad SP = PF.$

4. Draw SD, SD' perpendicular to the asymptotes.

Then DD' is the directrix.

5. If the asymptote meet the directrix in D, then DS drawn at right angles to CD meets the axis in the focus.

6. If PK, QL be the perpendiculars from the given points on the directrix $PS-SQ=PK-QL$ which is constant.

Therefore S lies on an hyperbola of which P and Q are foci.

7. If the circle inscribed in the triangle ABC touch the sides in D, E, F; B, C, D being given,
$$BA-CA=BF-EC=BD-DC.$$

Hence A lies on an hyperbola of which B and C are foci and D a vertex.

8. $PN : Pg :: SY : SP :: BC : CD$
$$:: PF : AC :: AC : Pg.$$
Therefore $\qquad PN=AC.$

9. Draw CD parallel to the given line and CP parallel to the tangent at D.

Then the tangent at P is parallel to CD and the given line.

10. Let $A'P$ and $P'A$ meet in Q, and draw the ordinate QM.

Then $\qquad QM : A'M :: PN : NA'$,

and $\qquad QM : AM :: P'N : NA$.

Therefore $\quad QM^2 : AM.MA' :: PN^2 : AN.NA'$
$$:: BC^2 : AC^2,$$
or Q lies on an hyperbola having the same axes.

11. Let the tangent at P, AP and $A'P$ meet the minor axis in t, E and E'.

Then $CE : PN :: CA : AN :: CA.A'N : AN.NA'$,

and $\qquad CE' : PN :: CA.AN :: AN.NA'$.

Hence $CE-CE' : PN :: 2CA^2 :: AN.NA'$.

Now $\qquad PN^2 : AN.NA' :: PN.Ct : AC^2$;

therefore $\qquad CE-CE'=2Ct$.

Therefore Pt bisects every line perpendicular to AA' terminated by $A'P$, AP.

12. SPT is an isosceles triangle since pST is parallel to $S'P$.

Therefore $$SP = ST.$$

13. Draw SD perpendicular to the asymptote and SK parallel to it.

If TS bisect the angle PSK, T being on the asymptote, TP is the tangent at P. Draw SY perpendicular to it.

Then CM which bisects DY at right angles will meet the asymptote in the centre C. DX drawn perpendicular to CS will be directrix.

14. If the tangent at P meet the tangents at A and A' in Z and Z', ZS and $Z'S$ are the internal and external bisectors of the angle ASP.

Hence the foci lie on a circle of which ZZ' is diameter.

15. Draw PK perpendicular to the directrix and DS at right angles to the asymptote.

Draw cxs at right angles to the directrix meeting it in x.

With centre P and radius PS such that $SP : PK :: cs : cD$, describe a circle meeting Ds in S.

Then S is the focus.

16. Draw Qq', Pp and Rr, Qq parallel to the asymptotes.

Then $CP : CQ :: Cp : Cq' :: Cq : Cr :: CQ : CR$.

17. $QC^2 - CB^2 : CN^2 - CA^2 :: BC^2 : AC^2$
$$:: PN^2 : CN^2 - CA^2.$$

Therefore $$PN^2 = QB . QB'.$$

18. $DN : NA :: QM : MA$ and $EN : NA'$
$$:: QM : MA'.$$

Therefore $ND . NE : AN . NA' :: QM^2 : AM . MA'$
$$:: PN^2 : AN . NA',$$

or $$PN^2 = ND . NE.$$

19. A, A', Y, Z lie on the auxiliary circle.

Therefore $$AT . TA' = YT . TZ.$$

20. If the tangent at P meet the asymptotes in L and L', $CD = PL = PL'$; therefore Q divides LL' in a constant ratio.

Draw QH. QK parallel to the asymptotes.

Then QH. QK varies as CL. CL' and is therefore constant.

Therefore Q lies on an hyperbola having the same asymptotes.

21. This is equivalent to the preceding.

22. Since $SK = S'K$, K lies on the circle passing through S, S' and P, and since KPt is a right angle, t lies on the same circle.

Therefore $GK : S'K :: SG : SP :: SA : AX$,

and $\qquad St : tK :: SY : SP :: BC : CD$.

23. Let P, P' be the points of trisection of the arc SS' and let XM bisect SS' at right angles, then $SP = 2PM$ and $S'P' = 2P'M$.

Hence P and P' lie on hyperbolas of which S and S' are foci and XM directrix.

If C, C' be the centres $CS = 4CX$ and $C'S' = 4C'X$.

Therefore C and C' are the points of trisection of the chord SS'.

24. Draw SZ parallel to the asymptote : the angle $STQ = TSZ = TSP$.

Therefore $\qquad\qquad SQ = QT$.

25. Since the hyperbolas have the same asymptotes the ratios $CS : BC : CA$ are constant.

Let NP be the fixed line parallel to an asymptote, and PQ proportional to an axis.

Then PQ^2 varies as CS^2, that is, as $CN . NP$, that is, as NP.

Hence Q lies on a parabola having NP for a diameter.

26. $PY.PY' = AC^2 - CP^2 = CD^2 - BC^2 = CS^2$
$$- \left(\frac{SP - S'P}{2}\right)^2 = CS^2 - CA'^2 = CB'^2.$$

27. Let TP meet the other asymptote in T', then $PT = PT'$.

Therefore $\qquad PQ = R'P = QR.$

28 Draw OrR parallel to PQ, meeting the ellipse and hyperbola in r and R.

Let Oa, Ob be the axes, then since OP, Or are conjugate in the ellipse, and OQ, OR in the hyperbola, if PN, QM, rl, RL be the ordinates,

$$ON.Ob = rl.Oa; \quad PN.Oa = Ol.Ob; \quad QM.Oa$$
$$= OL.Ob; \quad OM.Ob = RL.Oa.$$

Therefore $\qquad PN : ON :: QM : OM$

since $\qquad rl : Ol :: RL : OL,$

or OP and OQ are equally inclined to the axes.

29. Through S draw SC parallel to the bisector of the angle between the asymptotes meeting the asymptote which is given in position in C.

Draw SD perpendicular to that asymptote, and DX to CS. Then if A be taken in CS such that $CA^2 = CX.CS$, A is vertex.

30. If the tangents at P and Q meet in T, then since the perpendiculars from T on SP, SQ, HP, HQ are all equal, a circle can be described with centre T to touch SP, SQ, HP and HQ.

31. If CL, CL' be the asymptotes, S will lie in the bisector of the angle LCL'.

Draw PL, PL' parallel to the asymptotes to meet them in L and L';

and take S in CS such that $CS^2 = 4CL.CL'$,

then S is a focus.

32. If the conjugate diameters PCP', DCD' be given, complete the parallelogram $LML'M'$ formed by the tangents at D, P, D' and P'.

The diagonals LL', MM' are the asymptotes and the axes bisect the angles LCM, LCM'.

33. Let QT and RQ meet the asymptotes in L and M.

Then $QL : PH :: RH : TL :: CR : CT$

$$:: RM : TK :: PK : QM;$$

therefore $QL \cdot QM = PH \cdot PK$,

or Q is on the curve.

34. Let CL bisecting the angle ACB' meet PN in L, draw QM parallel to LC.

Then CL is proportional to CN;

therefore $CL \cdot CM$ is proportional to $CN \cdot NT + CN^2$, that is to CA^2.

Hence Q lies on an hyperbola of which CL and CB are asymptotes.

35. Draw SY perpendicular to the tangent and produce it to Z making $YZ = SY$.

Then if Q be the point of contact, and P the fixed point,

$$HP - PS = HQ - QS = HZ;$$

therefore $HP - HZ = PS$,

or the locus of H is an hyperbola of which P and Z are foci.

36. If PT, Pt the tangents to the ellipse and hyperbola meet BC in T and t, then since the curves have the same conjugate axis, for $CA^2 = CS^2 + CB^2$

$$Ct \cdot PN = BC^2 = CT \cdot PN,$$

or $CT = Ct$.

37. This problem is the converse of Ex. 3.

38. If G be the point of intersection
$$CG = \tfrac{2}{3}\, CP,$$
or G lies on an hyperbola having the same asymptotes.

39. The angle $CYY' = S'PY' = S'NY'$ since S', P, N
and Y' lie on a circle.
Therefore Y, Y', C and N lie on a circle.

40. If SY meet $S'P$ in Z, $SY = YZ$; therefore $S'Y$
bisects PG.
Similarly SY' bisects PG.

41. $BC^2 : AC^2 :: NG : CN :: CT.NG : CN.CT$;
therefore $\qquad\qquad CT.NG = BC^2.$

42. The angle $STP = TS'P + S'PT = TS'P + SPT$
$$= PS'S + SS't = \text{supplement of } PSt.$$

43. $\qquad\qquad Pt.PT = CD^2 = SP.S'P,$
or $\qquad\qquad SP : PT :: Pt : S'P$;
and the angles SPT and tPS' being equal, the triangles
SPT, tPS' are similar.

44. The circles SCE, $S'CE'$ stand upon equal chords
SC, $S'C$ and contain equal angles SEC, $S'E'C$, since CE is
parallel to the bisector of SPS'.

45. If the tangent at P meet the tangent at A, the
vertex of the branch on which P lies, in T, T is the centre of
the circle inscribed in the triangle SPS', since TS, TS'
bisect the angles ASP, $AS'P$.

46. $CT : CA :: CA : CN :: CP : CQ$, or AQ is
parallel to PT.

47. $\qquad\qquad CE : CA :: CS : CA$;
therefore $\qquad\quad CE = CS$; but $CD = CA$.
Therefore $\qquad\quad AD$ and SE are parallel.

48. If E be the centre of the circle and EK its radius,
$$EK : CE :: BC : SC,$$

or $\quad\quad EK : CA :: BC : SC + BC$

$\quad\quad\quad\quad\quad\quad\quad :: (SC - BC)\, BC : CA^2.$

And $\quad SR' : CS :: BC : AC :: SR : BC$;

therefore $\quad RR' : SR :: CS - BC : BC$,

or $\quad RR' : CS - BC :: SR : BC :: BC : CA.$

Therefore $\quad\quad\quad\quad EK = RR'.$

49. $\quad\quad\quad\quad\quad PM = PL$;

therefore $\quad\quad\quad\quad GL = GM.$

50. If Ca, Cb be the conjugate diameters and one hyperbola touch the ellipse in P, the tangent at P will meet Ca, Cb in T, t, such that $TP = Pt = CD$.

Hence PD is bisected by Ct,

and tD touches the other hyperbola and is parallel to CP.

51. If LL' and MM' be the tangents,

$\quad\quad\quad CL : CM :: CM' : CL'$,

or $\quad\quad\quad LM'$ and $L'M$ are parallel.

52. If the tangent at P meet the tangents at A and A' in F and F' and QM be the ordinate of the centre of the circle,

$\quad\quad\quad QM : MS :: SA : AF$,

and $\quad\quad QM : MS' :: S'A' : A'F'$.

Hence $QM^2 : SM . MS' :: SA^2 : AF . A'F' :: SA^2 : BC^2$

$\quad\quad\quad\quad\quad$ (Art. 126).

Hence the locus of Q is an hyperbola of which S and S' are vertices.

53. If PM be perpendicular to the directrix,

$PK : PM :: CS : CA :: SA : AX :: SP : PM$,

or $\quad\quad\quad\quad PK = SP.$

54. Let PD meet an asymptote in n, draw Pl, Dm parallel to Cn.

Then $$Dm \cdot Dn = Pn \cdot Pl \, ;$$
therefore $$Dn = Pn.$$

Therefore if LPL' is tangent at P, LD is tangent at D, and CP, CD are conjugate.

55. If QR meet the asymptotes in q and r, $qQ = rR$; therefore if EPe be the tangent at P,
$$CL : CN :: qQ : Pe :: PE : qR :: CN : CM.$$

56. If the circle intersect the axis in b, B,
$$CB \cdot Cb = CS^2$$
or $$CB^2 + CB \cdot Bb = CA^2 + CB^2 \, ;$$
therefore $$CB \cdot Bb = CA^2.$$

57. Let the straight line $q'Q'APQq$ meet the asymptotes in Q', Q.

Draw RCR' parallel to AP terminated by $A'q$, $A'q'$.
Then $$PQ' = AQ = CR = Qq,$$
and $$Q'q' = CR' = AQ' = PQ \, ;$$
therefore $$Pq' = Pq.$$

58. T is the centre of the circle inscribed in the triangle $PS'Q$, therefore the difference between PTQ and half $PS'Q$ is a right angle.

59. Draw CD, CE parallel to OA, OB and PH, PK parallel to and terminated by CE, CD.

Then $PH : OD :: CK : CE :: CP : CA :: CB : CP$
$$:: CD : CH :: OE : PK \, ;$$
therefore $$PH \cdot PK = OD \cdot OE,$$
or P lies on an hyperbola having CD, CE for asymptotes.

60. Draw PH, PK, QH', QK' parallel to the asymptotes.

Then $PL : QM :: PH : QH' :: QK' : PK :: QN : PR$,
or $$PL \cdot PR = QM \cdot QN.$$

61. If TK, TN be perpendicular to the directrix and SP, $TK = SN$.

Therefore $ST : TK :: ST : SN$ a constant ratio, and the angle between the asymptotes is double PST, that is, double the external angle between the tangents.

62. $$Q'V^2 - RV^2 = CD^2 = RV^2 - QV^2,$$

or $$QV^2 + Q'V^2 = 2RV^2.$$

Again $$CT.CV = CP^2 = CV.CT';$$

hence $$CT = CT'$$

63. If V be the middle point of PQ, then since R, V are the middle points of LRL' and $LPQl$, RV is parallel to the asymptote $CL'l$.

Hence $$PM + QN = 2RE.$$

64. If TP, TQ be the tangents, PTQ, STS' have the same bisector which passes through the point where the circle meets BCB'.

65. The tangents at P and Q intersect in t on the circle and BCB'.

Hence the angle $PtQ = PSQ$.

66. The angle $PNY = PSY = PS'Y' = $ supplement of PNY'.

67. The triangle YCY' is greatest when YCY' or SPS' is a right angle.

In that case PT meets BC in t such that $Ct = CS$;

therefore $$CS.PN = Ct.PN = BC^2.$$

68. The triangle SPS' : triangle StS' :: $SP.PS'$: St^2 :: $CD^2 : St^2$.

69. $S'P$ is parallel to CY and $S'Q$ to CZ.

Therefore $S'T$ is parallel to bisector of YCZ and is perpendicular to YZ.

70. If G be the centre of the circle, GL bisects SP;

therefore $$SP = 2PL = 2SR.$$

71. The tangent at Q is parallel to the normal at P, therefore the tangent at P is parallel to the normal at Q, or CP is conjugate to normal at Q.

72. If Y be the point from which the tangents are drawn, SP and $S'P'$ are both parallel to CY.

73. $SC^2 : AC^2 :: CG : CN :: CG.CT : CN.CT$,
or $$CG.CT = SC^2.$$

74. By Ex. 73, G the foot of the normal is a fixed point;

therefore P lies on the circle of which TG is diameter.

75. If TP, TQ be the tangents, CT will bisect PQ in V,
and $$CT.CV = CT^2,$$
or PQ is a tangent at V.

76. Let GQ meet the conjugate in G'.
Then $QG' : QG :: CN : NG :: AC^2 : BC^2$.
Therefore, by Art. 111, QG' is normal at Q.

77. If PM be drawn perpendicular to the directrix of the parabola the angle $PTQ = SPT - SQT = $ half $SPM - $ half $SQS' = $ half $SS'Q$.

78. If $abcd$ be the quadrilateral and S lie on the circle the angle $Hcd = Scb = Sab = Had$,
or H is on the circle.

79. If PP' be the chord of contact and CV bisect PP' then CV, PP' are parallel to a pair of conjugate diameters in both conics.

Hence if from a common point Q, a double ordinate QVQ' be drawn parallel to PP', Q' must lie on both curves.

Similarly RR' the line joining the other two common points is parallel to PP'.

80. If SD, SD' are perpendiculars from the common focus on the asymptotes, DD' is the tangent at the vertex of P and a directrix of H.

If P be a common point, and PM perpendicular to DD';

$$SP : PM :: SC : CA,$$

but $$SP = PM + SX.$$

Therefore $SP : SX :: CS : CS - CA :: CS : AS.$

Hence $\quad AS.SP = SX.CS = BC^2 = AS.SA',$

or $$SP = A'S.$$

Therefore $A'P$ touches the parabola at P.

81. With centre P, the given point and radius of the given length describe a circle meeting the other asymptote in p.

Then $pPQq$ is the line required.

82. Let CB, CA be semiaxes of the ellipse, Ca, Cb of the hyperbola.

Let PN meet the asymptote in Q,

then $$QN^2 : CN^2 :: Cb^2 : Ca^2,$$

or $$QN^2 + CN^2 : Ca^2 + Cb^2 :: CN^2 : Ca^2;$$

but $$SP + S'P = 2CA,$$

and $$SP - S'P = 2Ca.$$

Hence $4CA.Ca = SP^2 - S'P^2 = SN^2 - S'N^2 = 4CN.CS;$

therefore $CA^2 : CS^2 :: CN^2 : Ca^2 :: QN^2 + CN^2 :$
$$Ca^2 + Cb^2 :: CQ^2 : CS^2.$$

Therefore Q lies on the auxiliary circle of the ellipse.

83. Let Q be a common point.

Then $SQ - QH = AA'$ and $SQ - QP = SP - 2PH$
$$= AA' - PH.$$

Therefore $$QP = QH + PH,$$

or Q must be the other extremity of the focal chord PH.

84. If $A'K$ meet the directrix in F, then,
since $$SA' = 2A'X,$$
$FA'S$ is an isosceles triangle and FS is parallel to KD.

Also $\quad A'F : FP :: A'X : XN :: A'S : SP$
or FS bisects the angle $A'SP$;

therefore if SP and DK meet in Q, QSD is an isosceles triangle.

Therefore Q lies on the circle of which $A'D$ is diameter.

85. This problem is a particular case of Ex. 61.

86. $\quad\quad PL . PL' = PL^2 = CD^2 = PG . Pg$;
therefore G, g, L, L' lie on a circle of which Gg is diameter.

C is on this circle since GCg is a right angle.

The radius of this circle varies as Gg and therefore as CD and therefore inversely as the perpendicular from C on LL'.

87. If PCP', DCD' be conjugate diameters and Q any point on the curve,
$$QP^2 + QP'^2 = 2CP^2 + 2CQ^2; \quad QD^2 + QD'^2 = 2CD^2 + 2CQ^2.$$
Therefore
$$QP^2 + QP'^2 - QD^2 - QD'^2 = 2CP^2 - 2CD^2 = 2AC^2 - 2BC^2.$$

88. If $S'L'$, $S'M'$ be drawn parallel to the asymptotes LS', MS' bisect the angles $PS'L'$, $PS'M'$.
Hence $LS'M = $ half the angle between the asymptotes.

89. If PT meets the tangent at A in V, VS bisects the angle ASP;

therefore $\quad SP : ST :: PV : VT :: AN : AT.$

90. If P is a point of intersection, let the tangent and normal of the ellipse at P meet the transverse axis in T and G, and the conjugate axis in t and g.

Then, PT being the normal of the hyperbola, the semi-axes of which are $A'C$ and $B'C$,

$$CT : CN :: SC^2 : A'C^2, \qquad \text{Art. 111,}$$

$$\therefore AC^2 : CN^2 :: SC : A'C;$$

$$\therefore CN.SC = AC.A'C.$$

Again, Pg being the normal of the ellipse,

$$Cg : PN :: SC^2 : BC^2, \qquad \text{Art. 72,}$$

and $$Cg.PN = B'C^2,$$

$$\therefore B'C^2 : PN^2 :: SC^2 : BC^2$$

and $$'PN.SC = BC.B'C.$$

Hence, if PN meet the asymptote in Q,

$$QN : CN :: B'C : A'C,$$

and it is easily deduced that

$$QN : PN :: AC : BC.$$

91. Let $ABCD$ be the quadrilateral, A, B, and C being fixed points.

Then $$AB + CD = BC + AD,$$

or $$CD - DA = CB - AB.$$

Hence D lies on an hyperbola of which A and C are foci.

92. Since Q, S, C, t lie on a circle, the angle

$$tQC = tSS' = SPt,$$

hence CQ is parallel to SY and CY, SQ are equally inclined to SY;

therefore $$SQ = CY = CA.$$

93. Draw SY perpendicular to the tangent and produce to Z making $SY = YZ$.

Then if P be the point of contact $HZ = HP - SP = AA'$.

Hence the locus of H is a circle of which Z is centre.

94. RS and VS' bisect the angles PSQ and $PS'Q$; let QS, $S'P$ meet in Z.

Then $RSP + VS'Q = $ half $PSQ + $ half $PS'Q = $ half QSP
$+ $ half $SZS' - $ half $SQS' = QSP + $ half $SPS'' - $ half SQS'
$$= QTP + TQS - SQT = PTQ.$$

95. CgP is an isosceles triangle, and the angle
$$CGt = CPT = TCP;$$
therefore $PG = Ct$ and $CD^2 = PG . Pg = Cg . Ct = CS^2.$

96. Since the asymptote CD bisects BA, CD is parallel to the axis of the parabola and BA is parallel to the other asymptote.

If $QPVP'Q'$ parallel to BA meet CD in V,
$$QV = VQ' \text{ and } PV = VP';$$
therefore $\qquad QP = Q'P'.$

97. Let EL be the ordinate of E, and draw EF perpendicular to PN.

Then, CD being conjugate to CP, the triangles CDM and PFE are similar and equal.
$$\therefore CL = CN + EF = CN + DM,$$
$$\therefore CN : CL :: AC : AC + BC,$$
and similarly $\quad EL : PN :: BC : BC - AC;$
$$\therefore EL^2 : (BC - AC)^2 :: PN^2 : BC^2$$
$$:: CN^2 - AC^2 : AC^2$$
$$:: CL^2 - (AC + BC)^2 : (AC + BC)^2.$$

98. If PM be drawn from the centre perpendicular to BC
$AP : PM :: PC : PM$, a constant ratio;
therefore P lies on an hyperbola of which A is focus and BC directrix.

If S be the other focus and SP meet the circle in Q
$$SQ = SP - PA = \text{constant},$$
or, the envelope is a circle of which S is centre.

99. The conics will be confocal having their foci H and H' on PG,
such that $\qquad PH^2 = PT . Pt = CD^2.$
For their locus see Ex. 9 on the ellipse.

100. If SY, SZ, $S'Y'$, $S'Z'$ be perpendiculars on tangents at right angles

$$CT^2 - CA^2 = TY . TY' = SZ . SZ' = CB'^2.$$

If SYZ, $S'Y'Z'$ are perpendicular to parallel tangents and CWW' be the perpendicular through the centre

$$2CW = SY + S'Y' ; \quad 2CW' = SZ - S'Z',$$

and $$SY - SZ = S'Y' + S'Z' ;$$

$$\therefore 4CW^2 + 4CB^2 = 4CW^2 - 4CB^2.$$

Hence $$CW^2 - CW'^2 = CB^2 + CB'^2.$$

101. The chord QR is inclined to the axis at the same angle as the tangent at P and is therefore always parallel to a fixed line.

102. TP and the asymptote subtend equal angles at S';

$$\therefore PS'T = S'TC = STP.$$

103. $$SF^2 = FX^2 + SX^2 = CF^2 - CX^2 + SX^2$$
$$= CF^2 + CS^2 - 2CS . CX$$
$$= CF^2 + CS^2 - 2CA^2 = CF^2 - CA^2 + CB^2$$
$$= \text{square of tangent from } F$$
$$= FA . FB.$$

104. If S be the focus of the ellipse and S' of the hyperbola,

$$CS : CA :: CA : CS';$$

$\therefore S$ and S' coincide with the feet of the directrices.

The relation, $CN . CT = CA^2$, proves that S and S' are the feet of the ordinates, and the relation, $Ct . PN = BC^2$, proves that t and t' are on the auxiliary circle.

Also $$Ct' : CS = AC : CS = CS' : Ct;$$

$\therefore$ the tangent intersects at right angles.

105. For $PG . Pg = CD^2$, Art. 123, $= PL^2 - PL . PL'$, and the diameter of the circle is Gg, which varies as CD, Art. 123, and therefore inversely as CY.

CHAPTER V.

THE RECTANGULAR HYPERBOLA.

1. THE angle $PCL = CLP =$ complement of LCY.

2. $QV^2 = VP \cdot Vp$, hence VQ touches at Q the circle QPp.

3. $$LP = PM = CD = PG = Pg.$$
Hence LGM is a right angle.

4. If LM be the straight line, and C be the corner of the square, $CL \cdot CM$ is constant, hence LM touches an hyperbola of which CL, CM are asymptotes.

5. Let AP, $A'P'$ meet in Q, and draw the ordinate QM.

Then $$QM : MA :: PN : NA,$$
and $$QM : MA' :: P'N : NA'.$$
Hence $$QM^2 : AM \cdot MA' :: PN^2 : AN \cdot NA' \, ;$$
therefore $$QM^2 = AM \cdot MA'.$$
Hence Q lies on a rectangular hyperbola having AA' for transverse axis.

6. If P, P' be joined to Q meeting an asymptote in R and R',
the angle
$$QRL = CLP - QPL = LCP - PP'Q = CR'P' = QR'L.$$

7. Produce LP to M making $PM=PL$, then if MC be drawn perpendicular to the given asymptote CL, C is the centre. In CS the bisector of the angle LCM take S such that CS is a mean proportional between CL and CM.

Then S is focus, and X the middle point of CS is the foot of the directrix.

8. The angle
$$DCL = PCL,$$
and
$$D'CL = P'CL \, ;$$
hence
$$DCD' = PCP'.$$

9. A diameter is a mean proportional between the parallel focal chord and AA', therefore focal chords parallel to conjugate diameters are equal.

10. As in the preceding, focal chords at right angles are equal, since diameters at right angles are equal.

11. If CD, Cd be conjugate to CP in the ellipse and hyperbola,
$$CD^2 = SP \cdot PS' = Cd^2 = CP^2.$$

12.
$$PN = CM \, ; \quad DM = CN \, ;$$
and
$$CD = CP.$$

13.
$$SP \cdot PS' = CD^2 = CP^2.$$

14. $QV^2 = CV^2 - CP^2 = CV^2 - CT \cdot CV = CV \cdot VT.$
Hence QV touches the circle CTQ.

15. If D be the intersection of tangents at A and B,
$$CD^2 = AC^2 + BC^2 = SC^2.$$

Hence D lies on the circle of which SS' is diameter.

16. If LPM, $G'PG$ be the tangent and normal to one,
$$LP = PM = CD = CP = PG = PG' \, ;$$
therefore GG' is the tangent to the second hyperbola.

17. The angle $CRT = CQT + RTQ = 2CLQ + LTL'$
$$= CLM + CL'T = CLR' + L'CQ' = TR'Q'.$$
Hence C, T, R and R' lie on a circle.

18. $QR^3 = CN^2 = CA^2 + RN^2 = CA^2 + CQ^2 = AQ^2.$

19. Let AQ, AR be the fixed straight lines, and P the middle point of QOR.

Through C the middle point of AO draw CH, CK parallel to AQ, AR,

and through P draw PHM, PKN parallel to AR, AQ.

Then the complements AC, HK about the diagonal MN are equal.

Therefore $PH . PK$ is constant, and P lies on a rectangular hyperbola, having CH and CK for asymptotes.

20. Draw QB perpendicular to AB, and make AB equal to CD, then A is a fixed point.

Then $AD : AB :: BC : CD :: QB : DP,$

or $PD . DA = AB . BQ.$

Hence P lies on a rectangular hyperbola of which AB is one asymptote.

21. If D, E, F, O be the centres of the escribed and inscribed circles,

$$OC . CF = DC . CE,$$

since the triangles OCE, DCF are similar.

Hence the hyperbola is rectangular since diameters at right angles are equal.

22. If the diameters of the parallelogram $LML'M'$ meet in C, the angle $SLS' = SL'S'$.

Hence S and S' lie on a rectangular hyperbola circumscribing $LML'M'.$ (Art. 137.)

23. If PSq, SQq be the chords and D be a point on the directrix, such that DS bisects the angle QSp, then D will lie in pq.

But DS is perpendicular to the asymptote since Pp, Qq are equally inclined to it, therefore D lies on the asymptote.

Similarly Pq, Qp meet in D', the foot of the perpendicular from S on the other asymptote.

24. If V be the middle point of PQ and POP' be a diameter,

the angle $VNP = NPV = QP'P = POV.$

Hence O, P, V, N lie on a circle.

If OQ meet the given tangent in T, produce OQ to V, making OV a third proportional to OT and OQ; with centre V and radius, a mean proportional between VO and VT, describe a circle meeting the given tangent in P its point of contact. In the tangent measure off

$$PL = PM = OP,$$

then OL and OM are the asymptotes.

25. Draw PD perpendicular to the base QR,

then, since $PD^2 = DQ \cdot DR,$

DP is the tangent at P.

26. Let a circle on DE meet the hyperbola in P and Q, draw the diameters PCP', QCQ'.

Then, since the angle DPE is either equal or supplementary to $DP'E$ and DQE to $DQ'E$, the similar circle on the other side of DE, will meet the curve in P' and Q'.

27. Let OAD, OBC be the fixed straight lines, PM, PN, PL perpendiculars from the centre of the circle on BC, AD and the bisector of the angle AOB.

Let PM, PN meet OL in m, n.

Draw Ll, Ll', Pr, Pr' perpendicular to BC, AD, Ll, Ll' respectively.

Then, $BM^2 + MP^2 = AN^2 + NP^2,$

or $PN^2 - PM^2 = BM^2 - AN^2$

which is constant,

$PN^2 = (Ll' + Lr')^2 = (Ll - Lr)^2 + 4Ll' \cdot Lr' = PM^2 + 4Ll \cdot Lr.$

Hence the rectangle $Ll \cdot Lr$ is constant;

but $Ll : OL$ in a constant ratio and $Lr : PL$ is constant.

Therefore $PL \cdot LO$ is constant, or P lies on a rectangular hyperbola having OL for an asymptote.

28.   If P be the point of contact
$$CL = 2PN, \ CL' = 2CN \ ;$$
therefore $\qquad CL . CL' = 2Ca^2 = 4PN . CN.$

Hence $\qquad CN . CL' \ : \ Ca^2 \ :: \ Ca^2 \ : \ PN . CL,$

or $\qquad\qquad AC^2 \ : \ Ca^2 \ :: \ Ca^2 \ : \ CB^2.$

29.   Draw CV conjugate to PQ.

Then, the angle
$$TPQ = PP'Q = PCV = CPQ' \ ;$$
therefore the angles $CPQ, \ TPQ'$ are equal.

30.   Let $DB, \ DC$ meet the asymptotes in b and c: draw $AH, \ AK$ parallel to the asymptotes. Then OBb, OAK are similar triangles, also $OCc, \ OAK$.

Hence
$$OB : OA :: Ob : OK :: OH : Oc :: AK : Oc :: OA : OC \ ;$$
therefore A lies on the circle of which BC is diameter as does D.

31.   Let the tangents at P and Q meet the asymptote in L and M.

The angle
$$PCQ = PCL - QCM = PLC - CMT = LTM$$
$$= \text{supplement of } PTQ.$$

32.   Each hyperbola passes through the orthocentre of the triangle ABC.

Hence D is that orthocentre.

Now the line joining the middle point of AB to the middle point of CD is a diameter of the nine point circle.

And $AB^2 + CD^2 = $ square on diameter of circumscribed circle.

Hence the circles intersect at right angles.

33. $PN^2 = CN^2 - CA^2 = CN^2 - CN . CT = CN . NT.$

Hence the triangle CPN is similar to PTN, and therefore to tTC.

34. This problem is a particular case of **Ex.** 61 on the hyperbola.

35. *CM*, *CN* which bisect *PP'*, *PQ'* are conjugate being equally inclined to the asymptotes;

therefore *P'Q'* is a diameter.

36. If *AB* be a diameter of the hyperbola, *CD* subtends, at *A* and *B*, angles which are both equal and supplementary, and are therefore right angles.

37. If *AQ'*, *BQ* meet in *R*, the angle $QAQ' = $ supplement of $QBQ' = RBP$;

therefore *R* lies on the circle.

38. Let *CV*, *CV'* bisect *PQ*, *P'Q*.

The angle $PRQ = PQL = VCQ = CQV'$,

so $PR'Q = CQV.$

Draw *VM* perpendicular to *CQ*,

then $PQ : QR :: VM : CV,$

and $P'Q : R'Q :: VM : QV.$

Now $P'Q = 2CV,$

and $PQ = 2QV;$

therefore $QR = QR'.$

39. Let *PN* meet *CF* in *K*,

then *CF* varies as *CG*, and *PF* varies as *PK*, or *Ct*, or *CT*.

Hence *PF . FC* is proportional to *CG*, *CT* or CS^2.

Hence *P* lies on a rectangular hyperbola having *CF* for asymptote.

40. If the tangent at *Q* meet in *V* the line joining the fixed points *A* and *B*,

$$VQ^2 = VA . VB.$$

41. If the chord *QR* meet the tangent at *P* in *E*,

$$RPL = QPE = PRQ.$$

42. If D be the point,
$$CD \cdot CT = 2AC^2 \text{ and } CD = AC.$$

43. $CP = CD$ and are equally inclined to the asymptote.

44. Let BAD be the given difference, and draw CL parallel to AD, meeting BA in L. CAL, CBL are similar triangles;
$$CL^2 = AL \cdot BL.$$

45. If tQT, $t'QT'$ be the tangents, and SM perpendicular to the axis,
$$SQM = SQT - MQT = S'Qt' - CtT$$
$$= QS'M + QT'C - CtT = QS'M;$$
$$\therefore QM^2 = SM \cdot S'M.$$

46. If V is the middle point of OP,
$$OTV = VOT, \therefore OTP \text{ is a right angle,}$$
$$= OCV, \text{ Art. 136};$$
$$\therefore O, V, C, T \text{ are concyclic, and}$$
$$\therefore VCT = VOT = OCV.$$

Similarly, if U is the middle point of OQ,
$$OTU = UTQ.$$

CHAPTER VI.

THE CYLINDER AND THE CONE.

1. TAKE two points E and A on the generating line, and draw EX at right angles to the axis, making EX equal to EA.

Then the plane containing AX, and perpendicular to the plane EAX will cut the cylinder in an ellipse of the required eccentricity.

2. Take two points E and A on the generating line and with centre A, and radius twice EA, describe a circle meeting EF in X.

Then the plane through AX perpendicular to the plane EXA will intersect the cone in an ellipse of the required eccentricity.

3. Take two points EA on the generating line, the least angle of the cone will be, when EA is double the perpendicular from A on EF, that is when the semi-vertical angle is equal to the angle of an equilateral triangle.

4. A tangent plane to a cone touches it along a generating line OF, hence OF is parallel to all sections parallel to the tangent plane which are therefore parabolas.

If C be the centre of the sphere FES, the ratios $CS : CA$ and $CA : CO$ are constant, and the angle OCS is constant, therefore COS is constant, and S lies on a cone of which O is vertex and OC axis.

5. Through the flames of the candles which are treated as points, draw planes intersecting in the ceiling, in a straight line, since these fixed planes must always be tangent planes to the ball, the locus of the centre of the ball is a horizontal straight line.

6. The triangles AEX, $A'E'X'$ have all their sides parallel;

therefore

$$SA : AX :: EA : AX :: E'A' : A'X' :: S'A' : A'X'.$$

7. If C be the centre of the sphere FES, the angle OCS is constant.

And the ratios $CE : CA$, $CE : CV$, and $CS : CA$ are all constant ;

therefore $CS : CV$ is constant and the angle CVS is constant, or VS is a fixed straight line.

8. Take two points E and A on the generating line and with centre A and radius AX, such that $EA : AX$ in the ratio of the eccentricity, describe a circle intersecting FE in X, the section of which AX is axis will have the required eccentricity.

9. XS, XS' are tangents to the same sphere FES of which C is centre.

Let SS', EF meet the axis in V', L, and let CX meet SS' in M.

Then $\qquad CL . CV' = CM . CX = CE^2.$

Hence V and V' coincide.

10. Draw CN perpendicular to the axis,

then $\qquad\qquad 2CN = A'D' - AD,$

and $\qquad\qquad 2ON = OD + OD'.$

In CN take a point Q, such that

$$QN : CN :: DO^2 : DA^2;$$

Collins. **Easy Translations from Nepos, Cæsar, Cicero, Livy,**
&c., for Retranslation into Latin. With Notes. 2s.

Compton. **Rudiments of Attic Construction and Idiom.** By
the Rev. W. C. Compton, M.A., Assistant Master at Uppingham School. 3s.

Clapin. **A Latin Primer.** By Rev. A. C. Clapin, M.A. 1s.

Frost. **Eclogæ Latinæ;** or, First Latin Reading Book. With
Notes and Vocabulary by the late Rev. P. Frost, M.A. New Edition.
Fcap. 8vo. 1s. 6d.

———— **Analecta Græca Minora.** With Notes and Dictionary.
New Edition. Fcap. 8vo. 2s.

———— **Materials for Latin Prose Composition.** By the late Rev.
P. Frost, M.A. New Edition. Fcap. 8vo. 2s. **Key** (for Tutors only), 4s.

———— **A Latin Verse Book.** New Edition. Fcap. 8vo. 2s.
Key (for Tutors only), 5s.

———— **Materials for Greek Prose Composition.** New Edition.
Fcap. 8vo. 2s. 6d. **Key** (for Tutors only), 5s.

Harkness. **A Latin Grammar.** By Albert Harkness. Post 8vo.
6s.

Key. **A Latin Grammar.** By T. H. Key, M.A., F.R.S. 6th
Thousand. Post 8vo. 8s.

———— **A Short Latin Grammar for Schools.** 16th Edition.
Post 8vo. 3s. 6d.

Holden. **Foliorum Silvula. Part I.** Passages for Translation
into Latin Elegiac and Heroic Verse. By H. A. Holden, LL.D. 11th Edit.
Post 8vo. 7s. 6d.

———— **Foliorum Silvuia. Part II. Select Passages for** Translation into Latin Lyric and Comic Iambic Verse. 3rd Edition. Post 8vo.
5s.

———— **Foliorum Centuriæ. Select Passages for Translation**
into Latin and Greek Prose. 10th Edition. Post 8vo. 8s.

Jebb, Jackson, and Currey. **Extracts for Translation in Greek,**
Latin, and English. By R. C. Jebb, Litt. D., LL.D., H. Jackson, Litt. D.,
and W. E. Currey, M.A. 4s. 6d.

Mason. **Analytical Latin Exercises.** By C. P. Mason, B.A.
4th Edition. Part I., 1s. 6d. Part II., 2s. 6d.

Nettleship. **Passages for Translation into Latin Prose.** By
Prof. H. Nettleship, M.A. 3s. Key (for Tutors only), 4s. 6d.
'The introduction ought to be studied by every teacher of Latin.'
Guardian.

Paley. **Greek Particles and their Combinations according to**
Attic Usage. A Short Treatise. By F. A. Paley, M.A., LL.D. 2s. 6d.

Penrose. **Latin Elegiac Verse, Easy Exercises in.** By the Rev.
J. Penrose. New Edition. 2s. (Key, 3s. 6d.)

Preston. **Greek Verse Composition.** By G. Preston, M.A.
5th Edition. Crown 8vo. 4s. 6d.

Seager. **Faciliora. An Elementary Latin Book on a new**
principle. By the Rev. J. L. Seager, M.A. 2s. 6d.

Stedman (A. M. M.). **First Latin Lessons.** By A. M. M.
Stedman, M.A., Wadham College, Oxford. Second Edition, enlarged.
Crown 8vo. 2s.

———— **First Latin Reader.** With Notes adapted to the Shorter
Latin Primer and Vocabulary. Crown 8vo. 1s. 6d.

———— **Easy Latin Passages for Unseen Translation.** 2nd and
enlarged Edition. Fcap. 8vo. 1s. 6d.

———— **Easy Latin Exercises** on the Syntax of the Shorter and
Revised Latin Primers. With Vocabulary. 3rd Edition. Crown 8vo.
2s. 6d.

Stedman (A. M. M.) Notanda Quædam. Miscellaneous Latin
Exercises on Common Rules and Idioms. Fcap. 8vo. 1s. 6d.
———— Latin Vocabularies for Repetition : arranged according
to Subjects. 4th Edition. Fcap. 8vo. 1s. 6d.
———— First Greek Lessons. [*In preparation.*
———— Easy Greek Passages for Unseen Translation. Fcap.
8vo. 1s. 6d.
———— Easy Greek Exercises on Elementary Syntax.
[*In preparation.*
———— Greek Vocabularies for Repetition. Fcap. 8vo. 1s. 6d.
———— Greek Testament Selections for the Use of Schools.
2nd Edit. With Introduction, Notes, and Vocabulary. Fcap. 8vo. 2s. 6d.
Thackeray. Anthologia Græca. A Selection of Greek Poetry,
with Notes. By F. St. John Thackeray. 5th Edition. 16mo. 4s. 6d.
———— Anthologia Latina. A Selection of Latin Poetry, from
Nævius to Boëthius, with Notes. By Rev. F. St. J. Thackeray. 5th Edit.
16mo. 4s. 6d. .
Wells. Tales for Latin Prose Composition. With Notes and
Vocabulary. By G. H. Wells, M.A. Fcap. 8vo. 2s.

Donaldson. The Theatre of the Greeks. By J. W. Donaldson,
D.D. 10th Edition. Post 8vo. 5s.
Keightley. The Mythology of Greece and Italy. By Thomas
Keightley. 4th Edition. Revised by L. Schmitz, Ph.D., LL.D. 5s.
Mayor. A Guide to the Choice of Classical Books. By J. B.
Mayor, M.A. 3rd Edition. Crown 8vo. 4s. 6d.
Teuffel. A History of Roman Literature. By Prof. W. S.
Teuffel. 5th Edition, revised by Prof. L. Schwabe, and translated by
Prof. G. C. W. Warr, of King's College. 2 vols. medium 8vo. 15s. each.

CAMBRIDGE MATHEMATICAL SERIES.

Arithmetic for Schools. By C. Pendlebury, M.A. 5th Edition,
stereotyped, with or without answers, 4s. 6d. Or in two parts, 2s. 6d. each.
Part 2 contains the *Commercial Arithmetic*. A Key to Part 2 in preparation.
EXAMPLES (nearly 8000), without answers, in a separate vol. 3s.
In use at St. Paul's, Winchester, Wellington, Marlborough, Charterhouse,
Merchant Taylors', Christ's Hospital, Sherborne, Shrewsbury, &c. &c.
Algebra. Choice and Chance. By W. A. Whitworth, M.A. 4th
Edition. 6s.
Euclid. Newly translated from the Greek Text, with Supple-
mentary Propositions, Chapters on Modern Geometry, and numerous
Exercises. By Horace Deighton, M.A., Head Master of Harrison College,
Barbados. New Edition, Revised, with Symbols and Abbreviations.
Crown 8vo. 4s. 6d.
Book I. ... ... 1s. | Books I. to III. ... 2s. 6d.
Books I. and II. ... 1s. 6d. | Books III. and IV. 1s. 6d.
Euclid. Exercises on Euclid and in Modern Geometry. By
J. McDowell, M.A. 4th Edition. 6s.
Elementary Trigonometry. By J. M. Dyer, M.A., and Rev.
R. H. Whitcombe, M.A., Assistant Masters, Eton College. 4s. 6d.
Trigonometry. Plane. By Rev. T. Vyvyan, M.A. 3rd Edit. 3s. 6d.
Geometrical Conic Sections. By H. G. Willis, M.A. 5s.
Conics. The Elementary Geometry of. 7th Edition, revised and
enlarged. By C. Taylor, D.D. 4s. 6d.
Solid Geometry. By W. S. Aldis, M.A. 4th Edit. revised. 6s.

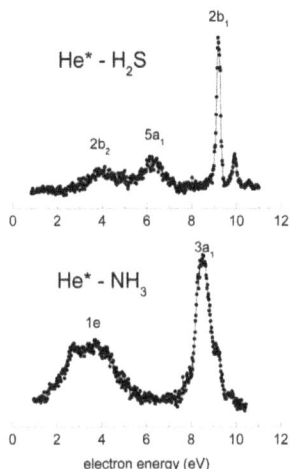

Figure 6. The obtained PIES spectra in Penning ionization studies of hydrogen sulfide (upper panel) and ammonia (lower panel) molecules by using $He^*(^3S_1, {}^1S_0,)$ metastable atoms. In the $2b_1$ band of the He^*-H_2S spectrum, the two peaks recorded at an electron energy value of 9.18 and 9.94 eV, and related to $He^*(^3S_1)$ and $He^*(^1S_0)$ collisional ionization events, respectively, appear well separated (see upper panel). The two $He^*(^3S_1, {}^1S_0,)$ metastable components are still present, even if less separated, in the wide $3a_1$ band of the He^*-NH_3 spectrum (see lower panel).

4. Conclusions

Considering that a basic step in the chemical evolution of planetary atmospheres and interstellar clouds (where 89% of atoms are hydrogen and 9% are helium) is the interaction of atoms and molecules with electromagnetic waves (γ and X rays, UV light) and cosmic rays [43], the formation of excited metastable species like He^* and Ar^* (argon is the third component of Earth and Mars' atmosphere) by collisional excitation with energetic target particles (electrons, protons, or alpha particles, coming from the solar wind) and the possible subsequent Penning ionization reactions, could be of importance in these environments. It has to be stressed that the effect of Penning ionization has not yet been fully considered in the modeling of terrestrial and extraterrestrial objects, even though metastable excited species can be formed in such environments. In this paper recent experimental results concerning production and characterization of simple ionic species of atmospheric interest are presented and discussed. Such results concern the formation of free ions in collisional ionization of H_2O, H_2S, and NH_3 induced by highly excited species, like He^* and Ne^* metastable atoms.

The analysis of the recorded electron spectra for the three presented systems clearly shows: (i) a very strong and anisotropic attractive behavior of Ne^*-NH_3

17

interaction promoting the Penning process; (ii) a still quite strong attractive behavior for Ne^*–H_2O collisions; and (iii) a corresponding attractive interaction in the Ne^*–H_2S collisions, which appear to be about 40% that of water. The analyzed experimental observables, in term of energy shifts, clearly indicate that H_2O and H_2S show similar behavior when these hydrogenated molecules are involved in Penning ionization induced by He^* and Ne^* metastable atoms, and in particular: (i) the interaction potential driving the formation of A $^2A_1$ excited state of H_2O^+ and H_2S^+ product ions exhibits an effective interaction potential well depth in the entrance channel of the ionization reaction, quite deeper (about 50% more in the case of water and about 25% more in the case of hydrogen sulfide) than that involved in the production of the same final ions in their X $^2B_1$ ground state (see the relative ε_{max} values reported in the Table 2); (ii) a quite pronounced anisotropic attraction controls the stereodynamics in the entrance channels of both (Ne^*–H_2O, H_2S) analyzed systems: in the case of Ne^*–H_2O autoionizing collisions a strong anisotropic interaction is operative, while in the Ne^*–H_2S system the attractive interaction appears to be weaker than the system involving water; and (iii) the entity of this smaller attraction, characterizing Ne^*–H_2S collisions, is dependent on the approach orientation between Ne^* atoms and the hydrogenated target molecules: it is about 43% with respect to the Ne^*–H_2O case for those collisions producing H_2X^+ parent ions (where X stands for O or S atom) in their X $^2B_1$ ground electronic state (this is the case for the Ne^*–H_2X collisions towards the orthogonal direction respect to the molecular plane), while it reaches about 60% of the Ne^*–H_2O interaction when the exchange mechanism of Penning ionization can exploit Ne^* collisions along the C_{2v} molecular axis in the opposite side with respect to hydrogen atoms, with the formation of H_2X^+ final ions in their first A $^2A_1$ excited electronic state. In the case of Ne^*–NH_3 collisions, a very strong and anisotropic attractive interaction is observed, characterizing the potential energy surface of the entrance channel with a much deeper well depth in the case of those collisions producing the [Ne ... NH_3]$^+$ intermediate ionic complex, leading product ion in its $X(^2A_2'')$ - $3a_1$ ground electronic state (which is an opposite situation to that observed in H_2O and H_2S systems). In this case the formation of [Ne ... $NH_3(X\ ^2A_2'')$]$^+$ occurs preferentially when the neon metastable atom approaches ammonia molecules towards the direction of the $3a_1$ non-bonding lone pair located on the nitrogen atom, removes one electron from this orbital, and gives rise to Penning ionization. Finally, the analysis of PIES spectra recorded for He^*–H_2S and –NH_3 collisions, confirms the attractive and anisotropic behavior of the potential energy surfaces already discussed for Ne^*–M interactions (with M = H_2O, H_2S, and NH_3) governing the entrance channel for Penning ionization of such hydrogenated molecules. A similar situation for metastable helium and metastable neon atoms interacting with N_2O molecules [19] has been explained by the use of a semiempirical method that accounts for the metastable atom orbital deformation because of the permanent dipole of the molecule. We have plans to extend such a model also to the systems presented here, and to test the potential energy surfaces so obtained in a comparative way. This work, is in progress in our laboratory, will allow us to achieve a deeper understanding of the stereodynamics of the collisional autoionization processes involving He* and Ne* with the hydrogenated molecules here discussed, and of the propensity to form ions.

Acknowledgments: Stefano Falcinelli is very grateful to Richard N. Zare of the Chemistry Department of Stanford University, CA, USA, for his constant encouragement and the useful suggestions during the period between June and September 2014 when Stefano Falcinelli was working in his laboratory as a visiting scholar and was conceiving the paper here presented.

Author Contributions: Stefano Falcinelli and Franco Vecchiocattivi conceived and designed the experiment; Stefano Falcinelli performed the experiment; Fernando Pirani developed the semiempirical method for potential energy surfaces calculations; all the authors analyzed the data, participated to the discussion on the obtained results, and contributed to write the paper.

Conflicts of Interest: The authors declare no conflict of interest.

References

1. Davies, K.; Rush, C.M. Reflection of high-frequency radio waves in inhomogeneous ionospheric layers. *Radio Sci.* **1985**, *20*, 303–309.
2. Hruska, F. Electromagnetic Interference and Environment. *Int. J. Circuits Syst. Signal Process.* **2014**, *8*, 22–29.
3. Qiu, H.; Chen, L.T.; Qiu, G.P.; Zhou, C. 3D visualization of radar coverage considering electromagnetic interference. *WSEAS Trans. Signal Process.* **2014**, *10*, 460–470.
4. Egano, F. *Radio Propagation Observatory*. Available online: http://www.qsl.net/ik3xtv (accessed on 14 August 2014).
5. Hotop, H. Detection of metastable atoms and molecules. In *Atomic, Molecular, and Optical Physics: Atoms and Molecules*; Dunning, F.B., Hulet, R.G., Eds.; Academic Press, Inc.: San Diego, CA, USA, 1996; pp. 191–216.
6. Brunetti, B.G.; Falcinelli, S.; Giaquinto, E.; Sassara, A.; Prieto-Manzanares, M.; Vecchiocattivi, F. Metastable-Idrogen-Atom scattering by crossed beams: Total cross sections for H*(2s)-Ar, Xe, and CCl$_4$ at thermal energies. *Phys. Rev. A* **1995**, *52*, 855–858.
7. Falcinelli, S. Penning ionization of simple molecules and their possible role in planetary atmospheres. In *Recent Advances in Energy, Environment and Financial Planning—Mathematics and Computers in Science and Engineering Series*; Batzias, F., Mastorakis, N.E., Guarnaccia, C., Eds.; WSEAS Press: Athens, Greece, 2014; pp. 84–92.
8. Rodger, C.J.; Nunn, D. VLF scattering from red sprites: Application of numerical modeling. *Radio Sci.* **1999**, *34*, 923–932.
9. Larsson, M.; Geppert, W.D.; Nyman, G. Ion chemistry in space. *Rep. Prog. Phys.* **2012**, *75*, 066901.
10. Stauber, P.; Doty, S.D.; van Dishoeck, E.F.; Benz, A.O. X-ray chemistry in the envelopes around young stellar objects. *Astron. Astrophys.* **2005**, *440*, 949–966.

11. Indriolo, N.; Hobbs, L.M.; Hinkle, K.H.; McCall, B.J. Interstellar metastable helium absorption as a probe of the cosmic-ray ionization rate. *Astrophys. J.* **2009**, *703*, 2131–2137.

12. Waldrop, L.S.; Kerr, R.B.; González, S.A.; Sulzer, M.P.; Noto, J.; Kamalabadi, F. Generation of metastable helium and 1083 nm emission in the upper thermosphere. *J. Geophys. Res.* **2005**, *110*, A08304.

13. Bishop, J.; Link, R. He (2^3S) densities in the upper thermosphere: Updates in modeling capabilities and comparison with midlatitude observations. *J. Geophys. Res.* **2005**, *104*, 17157–17172.

14. Siska, P.E. Molecular-beam studies of Penning ionization. *Rev. Mod. Phys.* **1993**, *65*, 337–412.

15. Brunetti, B.G.; Vecchiocattivi, F. Autoionization dynamics of collisional complexes. In *Ion Clusters*; Ng, C., Baer, T., Powis, I., Eds.; Springer: New York, NY, USA, 1993; pp. 359–445.

16. Penning, F.M. Über Ionisation durch metastabile Atome. *Naturwissenschaften. J.* **1927**, *15*.

17. Legon, A.C. The halogen bond: An interim perspective. *Phys. Chem. Chem. Phys.* **2010**, *12*, 7736–7747.

18. Biondini, F.; Brunetti, B.G.; Candori, P.; de Angelis, F.; Falcinelli, S.; Tarantelli, F.; Teixidor, M.M.; Pirani, F.; Vecchiocattivi, F. Penning ionization of N_2O molecules by $He^*(2^{3,1}S)$ and $Ne^*(^3P_{2,0})$ metastable atoms: A crossed beam study. *J. Chem. Phys.* **2005**, *122*, 164307:1–164307:10.

19. Biondini, F.; Brunetti, B.G.; Candori, P.; de Angelis, F.; Falcinelli, S.; Tarantelli, F.; Pirani, F.; Vecchiocattivi, F. Penning ionization of N_2O molecules by $He^*(2^{3,1}S)$ and $Ne^*(^3P_{2,0})$ metastable atoms: Theoretical considerations about the intermolecular interactions. *J. Chem. Phys.* **2005**, *122*, 164308:1–164308:11.

20. McClintock, W.E.; Lankton, M.R. The mercury atmospheric and surface composition spectrometer for the MESSENGER mission. *J. Phys. Chem. Ref. Data* **1989**, *21*, 1125–1499.

21. Atkinson, R.; Baulch, D.L.; Cox, R.A.; Hampson, R.F.; Kerr, J.A., Jr.; Troe, J. Evaluated kinetic and photochemical data for atmospheric chemistry: Supplement IV. IUPAC Subcommittee on kinetic data evaluation for atmospheric chemistry. *Space Sci. Rev.* **2007**, *131*, 481–521.

22. Brunetti, B.G.; Candori, P.; Cappelletti, D.; Falcinelli, S.; Pirani, F.; Stranges, D.; Vecchiocattivi, F. Penning ionization electron spectroscopy of water molecules by metastable neon atoms. *Chem. Phys. Lett.* **2012**, *539–540*, 19–23.

23. Balucani, N.; Bartocci, A.; Brunetti, B.G.; Candori, P.; Falcinelli, S.; Pirani, F.; Palazzetti, F.; Vecchiocattivi, F. Collisional autoionization dynamics of $Ne^*(^3P_{2,0})$-H_2O. *Chem. Phys. Lett.* **2012**, *546*, 34–39.

24. Brunetti, B.G.; Candori, P.; Falcinelli, S.; Pirani, F.; Vecchiocattivi, F. The stereodynamics of the Penning ionization of water by metastable neon atoms. *J. Chem. Phys.* **2013**, *139*, 164305:1–164305:8.

25. Čermák, V.; Yencha, A.J. Penning ionization electron spectroscopy of H_2O, D_2O, H_2S and SO_2. *J. Electron Spectr. Rel. Phenom.* **1977**, *11*, 67–73.

26. Brion, C.E.; Yee, D.S.C. Electron spectroscopy using excited atoms and photons IX. Penning ionization of CO_2, CS_2, COS, N_2O, H_2S, SO_2, and NO_2. *J. Electron Spectr. Rel. Phenom.* **1977**, *12*, 77–93.

27. Ohno, K.; Mutoh, H.; Harada, Y. Study of electron distributions of molecular orbitals by Penning ionization electron spectroscopy. *J. Am. Chem. Soc.* **1983**, *105*, 4555–4561.

28. Haug, B.; Morgner, H.; Staemmler, V. Experimental and theoretical study of Penning ionisation of H_2O by metastable Helium He (2^3S). *J. Phys. B At. Mol. Phys.* **1985**, *18*, 259–274.

29. Bentley, J. Potential energy surfaces for excited neon atoms interacting with water molecules. *J. Chem. Phys.* **1980**, *73*, 1805–1813.

30. Ishida, T. A quasi-classical trajectory calculation for the Penning ionization H_2O-He*(2^3S). *J. Chem. Phys.* **1996**, *105*, 1392–1401.

31. Falcinelli, S.; Candori, P.; Bettoni, M.; Pirani, F.; Vecchiocattivi, F. Penning ionization electron spectroscopy of hydrogen sulfide by metastable Helium and Neon atoms. *J. Phys. Chem. A* **2014**, *118*, 6501–6506.

32. Arfa, M.B.; Lescop, B.; Cherid, M.; Brunetti, B.; Candori, P.; Malfatti, D.; Falcinelli, S.; Vecchiocattivi, F. Ionization of ammonia molecules by collision with metastable neon atoms. *Chem. Phys. Lett.* **1999**, *308*, 71–77.

33. Brunetti, B.; Candori, P.; Falcinelli, S.; Vecchiocattivi, F.; Sassara, A.; Chergui, M. Dynamics of the Penning ionization of fullerene molecules by metastable Neon atoms. *J. Phys. Chem. A* **2000**, *14*, 5942–5945.

34. Brunetti, B.; Candori, P.; Falcinelli, S.; Kasai, T.; Ohoyama, H.; Vecchiocattivi, F. Velocity dependence of the ionization cross section of methyl chloride molecules ionized by metastable argon atoms. *Phys. Chem. Chem. Phys.* **2001**, *3*, 807–810.

35. Brunetti, B.; Candori, P.; Falcinelli, S.; Lescop, B.; Liuti, G.; Pirani, F.; Vecchiocattivi, F. Energy dependence of the Penning ionization electron spectrum of Ne*($^3P_{2,0}$) + Kr. *Eur. Phys. J. D* **2006**, *38*, 21–27.

36. Kraft, T.; Bregel, T.; Ganz, J.; Harth, K.; Ruf, M.-W.; Hotop, H. Accurate comparison of HeI, NeI photoionization and He($2^{3,1}S$), Ne($3s^3P_2$, $^3P_0$) Penning ionization of argon atoms and dimers. *Z. Phys. D* **1988**, *10*, 473–481.

37. Kimura, K.; Katsumata, S.; Achiba, Y.; Yamazaky, T.; Iwata, S. *Handbook of HeI Photoelectron Spectra of Fundamental Organic Molecules*; Japan Scientific Societies Press: Tokyo, Japan, 1981.

38. West, W.P.; Cook, T.B.; Dunning, F.B.; Rundel, R.D.; Stebbings, R.F. Chemiionization involving rare gas metastable atoms. *J. Chem. Phys.* **1975**, *63*, 1237–1242.

39. Arfa, M.B.; le Coz, G.; Sinou, G.; le Nadan, A.; Tuffin, F.; Tannous, C. Experimental study of the Penning ionization of the H_2O molecule by He* (2^3S, 2^1S) metastable atoms. *J. Phys. B At. Mol. Opt. Phys.* **1994**, *27*, 2541–2550.

40. Miller, W.H. Theory of Penning ionization. I. Atoms. *J. Chem. Phys.* **1970**, *52*, 3563–3571.

41. Hotop, H.; Niehaus, A. Reactions of excited atoms molecules with atoms and molecules. II. Energy analysis of penning electrons. *Z. Phys. D* **1969**, *228*, 68–88.

42. Falcinelli, S.; Rosi, M.; Candori, P.; Vecchiocattivi, F.; Bartocci, A.; Lombardi, A.; Lago, N.F.; Pirani, F. Modeling the intermolecular interactions and characterization of the dynamics of collisional autoionization processes. In *ICCSA 2013, Part I, LNCS 7971 Computational Science and its Applications*; Murgante, B., Misra, S., Carlini, M., Torre, C.M., Nguyen, H.Q., Taniar, D., Apduhan, B.O., Gervasi, O., Eds.; Springer-Verlag: Berlin Heidelberg, Germany, 2013; pp. 47–56.

43. Falcinelli, S.; Rosi, M.; Candori, P.; Vecchiocattivi, F.; Farrar, J.M.; Pirani, F.; Balucani, N.; Alagia, M.; Richter, R.; Stranges, S. Kinetic energy release in molecular dications fragmentation after VUV and EUV ionization and escape from planetary atmospheres. *Planet. Space Sci.* **2014**, *99*, 149–157.

Chemical Composition of Water Soluble Inorganic Species in Precipitation at Shihwa Basin, Korea

Seung-Myung Park, Beom-Keun Seo, Gangwoong Lee, Sung-Hyun Kahng and Yu Woon Jang

Abstract: Weekly rain samples were collected in coastal areas of the Shihwa Basin (Korea) from June 2000 to November 2007. The study region includes industrial, rural, and agricultural areas. Wet precipitation was analyzed for conductivity, pH, Cl^-, NO_3^-, SO_4^{2-}, Na^+, K^+, Mg^{2+}, NH_4^+, and Ca^{2+}. The major components of precipitation in the Shihwa Basin were NH_4^+, volume-weighted mean (VWM) of 44.6 $\mu eq \cdot L^{-1}$, representing 43% of all cations, and SO_4^{2-}, with the highest concentration among the anions (55%) at all stations. The pH ranged from 3.4 to 7.7 with a VMM of 4.84. H^+ was weakly but positively correlated with SO_4^{2-} ($r = 0.39$, $p < 0.001$) and NO_3^- ($r = 0.38$, $p < 0.001$). About 66% of the acidity was neutralized by NH_4^+ and Ca^{2+}. The Cl^-/Na^+ ratio of the precipitation was 37% higher than seawater Cl^-/Na^+. The high SO_4^{2-}/NO_3^- ratio of 2.3 is attributed to the influence of the surrounding industrial sources. Results from positive matrix factorization showed that the precipitation chemistry in Shihwa Basin was influenced by secondary nitrate and sulfate (41% $\pm$ 1.1%), followed by sea salt and Asian dust, contributing 23% $\pm$ 3.9% and 17% $\pm$ 0.2%, respectively. In this study, the annual trends of SO_4^{2-} and NO_3^- ($p < 0.05$) increased, different from the trends in some locations, due to the influence of the expanding power generating facilities located in the upwind area. The increasing trends of SO_4^{2-} and NO_3^- in the study region have important implications for reducing air pollution in accordance with national energy policy.

Reprinted from *Atmosphere*. Cite as: Park, S.-M.; Seo, B.-K.; Lee, G.; Kahng, S.-H.; Jang, Y.W. Chemical Composition of Water Soluble Inorganic Species in Precipitation at Shihwa Basin, Korea. *Atmosphere* **2015**, *6*, 1069–1101.

1. Introduction

Rain is the most effective process transporting soluble gases and particles from the atmosphere to the ground [1,2]. Precipitation chemistry plays an important role in understanding the air quality in a study area, because the concentrations and distribution of chemical components in rain depend on a variety of emission sources including sea spray, soil particles, and industrial pollutants [3–7].

In coastal areas, the Na^+/Cl^- ratio, SO_4^{2-}, Mg^{2+}, and Ca^{2+} are useful for evaluating both anthropogenic and natural influences [8]. Soil particles in northern China, Africa, and the Middle East play an important role in neutralizing the acidic components of rain [9]. The concentrations of anthropogenic components are related to energy consumption and the use of fertilizers [10,11].

Some anthropogenic acidic ions may greatly contribute to acidification and eutrophication of aquatic ecosystems [12]. Numerous long-term observations of precipitation have been carried out in Europe, North America, and Asia [13–17]. These studies have reported that SO_4^{2-} concentrations have decreased commensurate with reductions in SO_2 emissions. NO_3^- concentrations have also shown decreasing trends in regions where NO_2 emissions have decreased [18–20].

In many regions of Asia, rapid economic growth has led to an increase in pollutant emissions and seriously compromised air quality [21]. In addition, air quality has also been affected by wind-blown soil particles originating from the deserts of Mongolia and northern China [22–24]. Several studies conducted in this region have found that the rain is acidic, with H^+ concentration ranges similar to that reported for Europe [25–28].

In Korea, previous studies have shown that most precipitation events are acidic and even in background areas of Korea precipitation pH is <4.9 [29–31]. Some studies have suggested that air masses from the Asian continent increase the concentration of anthropogenic components of rainwater in Korea [31,32]. Choi *et al.* (2015) [33] studied precipitation chemistry in the area near Shihwa Lake (located in the mid-western area of the Korean peninsula), using both urban and background stations to compare precipitation chemistry. To date, precipitation chemistry in the industrial area has not been studied yet.

This study aimed to provide a detailed evaluation of the chemical composition of precipitation in the Shihwa Basin and to evaluate the relative contributions of various sources. Because the Shihwa Basin includes areas with industrial, agricultural, and coastal land use, the study are has experienced serious problems with air pollution [34–37].

2. Material and Methods

2.1. Sampling Sites

The Shihwa Basin is located 60 km southwest of Seoul, the capital of South Korea. The basin of Shihwa Lake has an area of about 475 km^2 and a population of more than 0.8 million people. The Shihwa and Banwol industrial complexes are located in Shihwa Basin. Three sampling stations were selected for precipitation monitoring, covering the southern (Hwasung), northern (Banwol), and western (Daeboo) areas of the Shihwa Basin (Table 1). Hwasung station is located inland about 20 km east of Shihwa Lake, in a residential and rural area. Banwol station is located 5 km east of Shihwa Lake and is important because the Banwol industrial complex is located there. This complex is a site for intensive chemical, leather, and metal processing industries, which have led to air pollution episodes, including pollutants such as ammonia, amines, hydrogen sulfide (H_2S) and mercaptans. Daeboo station is located on an island, about 2 km west of Shihwa Lake. The locations of the Shihwa Basin and the three sampling stations are shown in Figure 1.

Table 1. Characteristics of sampling stations in the Shihwa Basin.

Shihwa Basin	Latitude	Longitude	Elevation (m)	Land Type
Hwasung	37.236	126.918	31	Rural and residential
Banwol	37.315	126.750	56	Industrial
Daeboo	37.268	126.568	25	Agricultural

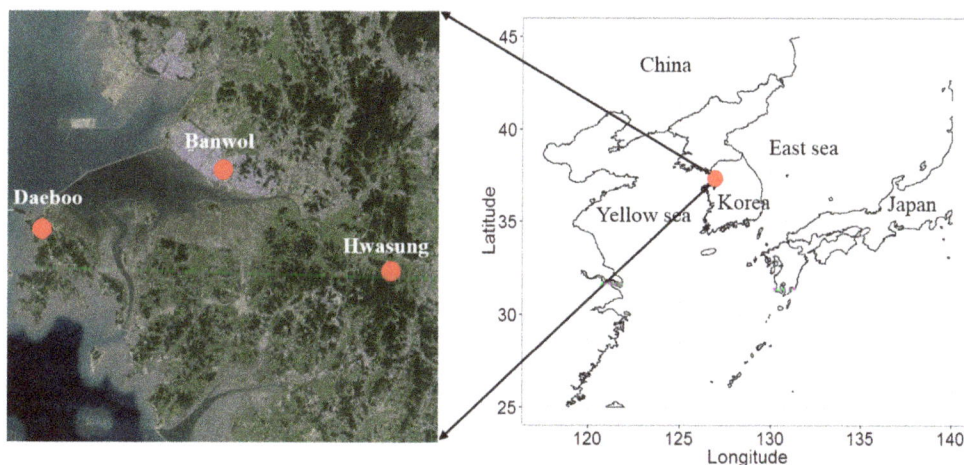

Figure 1. Locations of the study area and sampling stations.

The Shihwa Basin is strongly influenced by air masses driven by westerly winds. According to data from the Incheon meteorological administration (about 20 km north of Shihwa Lake), the average annual rainfall from June 2000 to November 2007 was 1187 mm·year^{-1}. The average temperature and wind speed were 15 °C and 2 m·s^{-1}, respectively. In the summer (June to August), the mean temperature was 24.5 °C and about 61% of the annual precipitation occurred during this season. In winter, the average temperature was 3.5 °C with only 5% of the annual precipitation occurring during winter. The spring season is characterized by the influence of Asian dust, with an average wind speed of 3.67 m·s^{-1}, 58% higher than that in winter. According to AWS data from Korea Meteorological Administration near the three sites, the prevailing wind directions were north and south at Hwasung station, east and west at Banwol station and northwest at Daeboo station. At Hwasung, the wind during spring and summer was north, but south during fall and winter. At Banwol and Daeboo stations, the prevailing wind was constant throughout the year.

2.2. Sample Collection

Precipitation samples were collected weekly using wet and dry collectors from June 2000 to November 2007. Butler and Likens (1998) [38] found that daily samples may provide a better estimate of actual rain chemistry than weekly samples. Gilliland et al. (2002) [39] also suggested that considerable monthly variation existed between daily and weekly data for H^+, NH_4^+ and SO_4^{2-}. All of the samplers were placed 1 m from the ground, and the wet sample collector was covered with a lid to prevent any contamination from dry deposition. The wet collector uses a conductivity sensor, which automatically opens at the onset of each precipitation event and closes again when a rain event stops. The automated samplers consisted of two Teflon buckets 28 cm in diameter. The average precipitation intensity for onset was 40 ± 59 mm/week during the study period. Before sampling, the buckets were carefully rinsed with deionized water several times until the conductivity of the water was <1.5 $\mu S \cdot cm^{-1}$. Precipitation samples were collected in a 30-mL high-density polyethylene bottle. The samples were unrefrigerated during sampling but stored in a freezer at $-20\,^\circ C$ until chemical analysis. Prior to analysis, the precipitation samples were filtered using 0.45-μm membrane filters (Millipore).

2.3. Sample Analysis

After the precipitation samples were collected, the weight of the bucket was measured to calculate the volume-weighted mean (VWM) concentrations of the atmospheric components. The pH and conductivity of the unfiltered precipitation were also measured immediately with a pH and conductivity meters (Fisher Scientific) [40]. The pH meter was calibrated before each measurement using standard 4.00 and 7.01 buffer solutions. The conductivity meter was also calibrated with a standard solution. For this study, 528 rain samples were analyzed. Major anion and cation concentrations were determined using ion chromatography (Waters, USA). For anions (Cl^-, NO_3^-, and SO_4^{2-}), AS14 or AS11 columns were used; CS12A or CS14 columns were used to measure cations (Na^+, NH_4^+, K^+, Mg^{2+}, and Ca^{2+}). The detection limits were 0.08 $\mu eq \cdot L^{-1}$ for Cl^-, 0.12 $\mu eq \cdot L^{-1}$ for NO_3^-, 0.11 $\mu eq \cdot L^{-1}$ for SO_4^{2-}, 0.54 for Na^+, 2.32 $\mu eq \cdot L^{-1}$ for NH_4^+, 0.2 $\mu eq \cdot L^{-1}$ for K^+, 0.44 $\mu eq \cdot L^{-1}$ for Mg^{2+} and 2.87 $\mu eq \cdot L^{-1}$ for Ca^{2+}. Detection limits were calculated by dividing the standard deviation of the response by the slope of the calibration and then multiplying by 3.3.

2.4. Quality Assurance

Of the 678 precipitation samples collected, about 7% were discarded because the amount of precipitation was insufficient to perform the chemical analyses. An additional 12% of the rain samples were discarded due to noticeable contamination by dry deposition (soil, leafs and insects) or because of sampler malfunction [41]. The remaining samples were subjected to a quality check based on ionic balance and conductivity balance. When the pH was above 5.6, the concentration of HCO_3^- (in $\mu eq \cdot L^{-1}$) was calculated using the formula $(HCO_3^-) = 10^{(pH - 11.24)}$ [42–44]. According to Okuda et al. (2005) [15], the acceptable ion range ($\sum$ cation/$\sum$ anion) and conductivity ($\sum$ measured conductivity/$\sum$ anion conductivity) ratio for a rain sample is 0.67–1.5. Data points outside this range were excluded. The percentage of samples excluded at each sampling station was 38% at Hwasung, 21% at Banwol, and 32% at Daeboo, averaging 30% overall. Linear regression of the relationship between cation sum vs. the anion sum showed an $r^2 = 0.95$ and slope of 0.95 (Figure 2a,b). Edmonds et al. (1991) [45] suggested that anion deficit means of the amount of unmeasured organic acids. Kim et al. (2013) [32] found that the organic acids (CH_3COO^-, $HCOO^-$) contributed to 12.4% of the acidity in precipitation at Jeju Island in Korea. The relationship between the calculated and measured conductivity was also highly correlated with a $r^2 = 0.96$ and slope of 0.92.

2.5. Positive Matrix Factorization (PMF)

Positive Matrix Factorization is a multivariable factor analysis tool to identify the contributions of various emission sources. PMF decomposes a speciated data matrix X of n by m dimensions (n number of samples and m chemical species) into factor profiles (g) and factor contributions (f) based on the correlation between the different components (Equation (1)). The objective of the PMF solution is to minimize the object function Q based on the uncertainties (u) as follows [46]:

$$Q = \sum_{i=1}^{n} \sum_{j=1}^{m} \left[\frac{X_{ij} - \sum_{k=1}^{p} g_{ik} f_{kj}}{u_{ij}} \right]^2 \tag{1}$$

where x_{ij} are the measured concentrations (in $\mu eq \cdot L^{-1}$), u_{ij} are the estimated uncertainty values (in $\mu eq \cdot L^{-1}$), n is the number of samples, m is the number of species and p is the number of factors included in the analysis [47].

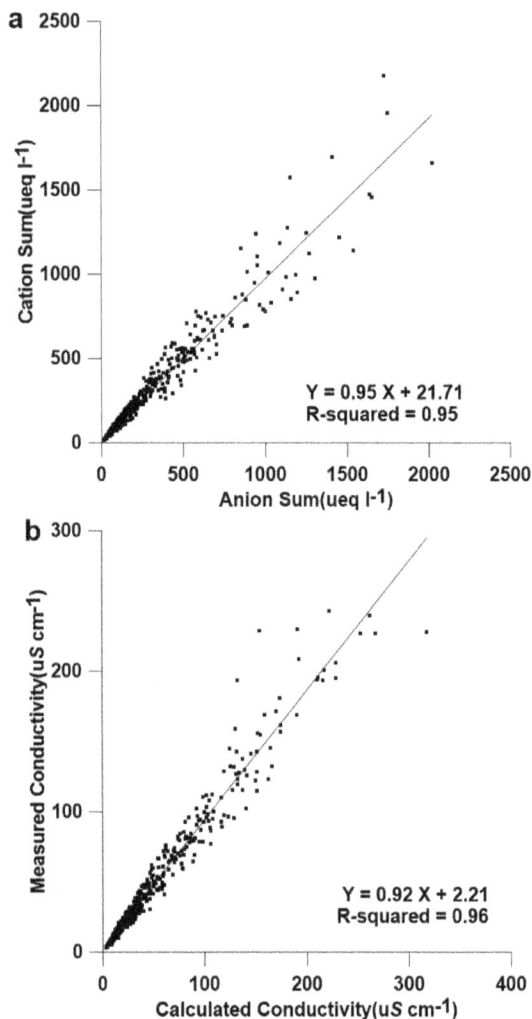

Figure 2. Linear regression (**a**) between the sum of cations and anions, and (**b**) between measured and calculated conductivity.

PMF needs two input files: one for the measured concentrations of the species and the other for the estimated uncertainties of the concentrations. The uncertainty greater than the detection limit was calculated using the concentration and detection limit (Equation (2)). If the concentration was less than or equal to the detection limit, the uncertainty was calculated with a fixed fraction (Equation (3)). Missing data and their uncertainty value were replaced with the value of the species-species median and four times the median value, respectively [48].

$$uncertainty = \sqrt{(\text{Error fraction} \times \text{Concentration})^2 + (0.5 \times \text{Detection limit})^2} \qquad (2)$$

$$uncertainty = \frac{5}{6} \times (Detection\ limit) \qquad (3)$$

A variable was defined to be weak if its S/N was between 0.2 and 2.0. Most of the scaled residuals were between −3.0 and 3.0. Fpeak values between −1 and 1 (in steps of 0.1) were examined to find out the most appropriate solution. Error estimate with bootstrap results showed 100% mapping for five factors at three sites.

3. Results and Discussion

3.1. Acidity

SO_2 and NO_x are the major precursors of acidity in precipitation. Wind-blown soil particles and sea spray species atmosphere play an important role in neutralizing acidic constituents. In uncontaminated precipitation, the equilibrium concentration of $CO_{2(aq)}$ generates a pH of approximately 5.6, which serves as a reference value. Figure 3 shows the frequency distribution for pH in the Shihwa Basin. Among the samples, 84% of the pH values were <5.6. The pH of uncontaminated precipitation is considered to be 5.0–5.6; this acidity originates from natural levels of atmospheric CO_2, NOx, and SO_x [49]. About 66% of the rain samples had a pH < 5.0, indicating that the rain samples were affected by additional acidic components. At Banwol station in the center of two industrial complexes, 69% of the samples had a pH of <5.0 and 14% had a pH < 4.0. At Hwasung station, 59% of the samples had a pH of <5.0 and 7% had a pH < 4.0. At Daeboo, 78% of the samples had a pH lower than 5.0 and 15% of the samples had a pH < 4.0.

Overall, the pH of the precipitation ranged from 3.4 to 7.7 (Table 2). The VWM pH was 4.84, indicating slight acidity in the study area. For Hwasung, located east of Banwol, the VWM pH was 4.96, higher than at the other sites ($p < 0.001$). At Banwol, the pH was 4.77, lowest among all the stations ($p < 0.001$), reflecting its proximity in the industrial complex. Daeboo had the highest VWM pH (4.81), similar to the VWM pH at Banwol.

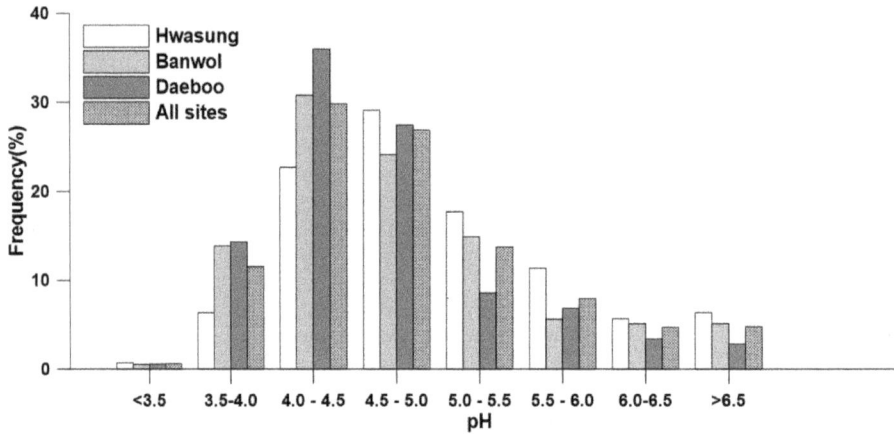

Figure 3. Frequency distribution for wet-only precipitation volume-weighted mean (VWM) pH.

Table 2. Major ionic compositions of precipitation in the Shihwa Basin ($\mu eq \cdot L^{-1}$ except pH).

Station	Parameter	Cl^-	NO_3^-	SO_4^{2-}	Na^+	NH_4^+	K^+	Mg^{2+}	Ca^{2+}	pH
	Min	0.5	3.8	6.8	0.6	5.9	0.02	0.3	0.7	7.5
	Max	472	482	1495	567	1758	163	297	781	3.4
Hwasung	Mean	45	68	127	33	117	9	15	57	4.5
	Standard deviation	82	101	72	79	122	20	32	134	0.8
	VWM	13.9	19.6	45.2	9.3	43.6	3.6	4.5	13.5	4.7
	Min	1.4	2.0	4.7	0.5	5.7	0.03	0.4	0.6	7.7
	Max	683	598	839	506	539	308	144	639	3.5
Banwol	Mean	58	71.3	145	34.7	116	9	14	65	4.3
	Standard deviation	93	83	160	66	104	12	23	108	0.8
	VWM	20	23	56	10	51	4.4	4.6	15	4.6
	Min	1.2	1.9	3.5	0.4	1.6	0.11	0.3	0.6	6.9
	Max	1505	594	753	1434	588	87	320	763	3.5
Daeboo	Mean	70	73	118	59	88	8.5	20	61	4.4
	Standard deviation	119	89	134	88	89	20	33	115	0.7
	VWM	25	22	40	17	40	5.1	6.6	15	4.6

3.2. Concentration of Inorganic Species

The minimum and maximum concentrations of chemical components in the individual precipitation samples and their VWMs are shown in Table 2. SO_4^{2-} and NH_4^+ were the most abundant ions in precipitation at the three stations. At Hwasung, the concentrations of anthropogenic species were lower than those at the

30

other stations. The VWM concentration of sea salts was also lowest in Hwasung, as it is located farthest from the seashore. The highest concentrations of SO_4^{2-}, NH_4^+ and NO_3^- were measured at Banwol, related to anthropogenic sources associated with two industrial complexes and motor vehicles. The VWM concentration of H^+ was also highest in Banwol. In Daeboo, the precipitation acidity was similar to that in Banwol due to the low concentration of NH_4^+. The VWM conductivities in Hwasung, Banwol, and Daeboo were 18, 24, and 23 $\mu S \cdot cm^{-1}$ with VWM total ions of 181, 253, and 235 $\mu eq \cdot L^{-1}$, respectively.

For the combined data collected in the Shihwa Basin, the ion concentrations fell in the order $SO_4^{2-} > NH_4^+ > H^+ > NO_3^- > Cl^- > Ca^{2+} > Na^+ > Mg^{2+} > K^+$. Among the anions, SO_4^{2-} had the highest concentrations at all stations (54%). The next most abundant anion was NO_3^- at nearly 24%. The precipitation chemistry in the Shihwa Basin was mainly dominated by NH_4^+ with an average VWM of $\mu eq \cdot L^{-1}$, representing 40% of all cations. Considering the sampling period, the concentration of NH_4^+ can be as high as 49.5 $\mu eq \cdot L^{-1}$ (10%). The next most abundant cations were H^+, Ca^{2+}, and Na^+, representing 24%, 15%, and 12% of the total, respectively. NH_4^+ in the Shihwa Basin is associated with agricultural activities and an industrial wastewater treatment facility. Ca^{2+} is related to resuspension of dust from the soil and Asian dust originating from China mainly in the spring. [50]. These alkali components neutralize and decrease the acidity of the rainwater [51,52]. The ratios of individual ions to Na^+ are higher than seawater ratios, which indicates that contributions from anthropogenic and soil sources are important, while the marine contribution is negligible [53,54]. The Cl^-/Na^+ ratio in the combined precipitation samples over six years was 45% higher than that of seawater Cl^-/Na^+ (1.165). This suggests that the Cl^- originated from industrial activities in the area such as coal combustion and incineration of polyvinyl chloride [55,56]. The SO_4^{2-}/Na^+, K^+/Na^+, Mg^2/Na^+, and Ca^{2+}/Na^+ ratios for precipitation in the Shihwa Basin were 4.02, 0.32, 0.45, and 1.25, respectively, higher than the seawater ratios (0.12, 0.23, 0.04, and 0.02, respectively). This suggests that anthropogenic influences were present from local (two industrial complexes) or regional (Asian continent) sources.

The SO_4^{2-}/NO_3^- ratio has been used as an index to evaluate anthropogenic characteristic sources in rainfall samples [54,57]. SO_4^{2-}/NO_3^- ratios are diverse globally and temporally: 0.67 in the southwestern United States during 1995~2010 [58], 1.5 in Belgium in 2003 [41], 1.6 in Turkey in 2002 [51], 2.3 in central Pennsylvania during 1993~2001 [59], 2.4 in Mexico from 1994 to 2000 [60], 8.7 in Brazil in 2002 [55] and 18.6 in Costa Rica in 2009 [61]. Kulshrestha *et al.* (2003) [62] reported that the SO_4^{2-}/NO_3^- ratio in India increased as the degree of urbanization or industrialization increased. In Korea, SO_4^{2-}/NO_3^- ratios ranging from 0.6 to 3.6 have been reported (1990 to 2013) [30,33,63,64]. The mean SO_4^{2-}/NO_3^- at three EANET sites located in the western area of Korea averaged 0.97 during 2001~2007

(EANET). The average SO_4^{2-}/NO_3^- ratio of 2.3 in this study was attributed to the close proximity to industrial areas in Korea [30]. The $H^+/(NO_3^- + SO_4^{2-})$ ratio can represent the relative contributions of H_2SO_4 and HNO_3 to the acidity of rain, and deviations from unity indicate the degree of neutralization [65]. Earlier studies in Korea showed that acidity due to H^+ was less than 7% in rural areas, while that in urban areas ranged from 18% to 34%. The combined contribution of H_2SO_4 and HNO_3 to acidity in the Shihwa Basin was 0.36 ± 0.26, indicating that about 64% of the acidity was neutralized by cations [66]. The $H^+/(NO_3^- + SO_4^{2-})$ ratios for China, Southeast Asia, Korea, and Japan are about 20%, 60%, 30%, and 37%, respectively. The mean pH for wet-only precipitation (4.84) in the Shihwa Basin is higher than or similar to the results from other major cities in Asia.

3.3. Temporal Variation

During the study period, there were 26 episodic events in which the total ion concentration precipitation was >2000 $\mu eq \cdot L^{-1}$. There were 14 events in winter, 11 in spring, and one in autumn. In summer, there were no events when the total ion concentration was >2000 $\mu eq \cdot L^{-1}$. Rastogi and Sarin (2007) [67] reported that low-solute events are associated with heavy amounts of rain or successive events. The concentrations of most of the chemical components in the wet-only samples were inversely related to precipitation amount. About 74% of the rain in the Shihwa Basin fell from June to September (Figure 4). Concentrations during the rainy period (June to August) were significantly lower ($p < 0.05$ for H^+ and $p < 0.01$ for other ions) than those in the dry period (September to May). The concentrations of the inorganic species decreased due to dilution during the rainy period, thus ion concentrations were lower during the rainy period than ion concentrations in the dry period. The concentrations of soil-derived species (Ca^{2+} and Mg^{2+}) in the dry season were about six times higher than those in the rainy season. In March, influenced by dust from Asian continental deserts such as the Taklamakan, Gobi, and Loess plateau, the concentrations of all chemical components were highest. The monthly VWM concentrations of SO_4^{2-} and NO_3^- in the wet-only samples were correlated with concentrations of SO_2 ($r = 0.466$, $p < 0.001$) and NO_2 ($r = 0.559$, $p < 0.001$) in atmosphere (Figure 5). The NO_2 and SO_2 data are available from the National Institute of Environmental Research (NIER) in the same administrative district (Ansan city). Tu et al. (2005) [19] also reported that the SO_4^{2-} concentration had a temporal trend corresponding to SO_2 in atmosphere in China. Similar results have been found in a number of studies [15,68].

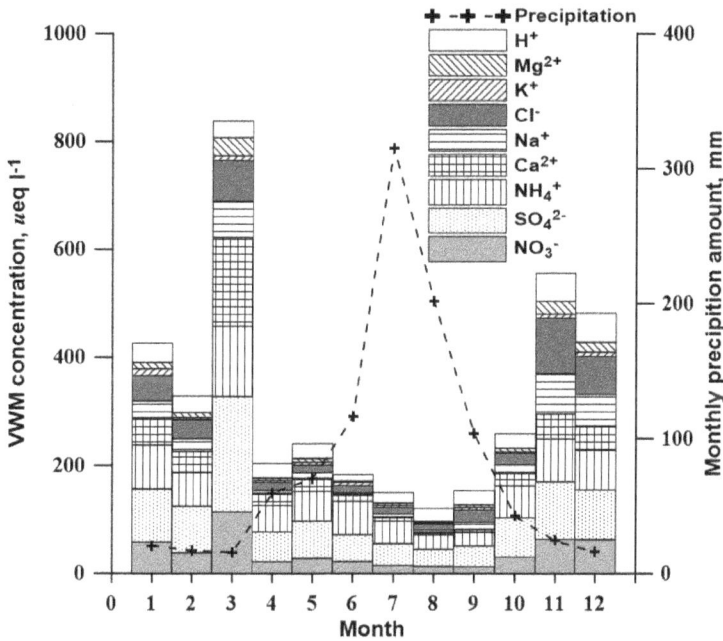

Figure 4. Monthly VWM concentrations of ions in rainfall and rainfall amount.

Figure 6 shows the annual VWM concentrations of the major components in precipitation. Anthropogenic SO_2 and NO_2 emitted from mobile and power plant sources contributed to the increasing of SO_4^{2-} and NO_3^- in the wet-only samples [69]. In 2002, power plants (450 MW) near the study area expanded to Boryeong thermal power plant located about 100 km south of Daebudo station. Thermal power facilities (1600 MW) were also added to the Yeungheung thermal power plant located 13 km west from Daebudo station in 2004. In 2005, the Incheon thermal power plants located 20 km north of the Banwol industrial complex expanded to a 342 MW power plant facility. Additional power generating capacity (5275 MW) has been added to these three power plants between 2008 and 2014. During the study period, the NH_4^+/SO_4^{2-} ratio decreased from 1.01 to 0.68. The NH_4^+/NO_3^- ratio also decreased from 2.39 to 1.20. Results from studies of wet-only precipitation in Austria, Brazil, Canada, USA, Japan, and China have shown significant decreasing trends in the concentrations of SO_4^{2-} and H^+ over long-term periods [15,16,70–72]. Table 3 shows VWM concentrations in recent precipitation studies in East Asia. The concentrations of water-soluble components in the Shihwa Basin were lower than those found in other Asian continental countries. The average concentrations of NO_3^- and SO_4^{2-} ranged from 1.8 to 2.0 and were 2.0 to 2.4 times higher than average concentrations in Seoul [33].

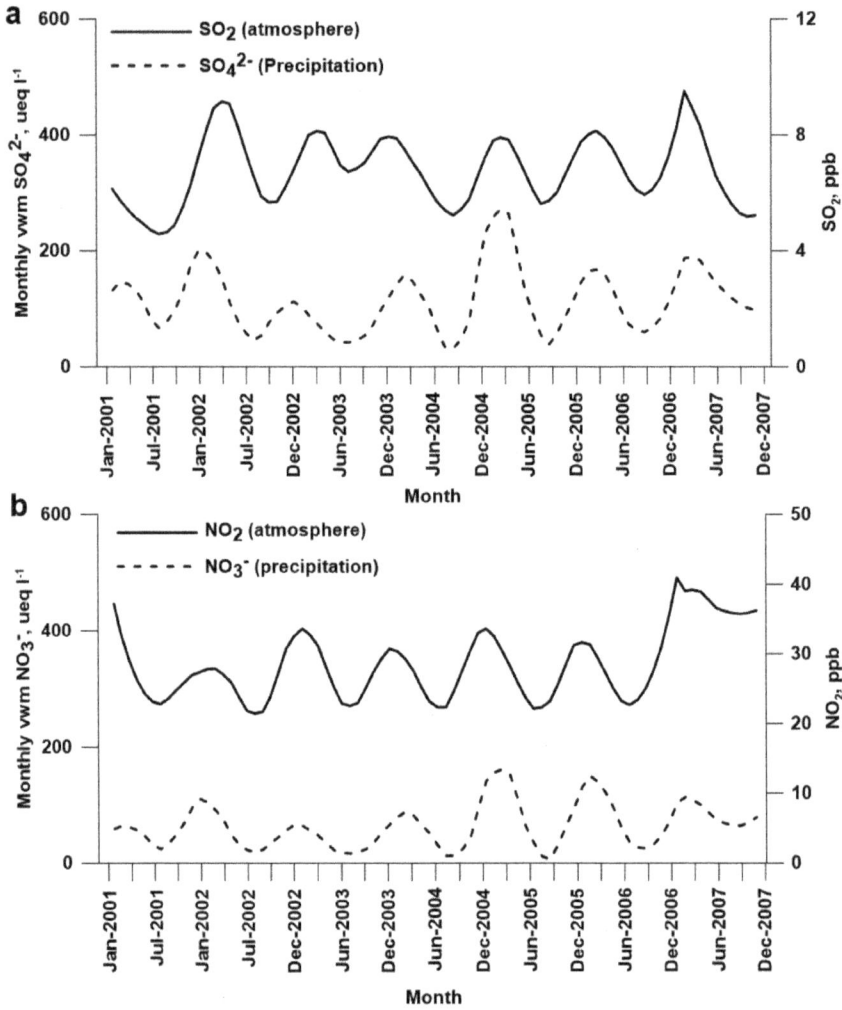

Figure 5. Trends with Lowess smoothing (span = 0.2): (**a**) monthly VWM $SO_4{}^{2-}$ concentration in wet-only precipitation and the monthly mean atmospheric SO_2 concentration and (**b**) monthly VWM $NO_3{}^-$ concentration in wet-only precipitation and the monthly mean atmospheric NO_2 concentration.

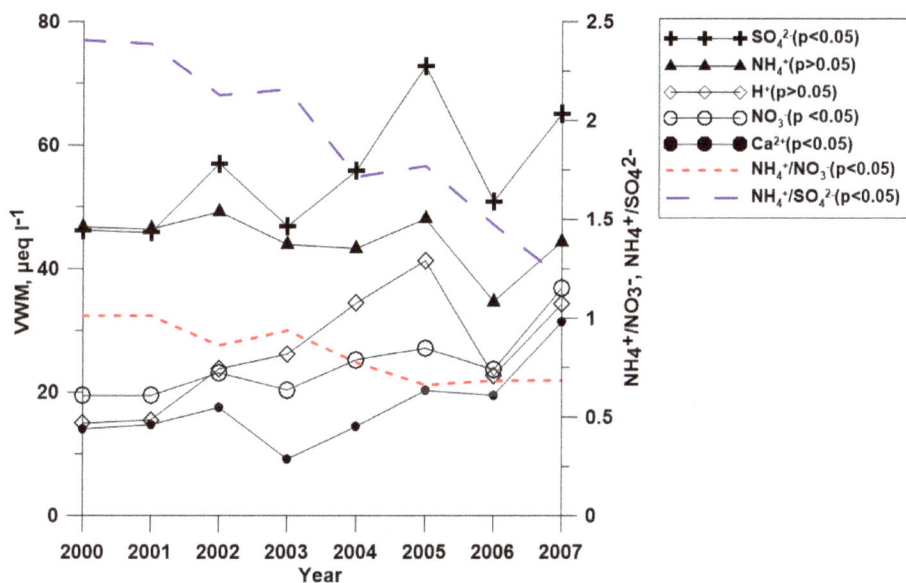

Figure 6. Annual variations of major inorganic VWM concentrations in the precipitation.

Table 3. Volume-weighted mean (VWM) concentrations of the chemical components in wet-only precipitation in East Asia ($\mu eq \cdot L^{-1}$ except pH).

East Asia	Year(s)	Cl^-	NO_3^-	SO_4^{2-}	Na^+	NH_4^+	K^+	Mg^{2+}	Ca^{2+}	pH
Beijing [a] (N China)	2003	-	118	380	-	211	-	-	159	6.48
China [b] (SE China)	2004	9	31	95	6	81	5	3	48	4.54
Cai Jia Tang [c] (S China)	2003	11	60	155	7	112	10	10	60	4.33
Hong Kong [d]	1999–2000	43	27	74	37	22	4	7	16	4.20
Singapore [e]	1999–2000	34	22	84	33	19	7	7	16	4.20
Japan [f] (Tokyo)	1990–2002	55	31	50	37	40	3	12	25	4.52
Jeju [g] (Korea)	2000–2007	40	23	22	24	40	21	3	9	5.37
This study	2000–2007	23	23	54	13	44	4	6	17	4.84

[a] Tang *et al.* (2005); [b] Zhang *et al.* (2007b); [e] Hu *et al.* (2003); [f] Okuda *et al.* (2005); [g] EANET(2013).

3.4. Statistical Analysis

Table 4 shows the correlation coefficients among chemical species for this study. Significant and strong correlations were found among Na^+, Cl^-, Mg^{2+} and Ca^{2+}, indicating that this region is heavily affected by sea salts and soil. Ca^{2+} and Mg^{2+} correlations with Cl^-, NO_3^-, and SO_4^{2-} indicate that $CaCl_2$, $MgCl_2$, $Ca(NO_3)_2$, $Mg(NO_3)_2$, $CaSO_4$, and $MgSO_4$ were the primary soil-derived components in the wet-only precipitation [73]. The correlation among NH_4^+, NO_3^-, and SO_4^{2-} indicate that they originated from similar anthropogenic sources. The significant correlation between NH_4^+ and NO_3^- ($r = 0.76$, $p < 0.001$) and between NH_4^+ and SO_4^{2-} ($r = 0.83$, $p < 0.001$) indicate that NH_4NO_3, $(NH_4)HSO_4$, and $(NH_4)_2SO_4$ were the major forms of NH_4^+ in precipitation. Although positive correlations between acidic anions and H^+ were observed, their correlations were weaker due to neutralization effect by basic soils components [74].

Table 4. Correlations between ionic components in wet-only precipitation over seven years in the Shihwa Basin ($p < 0.001$ for all correlation).

	Cl^-	NO_3^-	SO_4^{2-}	Na^+	NH_4^+	K^+	Mg^{2+}
NO_3^-	0.56						
SO_4^{2-}	0.62	0.82					
Na^+	0.89	0.58	0.64				
NH_4^+	0.52	0.76	0.83	0.48			
K^+	0.66	0.43	0.54	0.61	0.55		
Mg^{2+}	0.74	0.67	0.69	0.82	0.49	0.47	
Ca^{2+}	0.63	0.71	0.82	0.68	0.60	0.49	0.80

Additional details on the patterns in the chemical constituents were analyzed using PMF. Similar studies have also used this technique to identify sources of air pollutants. Anttila et al. (1995) [75] determined the sources of bulk wet deposition in Finland using PMF analysis. Recently, Kitayama et al. (2010) [76] used PMF to identify the sources of SO_2 in Japan. The number of factors was determined by the optimized values. Five factors were chosen (Figure 7). The factors were identified as sea salts (Na^+, ss-Cl^-), Asian dust (Mg^{2+}, Ca^{2+}), sulfuric acid (H^+, SO_4^{2-}), ammonium salt with nitrate (NH_4NO_3), and ammonium salt with sulfate (NH_4HSO_4 and $(NH_4)_2SO_4$)components. The first factor was characterized by high concentrations of sea salts and its contribution to ions in precipitation was 18%–26%. At the Banwol site, some anthropogenic components (NO_3^-, NH_4^+, and SO_4^{2-}) were also included in this factor. The second factor, accounting for 17% of the total ions, was dominated by Asian dust at all three sites, indicated by high concentrations of Mg^{2+} and Ca^{2+}. This factor explained 70%–79% of the variation in Ca^{2+} and 42%–74% of that in Mg^{2+}. The average contribution of the Asian dust factor was especially high during the spring. The third factor represents H^+ and SO_4^{2-}, accounting for 16%–23% of the total. The acidity of the precipitation is largely affected by the dissolution of SO_2 and NO_2 as precursors of the acidic components. As noted above, the high SO_4^{2-}/NO_3^- ratio in the Shihwa basin indicates that emissions of SO_2 from the industrial complex were dominant sources for acidity. SO_4^{2-} accounted for 93%, 52%, and 57% of this acidity factor at Hwasung, Banwol, and Daeboo, respectively. The fourth factor was characterized by ammonium nitrate, which explained 17%–21% of the total and originated from either local industrial or farmland sources. The presence of nitrate with ammonium ions indicates dissolutions of ammonium nitrate from aerosols. The fifth factor represents high loadings of NH_4^+ and SO_4^{2-}. The highest NH_4^+ concentrations were found in Hwasung where sizeable NH_3 sources, such as farmland, exist. The relative contributions of five factors to measured components varied largely by the month of year (Figure 8). The acidity and secondary aerosol factors were mainly associated with the temperature and precipitation amount. Their contribution was highest in summer. However, Asian dust influences were dominant during the spring.

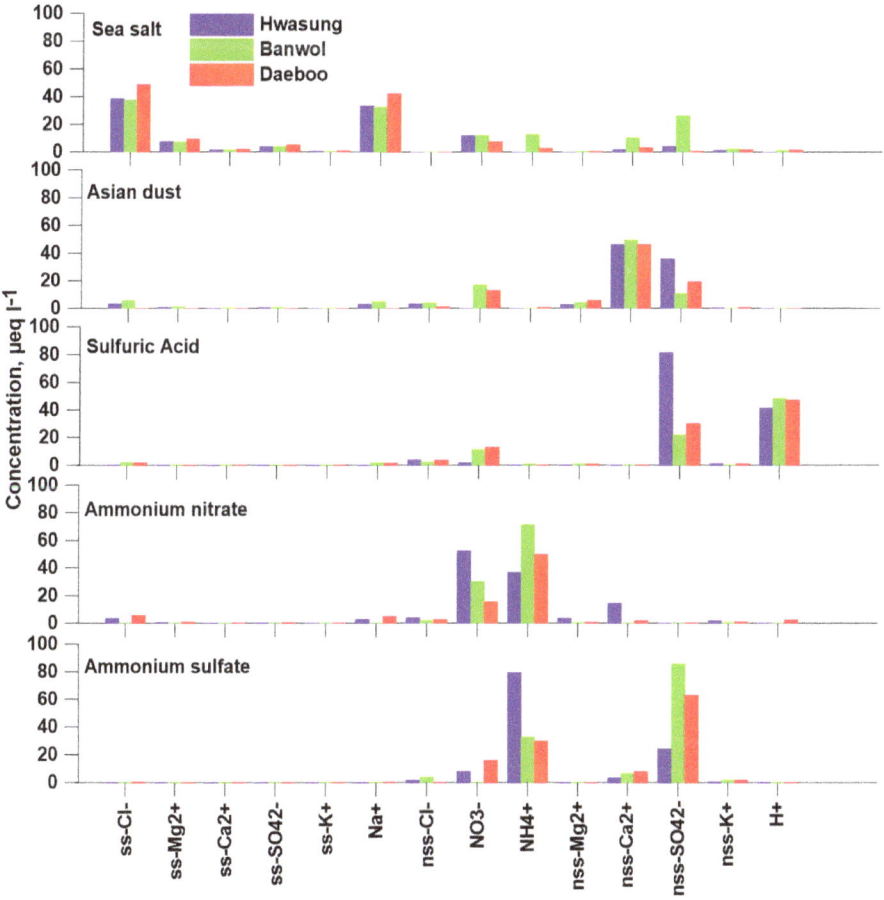

Figure 7. Source contributions of factors 1–5 at the three stations.

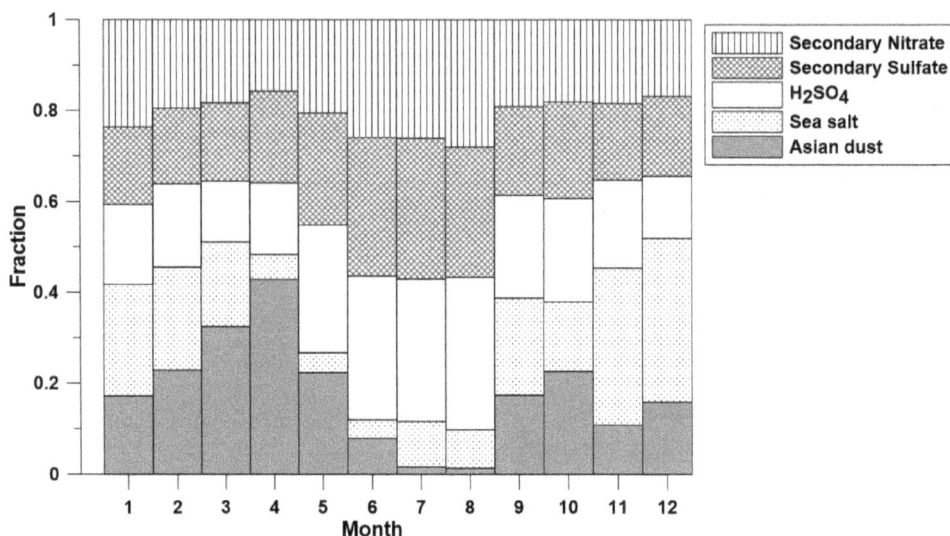

Figure 8. Temporal variations in the five source contributions.

4. Conclusions

The composition of precipitation at three sites in the Shihwa Basin was studied from June 2000 to November 2007. A total of 532 wet-only precipitation samples were used to characterize features of the precipitation and to evaluate the influence of anthropogenic sources.

The VWM for pH for all sites combined was 4.84, with pH < 5.0 in 66% of the precipitation samples, confirming that precipitation in the Shihwa Basin is acidic. The chemical composition of the precipitation was influenced by both natural (sea salt and soil components) and anthropogenic (acidic and alkali components) sources. SO_4^{2-} and NH_4^+ were the dominant ions, followed by NO_3^- and sea salts. Based on the $H^+/(NO_3^- + SO_4^{2-})$ ratio, 66% of the acidity in the Shihwa Basin was neutralized by alkaline particles. The SO_4^{2-}/NO_3^- ratio showed that the contribution of SO_4^{2-} to acidity was twice as high as that of NO_3^-. Seventy-four percent of the rain fell during the summer, when the concentrations of all of the ions were significantly lower than those in other seasons.

Although the mean concentrations of the anthropogenic components (NO_3^- + SO_4^{2-} + NH_4^+) in the Shihwa basin were 1.8~2.2 times higher than those in Seoul, the VWM concentrations of most of the chemical components were lower than those in other areas of East Asia. To identify the sources of the inorganic species in wet-only precipitation, the components were evaluated using statistical analysis. The moderate correlations of H^+ with NO_3^- and SO_4^{2-} and the strong correlations of NO_3^- and SO_4^{2-} with NH_4^+, Mg^{2+}, and Ca^{2+} suggest that neutralization reactions have a strong effect on acidity in the Shihwa Basin. Power generating facilities have expanded in the vicinity of the Shihwa basin from 2002 to 2013. This trend may have impacted the trend of increasing SO_4^{2-} concentrations during the study period. The results from PMF indicated that the main contributors to the wet-only precipitation chemical components in this study were secondary aerosols (40%–42%) followed by acidity (16%–23%), while sea salts and Asian dust contributed 19%–26% and 17%–18%, respectively.

Acknowledgments: This work was supported by the National Research Foundation of Korea (NRF-2009-413-B00004) and funded by the Korean Ministry of Education, Science and Technology.

Author Contributions: Seung-Myung Park: Sampling, Data analysis, Drafting of Manuscript
Beom-Keun Seo: Sampling, Data analysis
Gangwoong Lee: Study conception and design, Research performance, Critical revision
Sung-Hyun Kahang: Study conception and design
Yu Woon Jang: Sampling, Data analysis, Drafting of Manuscript

Conflicts of Interest: The authors declare no conflict of interest.

References

1. Draaijers, G.; Erisman, J.; Lövblad, G.; Spranger, T.; Vel, E. *Quality and Uncertainty Aspects of Forest Deposition Estimation using Throughfall, Stemflow and Precipitation Measurements*; TNO-MEP Report 98/093; TNO Institute of Environmental Sciences, Energy Research and Process Innovation: Apeldoorn, The Netherlands, 1998.
2. Duhanyan, N.; Roustan, Y. Below-cloud scavenging by rain of atmospheric gases particulates. *Atmos. Environ.* **2011**, *39*, 7201–7217.
3. Saxena, A.; Kulshrestha, U.C.; Kumar, N.; Kumari, K.M.; Srivastava, S.S. Characterization of precipitation at Arga. *Atmos. Environ.* **1996**, *30*, 3405–3412.
4. Hara, H. Acid deposition chemistry in Asia, Europe, and North America. *Progress Nuclear Energy* **1998**, *32*, 331–338.
5. Akkoyunlu, B.O.; Tayanç, M. Analyses of wet and bulk deposition in four different regions of Istanbul, Turkey. *Atmos. Environ.* **2003**, *37*, 3571–3579.
6. Vázquez, A.; Costoya, M.; Peña, R.M.; García, S.; Herrero, C. A rainwater quality monitoring network: A preliminary study of the composition of rainwater in Galicia (NW Spain). *Chemosphere* **2003**, *51*, 375–386.
7. Wai, K.M.; Wang, S.H.; Taner, P.A.; Lin, N.H. A dual site study of the rainwater chemistry within the western Pacific region. *J. Atmos. Chem.* **2007**, *57*, 85–103.
8. Santos, M.A.; Illanes, C.F.; Fornaro, A.; Pedrotti, J.J. Acid rain in downtown São Paulo City, Brazil. *Water Air Soil Pollut. Focus* **2007**, *7*, 85–92.
9. Avila, A.; Alarcó, M.; Queralt, I. The chemical composition of dust transported in red rains-its contribution to the biogeochemical cycle of a holm oak forest in Catalonia (Spain). *Atmos. Environ.* **1997**, *32*, 179–191.

10. Galloway, J.N. Acidification of the world: Natural and anthropogenic. *Water Air Soil Pollut.* **2001**, *130*, 17–24.

11. Bashkin, V.N.; Radojevic, M. Acid rain and its mitigation in Asia. *Int. J. Environ. Stud.* **2003**, *60*, 205–214.

12. Luo, J.; Wang, X.; Yang, H.; Yu, J.Z.; Yang, L.; Qin, B. Atmospheric phosphorous in the northern part of Lake Taihu, China. *Chemosphere* **2011**, *84*, 785–791.

13. Munger, J.W.; Eisenreich, S. Continental-scale variations in precipitation chemistry. *Environ. Sci. Technol.* **1983**, *17*, 32A–42A.

14. Downing, C.E.H.; Vincent, K.J.; Campbell, G.W.; Fowler, D.; Smith, R.I. Trends in wet and dry deposition of sulphur in the United Kingdom. *Water Air Soil Pollut.* **1995**, *85*, 659–664.

15. Okuda, T.; Iwase, T.; Ueda, H.; Suda, Y.; Tanaka, S.; Dokiya, Y.; Fushimi, K.; Hosoe, M. Long-term trend of chemical constituents in precipitation in Tokyo metropolitan area, Japan, from 1990 to 2002. *Sci. Total Environ.* **2005**, *339*, 127–141.

16. Tu, J.; Wang, H.; Zhang, Z.; Jin, X.; Li, W. Trends in chemical composition of precipitation in Nanjing, China, during 1992–2003. *Atmos. Res.* **2005**, *73*, 283–298.

17. Noguchi, I.; Hara, H. Ionic imbalance due to hydrogen carbonate from Asian dust. *Atmos. Environ.* **2004**, *38*, 6969–6976.

18. Fowler, D.; Smith, R.I.; Muller, J.B.A.; Hayman, G.; Vincent, K.J. Changes in the atmospheric deposition of acidifying compounds in the UK between 1986 and 2001. *Environ. Pollut.* **2005**, *137*, 15–25.

19. Sicard, P.; Coddeville, P.; Sauvage, S.; Galloo, J.C. Trends in chemical composition of wet-only precipitation at rural French monitoring stations over the 1990–2003 period. *Water Air Soil Pollut. Focus* **2007**, *7*, 49–58.

20. Zbieranowski, A.; Aheme, J. Long-term trends in atmospheric reactive nitrogen across Canada: 1899–2007. *Atmos. Environ.* **2011**, *45*, 5853–5862.

21. Larssen, T.; Seip, H.M.; Semb, A.; Mulder, J.; Muniz, I.P.; Vogt, R.D.; Lydersen, E.; Angell, V.; Dagang, T.; Eilertsen, O. Acid deposition and its effects in China: An overview. *Environ. Sci. Policy* **1999**, *2*, 9–24.

22. Park, S.U.; Kim, J.W. Aerosol size distributions observed at the Seoul National University campus in Korea during the Asian dust and non-Asian dust periods. *Atmos. Environ.* **2006**, *40*, 1772–1730.

23. Fu, P.; Huang, J.; Li, C.; Zhong, S. The properties of dust aerosol and reducing tendency of the dust storms in northwest China. *Atmos. Environ.* **2008**, *42*, 5896–5904.

24. Kim, J. Transport routes and source regions of Asian dust observed in Korea during the past 40 years (1965–2004). *Atmos. Environ.* **2008**, *42*, 4778–4789.

25. Zhao, D.; Xiong, J.; Xu, Y.; Chan, W.H. Acid rain in southwestern China. *Atmos. Environ.* **1988**, *22*, 349–358.

26. Herut, B.; Starinsky, A.; Katz, A.; Rosenfeld, D. Relationship between the acidity and chemical composition of rainwater and climatological conditions along a transition zone between large deserts and Mediterranean climate, Israel. *Atmos. Environ.* **2000**, *34*, 1281–1292.

27. Basak, B.; Alagha, O. The chemical composition of rainwater over Büyüqkçekmece Lake, Istanbul. *Atmos. Res.* **2004**, *71*, 275–288.

28. Wei, H.; Wang, J.L. Characteristics of acid rain in Jinyun Mountain, Chongqing, China. *Appl. Ecol. Environ. Res.* **2005**, *3*, 29–37.

29. Park, C.J.; Noh, H.R.; Kim, B.G.; Kim, S.Y.; Jung, I.U.; Cho, C.R.; Han, J.S. Evaluation of acid deposition in Korea. *Water Air Soil Pollut.* **2001**, *130*, 445–450.

30. Kang, G.; Collettjr, J.L.; Shin, D.Y.; Fujita, S.I.; Kim, H.K. Comparison of the chemical composition of precipitation on the western and eastern coasts of Korea. *Water Air Soil Pollut.* **2004**, *151*, 11–34.

31. Kim, S.B.; Choi, B.C.; Oh, S.Y.; Kim, S.; Kang, G.U. Acidity and chemical composition of precipitation at background area of the Korean Peninsula (Anmyeon, Uljin, Gosan). *J. Korean Soc. Atmos. Environ.* **2006**, *22*, 15–24.

32. Kim, K.J.; Bu, J.-O.; Kim, W.H.; Lee, Y.S.; Hyeon, D.R.; Kang, C.H. Pollution characteristics of rainwater at Jeju Island during 2009~2010. *J. Korean Soc. Atmos. Environ.* **2013**, *29*, 818–829.

33. Choi, J.S.; Park, J.S.; Ahn, J.Y.; Oh, J.; Son, J.S.; Kim, H.J.; Sung, M.Y.; Lee, Y.H.; Lee, S.D.; Hong, Y.D.; *et al.* The characteristics of long-term variation of acid deposition in Korea. *J. Korean Soc. Environ. Anal.* **2015**, *1*, 26–37.

34. Lee, S.I.; Kim, B.C.; Oh, H.J. Evaluation of lake modification alternatives for Lake Sihwa. *Korea Environ. Manag.* **2002**, *29*, 57–66.

35. Kim, K.T.; Kim, E.S.; Cho, S.R.; Chung, K.H.; Park, J.K. Distribution and pollution of heavy metals in the environmental samples of the Lake Sihwa. *J. Korean Soc. Marine Environ. Eng.* **2005**, *8*, 148–157.

36. Im, M.S.; Kim, K.H.; Choi, Y.J.; Jeon, E.C. Emission characteristics of VOC due to major industrial activities in the Banwol industrial complex. *J. Korean Soc. Atmos. Environ.* **2006**, *22*, 325–336.

37. Jung, J.; Jang, Y.W.; Arimoto, R.; Uematsu, M.; Lee, G. Atmospheric nitrogen deposition and its impact to Lake Sihwa in South Korea. *Geochem. J.* **2009**, *43*, 305–314.

38. Butler, T.J.; Likens, G.E. Weekly and daily precipitation chemistry network comparisons in the eastern U.S.: NADP/NTN *vs.* MAP3S/AIRMoN. *Atmos. Environ.* **1998**, *32*, 3749–3765.

39. Gilliland, A.B.; Butler, T.J.; Likens, G.E. Monthly and annual bias in weekly (NADP/NTN) *vs.* daily (AIRMoN) precipitation chemistry data in the Eastern USA. *Atmos. Environ.* **2002**, *36*, 5197–5206.

40. Sequeira, R.; Lung, F. A critical data analysis and interpretation of the pH, ion loadings and electrical conductivity of rainwater from the territory of Hong Kong. *Atmos. Environ.* **1995**, *29*, 2439–2447.

41. Staelens, J.; Schrijver, A.D.; Avermaet, P.V.; Genouw, G.; Verhoest, N. A comparison of bulk and wet-only deposition at two adjacent sites in Melle (Belgium). *Atmos. Environ.* **2005**, *39*, 7–15.

42. Satsangi, G.S.; Khare, L.P.; Singh, S.P.; Kumari, K.M.; Srivastava, S.S. Composition of rain water at a semi-arid rural site in India. *Atmos. Environ.* **1998**, *32*, 3783–3793.

43. Jain, M.; Kulshrestha, U.C.; Sarkar, A.K.; Parashar, D.C. Influence of crustal aerosols on wet deposition at urban and rural sites in India. *Atmos. Environ.* **2000**, *34*, 5129–5137.

44. Al-Khashman, O.A. Ionic composition of wet precipitation in the Petra Region, Jordan. *Atmos. Environ.* **2005**, *78*, 1–12.

45. Edmonds, R.L.; Thomas, T.B.; Rhodes, J.J. Canopy and Soil Modification of Precipitation Chemistry in a Temperate Rain Forest. *Soil Sci. Soc. Am. J.* **1991**, *55*, 1685–1693.

46. Environmental Protection Agency (EPA). *EPA Positive Matrix Factorization (PMF) 3.0 Fundamentals and User Guide*; USEPA Office of Research and Development: Research Triangle Park, NC, USA, 2008.

47. Gugamstty, B.; Wei, H.; Liu, C.N.; Awasthi, A.; Hsu, S.C.; Tsai, C.J.; Roam, G.D.; Wu, Y.C.; Chen, C.F. Source characterization and apportionment of PM_{10}, $PM_{2.5}$ and $PM_{0.1}$ by using positive matrix factorization. *Aerosol Air Qual. Res.* **2012**, *12*, 476–491.

48. Environmental Protection Agency (EPA). *EPA Positive Matrix Factorization (PMF) 5.0 Fundamentals and User Guide*; Environmental Protection Agency: Washington, DC, USA, 2014.

49. García, R.; Ma, D.T.; Padilla, H.; Belmont, R.; Azpra, E.; Arcega-Cabrera, F.; Báez, A. Measurement of chemical elements in rain from Rancho Viejo, a rural wooded area in the State of Mexico, Mexico. *Atmos. Environ.* **2006**, *40*, 6088–6100.

50. Kai, Z.; Huiwang, G. The characteristics of Asian-dust storms during 2000–2002: From the source to the sea. *Atmos. Environ.* **2007**, *41*, 9136–9145.

51. Topcu, S.; Incecik, S.; Atimtay, A.T. Chemical composition of rainwater at EMEP station in Ankara, Turkey. *Atmos. Res.* **2002**, *65*, 77–92.

52. Hegde, P.; Sudheer, A.K.; Sarin, M.M.; Manjunatha, B.R. Chemical characteristics of atmospheric aerosols over southwest coast of India. *Atmos. Environ.* **2007**, *41*, 7751–7766.

53. Das, R.; Das, S.N.; Misra, V.N. Chemical composition of rainwater and dustfall at Bhubaneswar in the east coast of India. *Atmos. Environ.* **2005**, *39*, 5908–5916.

54. Tanner, P.A. Analysis of Hong Kong daily bulk and wet deposition data from 1994 to 1995. *Atmos. Environ.* **1999**, *33*, 1757–1766.

55. Migliavacca, D.M.; Teixeira, E.C.; Wiegand, F.; Machado, A.C.M.; Sanchez, J. Atmospheric precipitation and chemical composition of an urban site, Guaíba hydrographic basin, Brazil. *Atmos. Environ.* **2005**, *39*, 1829–1844.

56. Sanusi, A.; Wortham, H.; Millet, M.; Mirabel, P. Chemical composition of rainwater in eastern France. *Atmos. Environ.* **1996**, *30*, 59–71.

57. Al-Khashman, O.A. Study of chemical composition in wet atmospheric precipitation in Eshidiya area, Jordan. *Atmos. Environ.* **2005**, *39*, 6175–6183.

58. Sorooshian, A.; Shingler, T.; Harpold, C.; Feagles, C.W.; Meixner, T.; Brooks, P.D. Aerosol and precipitation chemistry in the southwestern United States: Spatiotemporal trends and interrelationship. *Atmos. Chem. Phys.* **2013**, *13*, 7361–7379.

59. Dayan, U.; Lamb, D. Meteorological indicators of summer precipitation chemistry in central Pennsylvania. *Atmos. Environ.* **2003**, *37*, 1045–1055.

60. Báez, A.P.; Belmont, R.D.; García, R.M.; Torres, M.C.B.; Padilla, H.G. Rainwater chemical composition at two sites in Central Mexico. *Atmos. Res.* **2006**, *80*, 67–85.

61. Herrera, J.; Rodriguez, S.; Baez, A.P. Chemical composition of bulk precipitation in the metropolitan area of Costa Rica, Central America. *Atmos. Res.* **2009**, *94*, 151–160.

62. Kulshrestha, U.C.; Kulshrestha, M.J.; Sekar, R.; Sastry, G.S.R.; Vairamani, M. Chemical characteristics of rainwater at an urban site of south-central India. *Atmos. Environ.* **2003**, *37*, 3019–3026.

63. Acid Deposition Monitoring Network in East Asia (EANET). Data report on the acid deposition in the East Asian region 2001–2012; Network Center for EANET: Nigata-si, Japan, 2013.

64. Kang, G.U.; Lim, J.H.; Kim, H.K. An analysis of long-term trends in precipitation acidity of Seoul, Korea. *J. Korea Air Pollut. Res. Assoc.* **1997**, *13*, 9–18.

65. Safai, P.D.; Rao, P.S.P.; Momin, G.A.; Ali, K.; Chate, D.M.; Praveen, P.S. Chemical composition of precipitation during1984–2002 at Pune, India. *Atmos. Environ.* **2004**, *38*, 1705–1714.

66. Zhang, G.S.; Zhang, J.; Liu, S.M. Chemical composition of atmospheric wet depositions from the Yellow Sea and East China Sea. *Atmos. Res.* **2007**, *85*, 84–97.

67. Rastogi, N.; Sarin, M.M. Chemistry of precipitation events and inter-relationship with ambient aerosols over a semi-arid region. *J. Atmos. Chem.* **2007**, *56*, 149–163.

68. Aikawa, M.; Hiraki, T.; Tamaki, M. Comparative field study on precipitation, throughfall, stemflow, fog water, and atmospheric aerosol and gases at urban and rural sites in Japan. *Sci. Total Environ.* **2006**, *366*, 275–285.

69. Hand, J.L.; Gebhart, K.A.; Schichtel, B.A.; Malm, W.C. Increasing trends in wintertime particulate sulfate and nitrate ion concentrations in the great Plains of the United States (2000–2010). *Atmos. Environ.* **2012**, *55*, 107–110.

70. Aherne, J.; Farrell, E.P. Deposition of sulphur, nitrogen and acidity in precipitation over Ireland: Chemistry, spatial distribution and long-term trends. *Atmos. Environ.* **2002**, *36*, 1379–1389.

71. Fornaro, A.; Gutz, I.G.R. Wet deposition and related atmospheric chemistry in the São Paulo metropolis, Brazil, Part 3: Trends in precipitation chemistry during 1983–2003. *Atmos. Environ.* **2006**, *40*, 5893–5901.

72. Puxbaum, H.; Simeonov, V.; Kalina, M.; Tsakovski, S.; Löffler, H.; Heimburger, G.; Biebl, P.; Weber, A.; Damm, A. Long-term assessment of the wet precipitation chemistry in Austria (1984–1999). *Chemosphere* **2002**, *48*, 733–747.

73. Zhang, M.; Wang, S.; Wu, F.; Yuan, X.; Zhang, Y. Chemical compositions of wet precipitation and anthropogenic influences at a developing urban site in southeastern China. *Atmos. Res.* **2007**, *84*, 311–322.

74. Báez, A.; Belmont, R.; García, R.; Padilla, H.; Torres, M.C. Chemical composition of rainwater collected at a southwest site of Mexico City, Mexico. *Atmos. Res.* **2007**, *86*, 61–75.

75. Anttila, P.; Paatero, P.; Tapper, U.; Järvinen, O. Source identification of bulk wet deposition in Finland by positive matrix factorization. *Atmos. Environ.* **1995**, *29*, 1705–1718.

76. Kitayama, K.; Murao, N.; Hara, H. PMF analysis of impacts of SO2 from Miyakejima and Asian continent on precipitation sulfate in Japan. *Atmos. Environ.* **2010**, *44*, 95–105.

Airborne Aerosol *in Situ* Measurements during TCAP: A Closure Study of Total Scattering

Evgueni Kassianov, Larry K. Berg, Mikhail Pekour, James Barnard, Duli Chand, Connor Flynn, Mikhail Ovchinnikov, Arthur Sedlacek, Beat Schmid, John Shilling, Jason Tomlinson and Jerome Fast

Abstract: We present a framework for calculating the total scattering of both non-absorbing and absorbing aerosol at ambient conditions from aircraft data. Our framework is developed emphasizing the explicit use of chemical composition data for estimating the complex refractive index (RI) of particles, and thus obtaining improved ambient size spectra derived from Optical Particle Counter (OPC) measurements. The feasibility of our framework for improved calculations of total scattering is demonstrated using three types of data collected by the U.S. Department of Energy's (DOE) aircraft during the Two-Column Aerosol Project (TCAP). Namely, these data types are: (1) size distributions measured by a suite of OPC's; (2) chemical composition data measured by an Aerosol Mass Spectrometer and a Single Particle Soot Photometer; and (3) the dry total scattering coefficient measured by a integrating nephelometer and scattering enhancement factor measured with a humidification system. We demonstrate that good agreement (~10%) between the observed and calculated scattering can be obtained under ambient conditions (RH < 80%) by applying chemical composition data for the RI-based correction of the OPC-derived size spectra. We also demonstrate that ignoring the RI-based correction or using non-representative RI values can cause a substantial underestimation (~40%) or overestimation (~35%) of the calculated scattering, respectively.

Reprinted from *Atmosphere*. Cite as: Kassianov, E.; Berg, L.K.; Pekour, M.; Barnard, J.; Chand, D.; Flynn, C.; Ovchinnikov, M.; Sedlacek, A.; Schmid, B.; Shilling, J.; Tomlinson, J.; Fast, J. Airborne Aerosol *in Situ* Measurements during TCAP: A Closure Study of Total Scattering. *Atmosphere* **2015**, *6*, 1069–1101.

1. Introduction

Although the importance of atmospheric aerosol in modifying the Earth's radiation budget has been recognized by many studies [1,2], the extent to which aerosol shapes the regional and global climate is still ambiguous [3,4]. The magnitude and sign of the aerosol-induced changes of the radiation budget at the regional and global scales are highly uncertain, since these changes are influenced substantially by strong temporal and spatial variations of aerosol loading, chemical composition

44

and mixing state [5–7]. Since the advent of observational techniques for monitoring these variations from surface, air and space, the diversity of sensors with improved precision and accuracy has increased and corresponding innovative methods have been developed [8–11]. Aircraft measurements are becoming increasingly important for model validation studies because they can document aerosol variations in remote regions where access to ground-based observations is difficult or unavailable, and offer observations with higher temporal resolution than can typically be attained with satellites [12–14].

Comprehensive and integrated measurements of aerosol properties provide an important observational basis for evaluations of climate model predictions and necessarily involve combining data collected by several instruments with different designs and uncertainties. To determine whether these data are consistent and reasonable, a special kind of quantitative comparison experiment is commonly performed. Such an experiment, traditionally referred to as a closure study, compares the measured values of a selected aerosol property with those calculated from independent measurements [15–18]. For example, an optical closure experiment compares the measured values of an aerosol optical property, such as total scattering coefficient, with those calculated from independently measured size distributions and chemical composition under a variety of conditions [19–21]. Good agreement between the measured and calculated aerosol properties (within error bars) indicates consistency of the observational dataset, and bolsters its relevance for further use in global and regional climate model evaluations.

Optical closure studies have become an essential part of testing integrated datasets where simultaneous measurements of the optical, microphysical and chemical properties of aerosol at dry and ambient conditions are available [21–23]. Compared to the ground-based instrumentation suites, instrumentation on board aircraft platforms requires particular attention to its design and operation [24] mainly due to payload restrictions (requiring instruments with smaller dimensions and less weight; [25,26]) and abrupt changes in atmospheric and aerosol characteristics during the aircraft's rapid (about 100 m/s) motion (requiring instruments with faster response time and data acquisition speeds; [27]), which directly impact spatial resolution. While airborne instrumentation and associated data synergy are continuing to evolve on many fronts, rigorous scrutiny of airborne integrated measurements has not always been achieved. Moreover, demands to assess the consistency and reasonableness of integrated airborne data sets have been growing, given the increasingly heavy reliance of process-oriented model evaluations on aircraft measurements [6].

Optical Particle Counters (OPCs) are a common type of airborne instrument for deriving size spectra [24,28,29]. The fundamental quantity measured by OPCs is the amount of light scattered by individual particles over a large solid angle. The amount of scattered light depends on aerosol characteristics, such as size, shape and complex refractive index (RI), which is a function of the particle's chemical composition. The measured scattered light is converted into particle size using an appropriate scattering theory (e.g., Mie theory for spherical particles) and an assumed or estimated refractive index. For example, several parameterizations have been developed for correcting OPC-derived size distributions for weakly absorbing aerosol using Mie calculations and assuming that the RI-based correction depends on the real part of the complex RI only [30,31]. It is important to note that the assumptions employed for the refractive index may or may not be representative of the observed ambient conditions.

Although Mie theory does allow for the RI-based corrections associated with both the real and imaginary parts of the complex RI [32–34], airborne measurements of aerosol chemical composition and absorbing components that are required for the RI estimation are demanding and not always available. The mass loading of black carbon (BC) is an example of one of these absorbing components with relatively sparse relevant measurements [4,35]. As a result, iterative schemes that use assumed values of complex RI in combination with other assumptions are commonly applied to minimize differences between the measured and calculated aerosol properties of interest, such as PM_{10} [34]. For humid conditions, the particle size distributions exhibit a sensitivity to water uptake by particles [36], and therefore the hygroscopic growth factor (HGF) and its dependence on particle chemical composition must be considered in closure-related studies.

While closure studies using the microphysical, optical, chemical components, and ambient relative humidity (RH) are a well-known framework that has been used intensively for decades [37–39], its successful applications are mainly limited to the ground-based observations [21,40,41]. Given the complexity of conducting airborne measurements and the growing demand to use these measurements for process-oriented model validation and climate model assessments, there is a strong need to extend this framework to comprehensive airborne datasets [22,42,43]. The primary purpose of our work is to attempt to formally extend the framework for ground-based optical closure studies to airborne data sets by answering the following three main questions:

(1)   *What level of agreement can be achieved between the in-flight measured and calculated values of total scattering coefficient at ambient RH?*

(2)   *What is the effect of ignoring the influence of chemical composition data on this agreement?*

(3)   *How sensitive is this agreement to the assumed RI value, particularly if the assumed RI is non-representative of the ambient aerosol?*

The first question is associated with the consistency of the airborne measurements of the particle size distributions and optical properties when aerosol chemical composition data are available (a preferred "complete" dataset). The second and third questions can be considered as "practical-oriented" because they are focused mostly on practical situations when information on the chemical composition is not available (an "incomplete" dataset) and assumptions about aerosol composition are required. Given that the dimension/weight of several instruments commonly deployed to measure the aerosol chemical composition, such as the miniaturized version of the aircraft-compatible single particle mass spectrometer (miniSPLAT; [44]) and the Aerodyne Aerosol Mass Spectrometer (AMS), is substantial, they are deployed less frequently on mid-to-large size aerial platforms. Moreover, they cannot be deployed on small aerial platforms, such as small or unmanned aircraft. To assess the extended framework and evaluate the relevant assumptions through answering these three important questions, we use integrated airborne data collected during the recent Two-Column Aerosol Project (TCAP; http://campaign.arm.gov/tcap/) over the North Atlantic Ocean and US coastal region (Cape Cod, MA, USA).

We outline in the next section our approach for extending the ground-based framework for conducting optical closure experiments to airborne data. In Section 3 we briefly describe the TCAP data, which represent mainly non-absorbing aerosol and include measured size spectra, chemical composition and total scattering [45]. The complementary model components of our approach and the corresponding assumptions are discussed in Section 4. These assumptions, such as the homogeneous internal mixture, are reasonable and permit us to calculate the HGF and complex RI at ambient RH when additional information is missing. In Section 5, the calculated and measured total scattering coefficients are compared for the wide range of atmospheric conditions observed during TCAP flights conducted in July of 2012, including conditions with low and high ambient RH. The sensitivity of the calculated scattering to the RI and related issues are further discussed in Section 6. In particular, this section emphasizes that the ability to make complementary measurements of chemical composition holds promise for properly specifying the RI, and thus for improving the accuracy of total scattering calculations. The last section presents a summary of key findings.

2. Approach

Figure 1 outlines the major components and main steps in conducting our optical closure experiment by obtaining total scattering coefficients at ambient RH from airborne measurements and Mie calculations. Although our approach relies heavily on several important components and assumptions, such as homogeneous internal mixture and spherical geometry of particles, introduced earlier by previous studies [21,36,46], well-known challenges experienced in collection and interpretation of airborne data make evaluation of this unified approach mandatory. Our approach, as displayed in Figure 1, involves three major components: (1) integrated measurements of aerosol properties (Figure 1; top panel); (2) calculations of the ambient scattering coefficient using Mie theory and estimated hygroscopic growth factor (HGF) (Figure 1; middle panel), and (3) comparison of the scattering coefficients observed and calculated at ambient conditions (Figure 1; bottom panel). Note in Figure 1 how chemical composition information becomes encoded into the improved size spectra, making calculations of scattering more accurate. The left-hand part of diagram (Figure 1; top and bottom panels) illustrates the process of obtaining the observed total scattering at ambient conditions. For the TCAP data set used in this work, this step involves analyzing the total scattering measured by an airborne nephelometer at low RH and the light scattering hygroscopic growth f(RH) measured using a humidification system. The central part of diagram (Figure 1; top and middle panels) illustrates how the complex RI at dry conditions, derived from the chemical composition data, is used to adjust the particle size distributions. Note that the chemical composition data are also used to estimate the RH-dependent HGF. The latter is required for converting the dry complex RI and dry size spectra into their ambient counterparts. The right-hand part of diagram (Figure 1; top and middle panels) demonstrates how the ambient size spectra are obtained from the original (without RI-based correction) and adjusted (with RI-based correction) OPC-derived size distributions. The ambient complex RI and size distributions (both original and adjusted) are used as input for the Mie calculations. The output is the corresponding model total scattering coefficients calculated at ambient conditions. As we shall see, differences between model coefficients calculated using the original and the adjusted size distributions illustrate the importance of the RI-based correction for the total scattering calculation.

3. Data

This section is meant to survey the observational components of our framework (Figure 1; top panel) by reviewing the TCAP airborne data used in this investigation. The TCAP field campaign was designed to provide a comprehensive data set that can be used to investigate important climate science questions, including those related to aerosols. Conducted from June 2012 through June 2013, TCAP involved summer and winter periods of intensive aircraft observations that included the U.S. Department of Energy (DOE) Gulfstream-159 (G-1) aircraft. The G-1 typically sampled at multiple altitudes within two atmospheric columns, one located over Cape Cod, MA and a second over the Atlantic Ocean several hundred kilometers from the coast. Details of TCAP are given in Berg *et al.* [45]. To illustrate performance of our unified approach (Figure 1), we focus on the TCAP summertime data.

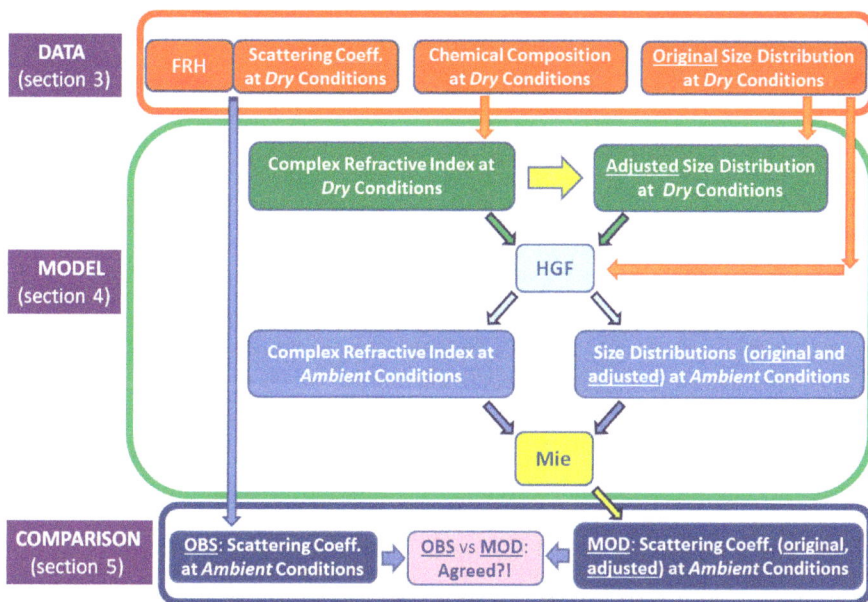

Figure 1. Schematic diagram summarizing the framework for an optical closure experiment using airborne data. The figure illustrates the link between the measured dry and ambient scattering coefficients (**left** part of diagram; top and bottom panels) and the connection between the measured dry chemical composition/size spectra and the calculated ambient scattering coefficient (**center** and **right** parts of diagram). Ambient size spectra (light blue, right) are obtained from the dry size distributions without (orange) and with (green) the RI-based correction, respectively. The estimated ambient size spectra (light blue, right) together with the ambient RI (light blue, left) are required as input for Mie calculations (yellow) of the corresponding total scattering coefficients (navy blue, right). See indicated text section for details of each component (Data, Model, Comparison).

An *in situ* instrumentation suite on board the DOE G-1 aircraft [24] together with airborne remote sensing sensors, such as the Spectrometer for Sky-Scanning, Sun-Tracking Atmospheric Research (4STAR; [26]) and the High Spectral Resolution Lidar (HSRL-2; [8]), were deployed during TCAP. Note that the HSRL-2 was operated aboard a NASA B-200 aircraft. Here, we focus on the instruments relevant to our study to improve calculations of the total scattering coefficient at ambient conditions. These calculations are based on Mie theory and require the ambient aerosol size distribution and complex RI. Aerosol size spectra, chemical composition and total scattering data were collected with high temporal resolution (<1 min) during the TCAP flights. We use these data to compute the corresponding averaged

49

characteristics for each flight leg (FL), which is defined as a straight level run at different altitudes with variable duration of approximately 5–15 min. We employ these FL-averaged aerosol characteristics in our investigation, consistent with earlier studies [28,29,46].

Particle size distributions were measured simultaneously by three airborne OPC instruments: an Ultra-High Sensitivity Aerosol Spectrometer (UHSAS, size range 0.06–1 μm), a Passive Cavity Aerosol Spectrometer (PCASP; size range 0.13–3 μm) and a Cloud and Aerosol Spectrometer (CAS; size range 0.6–>10 μm). These three instruments were mounted within PMS canisters on the same pylon underneath the right wing of the G-1 aircraft. The UHSAS and PCASP were operated with anti-ice heaters enabled and therefore the measured aerosol distributions are assumed to be dry. However, the CAS measured particle size distributions at ambient conditions. All three instruments were calibrated using polystyrene latex sphere (PSL) beads. The measured size distributions from the aforementioned probes are recovered from the raw counts (taking into account collection efficiencies and sampling volumes), merged, and smoothed (Appendix A) using a kernel based on Twomey's algorithm [42,47,48]. Figure 2 shows the resulting size distributions averaged over each FL on 21 July 2012. The altitude of each FL is shown in Figure 3.

Particle size spectra measured by these three OPC instruments (UHSAS, PCASP, and CAS) cover different particle size ranges. However, these instruments employ a similar underlying operating principle for determining particle size, namely light scattering by individual particles. The conversion of the scattered light into particle size requires the RI, which depends on the chemical composition of particles, thus demanding an accurate RI estimation. During TCAP, information on the organic and inorganic species mass loading and black carbon (BC) mass in individual aerosol particles came from complementary Aerodyne AMS and Droplet Measurement Technology Single Particle Soot Photometer (SP2; 0.06–0.6 μm range of mass-equivalent diameter; [49,50]) measurements, respectively (Figure 3). The AMS has a near unity transmission efficiency for particles with vacuum aerodynamic diameters between 0.06 and 0.6 μm, which falls to 50% at 1 μm, and is negligible above 1.5 μm and below 0.06 μm. The RH inside the SP2 does not exceed 10% and the AMS can differentiate particle-phase water from other species; therefore, the acquired chemical composition data represent dry conditions. Figure 3 illustrates that the aerosol chemical composition on 21 July 2012 was dominated by organic matter (OM); dominance of OM (generally greater than 70%) was observed for all TCAP flights [45].

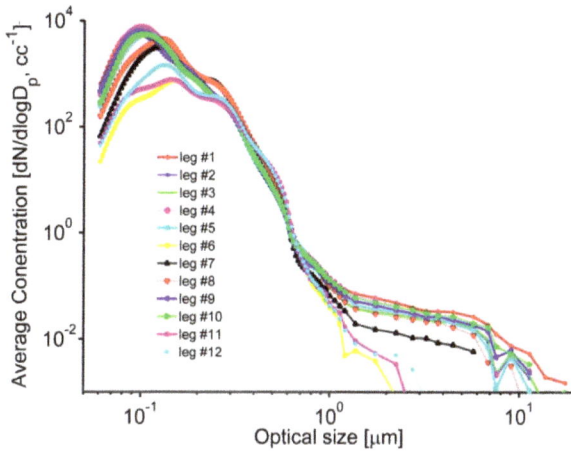

Figure 2. Example of combined size distributions generated for each FL during a given day (21 July 2012). Here and in the following plots, aerosol characteristics represent FL-averaged values. Elevation and time of each FL are shown in Figure 3.

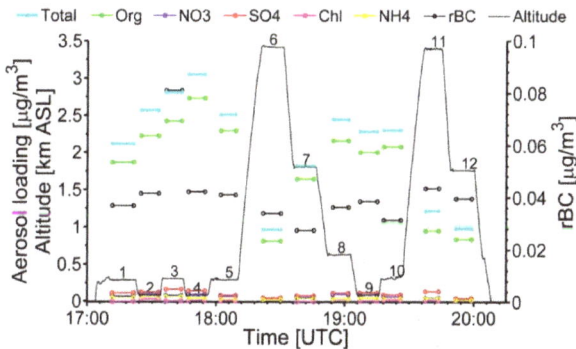

Figure 3. Example of FL-dependent chemical compositions (colored lines) and BC (thick black lines) mass measured by the AMS and SP2, respectively (21 July 2012). Additionally, altitude (thin black line) as a function of FL is included. FLs are labeled with numbers 1 through 12 on top of the thin black altitude line.

The total scattering coefficient at three wavelengths (0.45, 0.55, 0.7 μm) was measured at dry (RH < 20%) conditions using a TSI integrating nephelometer (Figure 4), while the light scattering hygroscopic growth, known as f(RH), was measured using a humidification system at three defined RHs (near 45%, 65% and 90%) at a single wavelength (0.525 μm) [51]. Similar to Shinozuka *et al.* [52], we adjust the f(RH) obtained at the 0.525 μm wavelength to the nephelometer wavelengths (0.45, 0.55, 0.7 μm) by multiplying the obtained f(RH) by 0.98, 1.01 and 1.04, respectively. The conventional truncation error correction [53] has been applied to the total scattering measured by the integrating nephelometer. We obtain the total scattering at ambient conditions (σ_{obs}) at three wavelengths (0.45, 0.55, 0.7 μm) using both the adjusted f(RH) and spectrally-dependent measured dry total scattering. It is to be noted that measurement uncertainties in the reported total scattering by the integrating nephelometer are quite small (~10%) for sub-μm, but can be considerable for super-μm particles (~50%) [53,54]. For a given FL, we assume that the combined uncertainty of σ_{obs} depends on its variability within FL (defined here as the standard deviation) and the measurement uncertainty [46]. In other words, the FL-dependent combined uncertainty in the ambient total scattering coefficient is comparable with the measurement uncertainty (10%) for a homogeneous FL and can exceed it substantially for an inhomogeneous FL where the standard deviation is large.

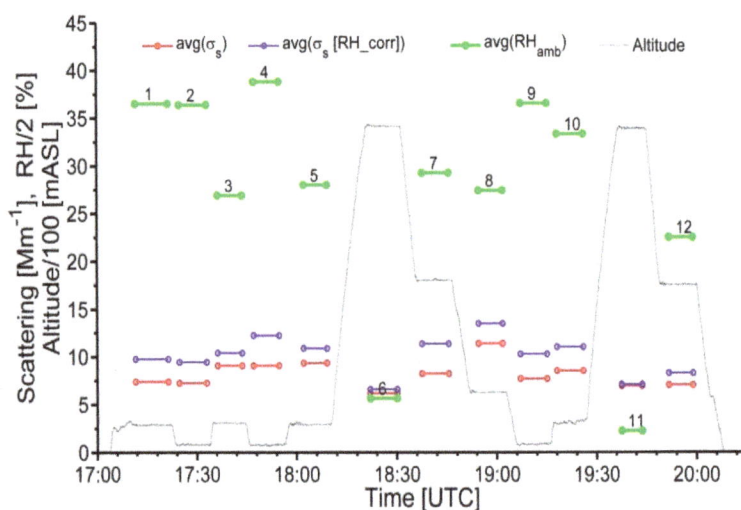

Figure 4. The same as Figure 3, except for the dry scattering measured by nephelometer (red lines) and ambient scattering obtained with measured f(RH) (blue lines) at 0.55 μm wavelength and ambient RH (green lines). FLs are labeled with numbers 1 through 12 on top of green lines.

TCAP aircraft data were screened prior to use in our study. To ensure that only high quality data are used, two quality assurance screening criteria are applied. First, we check the data streams for consistency between the size spectra, chemical composition and total scattering to prevent invalid data entry. Second, we consider only periods in which all data streams are available. For example, PCASP-measured size distributions were not available for several flights (e.g., 14 and 15 July) and the corresponding combined size distributions were generated from spectra measured by two instruments (UHSAS and CAS) only. These combined distributions obtained without PCASP data are excluded from our analysis. Similar to the size distributions, ambient scattering coefficients were not available for some episodes as well. Given the large uncertainties of the obtained f(RH) at very humid conditions (RH > 80%), the corresponding cases are also excluded from our analysis. Using the quality assurance screening criteria reduces the size of the original dataset by about 30%. A total of 45 TCAP FLs with the good quality data are included in our analysis. Further discussion of TCAP flight data quality can be found in Berg *et al.* [45].

4. Model and Adjustments

This section outlines the model components of our framework for an airborne optical closure experiment (Figure 1; middle panel) and describes the major assumptions required for estimating the hygroscopic growth factor (Section 4.1), ambient values of complex refractive index (Section 4.2) and correcting the OPC-derived size distributions (Section 4.3). The RI-based corrections of the size spectra together with estimates of HGF and complex RI form the basis for calculating the ambient total scattering (Section 4.4) with improved accuracy.

4.1. Hygroscopic Growth Factor

Water uptake by aerosol particles results in increased particle size and modifies the complex RI [36]. To account for changes associated with water absorption, information on aerosol chemical composition and hygroscopicity is needed. Typically, aerosol particles are a mixture of organic and inorganic substances. We estimate the hygroscopic growth factor of the mixture (HGF_{mix}) from the volume fractions of individual components (ε) and their growth factors as the volume-weighted average [46]:

$$HGF_{mix} = \left(\sum_i \varepsilon_i HGF_i^3 \right)^{1/3} \tag{1}$$

We convert the mass fractions measured by the AMS and SP2 instruments (Section 2) into the required volume fractions using densities reported in literature and listed in Table 1. We emphasize that the AMS and SP2 data are capturing a limited size range (sub-micron particles; Section 2) only. However, we use the AMS/SP2 data to infer the chemical composition of the entire size range (both sub- and super-micron particles). The application of the AMS/SP2 data for the entire size range should be appropriate for cases where the relative contribution of sub-micron particles to the scattering coefficient is dominant. Given that the aircraft data collected during the TCAP data represent such a favorable case with large contributions of sub-micron particles (Appendix B), application of the AMS/SP2 data for the entire size range is appropriate.

53

Table 1. Assumed size-independent density, real and imaginary parts of complex refractive index (RI) at 0.55 µm wavelength, and hygroscopic growth factor (HGF) values used in this study. Values are taken from [46,55,56].

	OM	SO$_4$	NO$_3$	Chl	NH$_4$	BC	Water
Density (g/cm^3)	1.4	1.8	1.8	1.53	1.8	1.8	1.0
RI (real)	1.45	1.52	1.5	1.64	1.5	1.85	1.33
RI (imag)	0.0	0	0	0	0	0.71	0
HGF (RH = 80%)	1.07	1.50	1.50	1.9	1.50	1.0	-

Although the growth factor can be quite sensitive to the particle size [21,57], we assume that particles with different sizes have the same HGF$_{mix}$. HGF$_{mix}$ from Equation (1) represents the hygroscopic growth factor at a specified relative humidity (RH$_{wet}$ = 80%). To obtain HGF$_{mix}$ for a different humidity, we assume that HGF$_{mix}$ follows the power law form [19,21]:

$$HGF_{mix}(RH) = \left(\frac{100 - RH_{dry}}{100 - RH} \right)^{\gamma} \tag{2}$$

where RH$_{dry}$ is 30% . The dimensionless exponent γ is calculated as:

$$\gamma = \frac{\log\left(HGF_{mix}(RH_{wet}) \right)}{\log\left[\left(100 - RH_{dry}\right) / (100 - RH_{wet}) \right]} \tag{3}$$

Recall that OM is the dominant component of aerosol sampled during the TCAP flights (Figure 2 and [45]). Given that the hygroscopic growth factor of the OM is relatively small compared to other chemical components (Table 1), the calculated RH-dependent HGF$_{mix}$ does not exceed 1.3 and approaches 1 as RH decreases (Figure 5).

4.2. Dry and Wet Refractive Indices

We apply a volume weighting approach to calculate the real (n_{dry}) and imaginary (k_{dry}) parts of the complex RI (m_{dry}) of particles at dry conditions:

$$m_{dry} = \sum_i \varepsilon_i m_{dry,i} \tag{4}$$

where m$_{dry,i}$ represents the real ($n_{dry,i}$) or imaginary ($k_{dry,i}$) part for each measured chemical component (Table 1). The underlying assumption of this popular approach is that the contribution of each chemical component to the light scattering and absorption is proportional to its volume fraction (ε_i). It should be emphasized that chemical composition, in general, depends on the particle size [57]. The same is true for the real (n_{dry}) and imaginary (k_{dry}) parts of the complex RI [40]. Here we assume that particles with different sizes have the same chemical composition, and therefore the same RI.

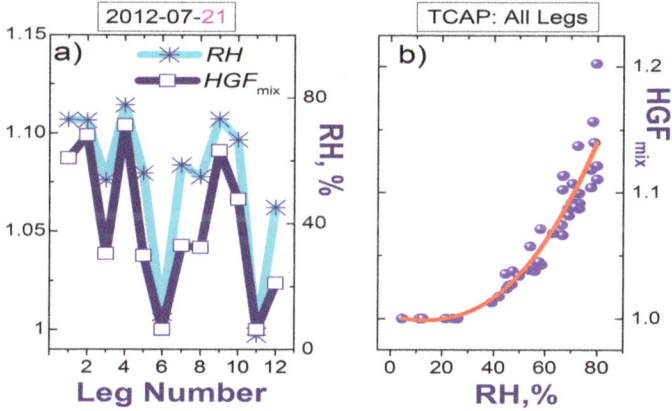

Figure 5. Example of ambient RH and RH-dependent HGF_{mix} calculated for each FL during a given day (21 July 2012) (**a**); scatterplot of RH-dependent HGF_{mix} (blue dots) for all TCAP FLs used in this study (Section 3) with polynomial fit (red line) (**b**).

The calculated dry RIs are applied to compute the corresponding ambient values [21,58]:

$$m_{wet} = \frac{m_{dry} + m_{water}\left(HGF_{mix}^3 - 1\right)}{HGF_{mix}^3} \tag{5}$$

where HGF_{mix} is the RH-dependent parameter defined in the previous section. As HGF_{mix} increases noticeably (near 1.1; Figure 5a), water becomes an influential component and the ambient RI decreases toward the water RI (Figure 6). In contrast, when HGF_{mix} is quite small (near 1; Figure 5a), the ambient RI increases toward the dry RI (Figure 6). Note that large values of the imaginary part of the RI (e.g., FL numbers 3, 6, 11 and 12) represent conditions where the relative contribution of BC to the total loading is substantial (>3%; Figure 3) and exceeds those for other FLs roughly by a factor of two.

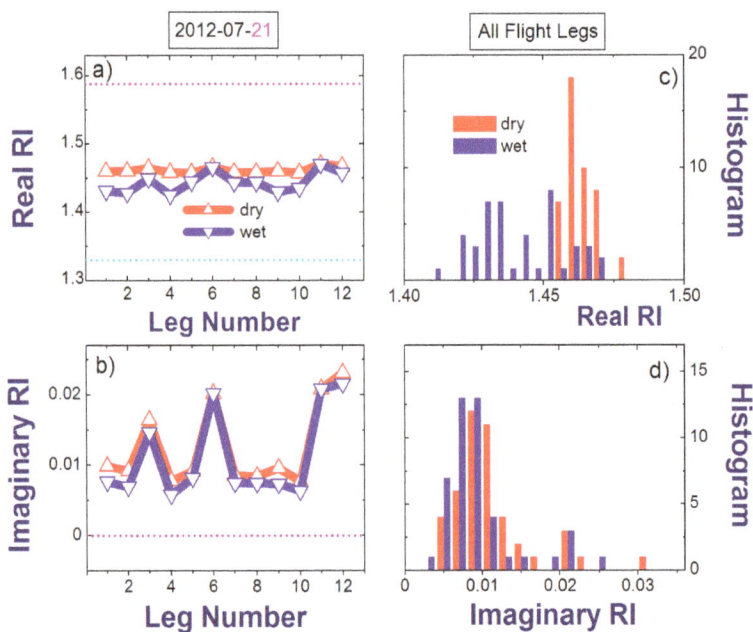

Figure 6. Example of RH-dependent dry and ambient values of the real and imaginary parts of the complex RI calculated for each FL during a given day (21 July 2012) (**a,b**); the corresponding histograms obtained for all TCAP FLs (**c,d**). The real (n_{OPC} = 1.588) and imaginary (k_{OPC} = 0) parts of the complex RI used for OPC calibration are shown in (a) (magenta) and (b) (magenta), respectively. The real RI of water (n_{water} = 1.33) is also shown (a, cyan). The imaginary RI of water (k_{water} = 0) is equal to the imaginary RI used for OPC calibration (k_{OPC} = 0).

4.3. Size Distribution

There is a substantial difference between the RI used for OPC calibration (n_{OPC} = 1.588; k_{OPC} = 0) and the FL-dependent dry RI calculated from chemical composition data (Figure 6). To take into account this difference, we apply the RI-based correction to adjust the OPC-derived size distributions (see Appendix C for further discussion). It should be emphasized that the PSL-based OPC calibration applies a RI that very likely overestimates those for the typical sub-micron aerosol. Since a higher RI produces a larger scattering signal, a particle with smaller size and higher RI scatters the same as a particle with larger size and smaller RI. As a result, all of the adjustments of the OPC calibration to the actual RI will increase the size of measured particles, and consequently will increase the total scattering for almost any reasonable size distribution. Moreover, adding an absorbing component further increases the size of the measured particles—after adjustment—due to the reduction of the particle's scattering and consequent decrease of its optical size. We will illustrate such increases of particle size and scattering in the following two sections.

The size spectra adjustment involves modification of the OPC-measured dry diameter of particles:

$$D_{dry,adj} = f\left(D_{dry}\right) D_{dry} \tag{6}$$

where $f(D_{dry})$ is the *size-dependent* scaling factor obtained from theoretical response calculations by extending well-established approaches for correcting the OPC-derived size distributions [30,31,34]. In our work we take advantage of available AMS and SP2 measurements (Section 3) for estimating the complex RI (Section 4.2), and thus our approach can be applied to both non-absorbing and absorbing aerosol.

The original OPC-derived and corrected dry size distributions are related as:

$$\frac{dN}{dlog\ D_{dry,adj}} = \frac{dN}{dlog\ D_{dry}} \frac{dlog\ D_{dry}}{dlog\ D_{dry,adj}} \tag{7}$$

In other words, the *size-dependent* correction (Equation (6)) modifies the original OPC-derived size spectra in two ways by (1) changing bin boundaries (horizontal shifting; replacement of D_{dry} with $D_{dry,adj}$) and (2) scaling of the normalized number concentration (vertical shifting; term d log D_{dry} /d log $D_{dry,adj}$).

The increase of particle diameter due to the water uptake is expressed as:

$$D_{wet} = HGF_{mix}\left(RH\right) D_{dry} \tag{8a}$$

$$D_{wet,adj} = HGF_{mix}\left(RH\right) D_{dry,adj} \tag{8b}$$

where the *size-independent* $HGF_{mix}(RH)$ is calculated from the chemical composition measurements (Section 4.1).

The corresponding dry and wet size distributions are related by the following equations:

$$\frac{dN}{dlog\ D_{wet}} = \frac{dN}{dlog\ D_{dry}} \frac{dlog\ D_{dry}}{dlog\ D_{wet}} = \frac{dN}{dlog\ D_{dry}} \tag{9a}$$

$$\frac{dN}{dlog\ D_{wet,adj}} = \frac{dN}{dlog\ D_{dry,adj}} \frac{dlog\ D_{dry,adj}}{dlog\ D_{wet,adj}} = \frac{dN}{dlog\ D_{dry,adj}} \tag{9b}$$

Note that functions in Equation (9) (left-hand *versus* right-hand side) have different arguments; for example, dN (D_{wet})/d log D_{wet} (Equation (9a); left-hand side) and dN (D_{dry})/d log D_{dry} (Equation (9a); right-hand side). In comparison with the *size-dependent* scaling factor (Equation (6)), the *size-independent* adjustment (Equation (8)) associated with water uptake modifies the original OPC-derived size spectra by changing bin boundaries only (horizontal shifting; e.g., replacement of

D_{dry} with D_{wet}; Equation (9a)); this adjustment does not cause the vertical scaling of the normalized number concentration (Equation (7) *versus* Equation (9)).

To illustrate the conversion of the measured dry size distribution into the corresponding wet size spectra (Equations (8a) and (9a)), we select two flight legs with high (RH~78%) and low (RH~5%) values of relative humidity (Figure 5a). The RH-related increase of particle size results in the horizontal shifting of the size distribution (wet size spectra *versus* dry size spectra) and this shifting is seen for the humid conditions (Figure 7a *versus* c). Since the theoretical response curves are multivalued for a resonance region (particle diameter > 0.5 μm) (Appendix C), the adjusted size distributions obtained at ambient RH with the *size-dependent* scaling factor (Equations (8b) and (9b)) exhibit a "bumpy" behavior for this region as well (Figure 7b,d).

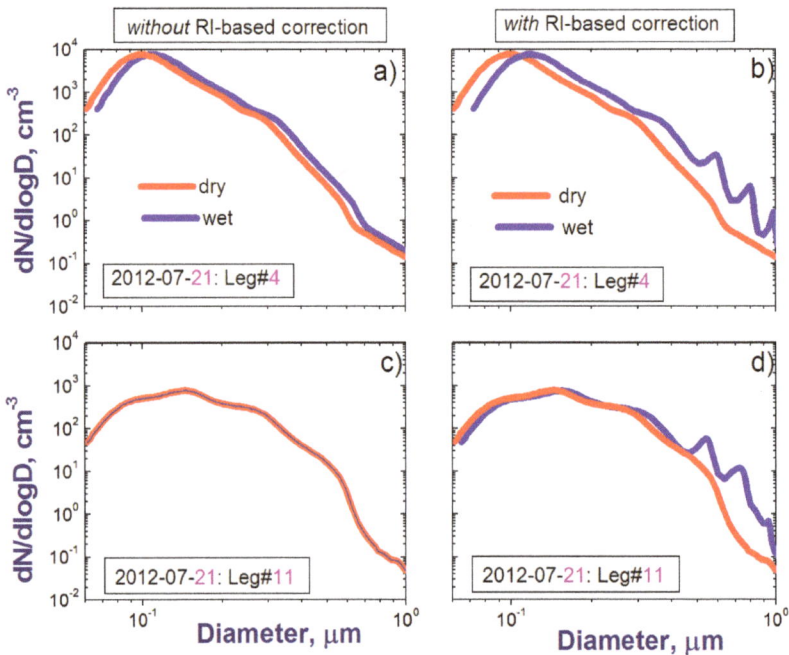

Figure 7. Example of size distributions obtained for two FLs (21 July 2012) with high (**a,b**) and low (**c,d**) values of ambient RH, respectively. Measured dry size distributions (red) are converted into their wet counterparts (blue) without (a,c) and with (b,d) the size-dependent scaling factor.

4.4. Scattering Coefficient Calculations

We calculate the ambient total scattering coefficient using both the ambient complex RI with real (n_{wet}) and imaginary (k_{wet}) parts (e.g., Figure 6a,b) and the ambient size distributions (e.g., Figure 7b,d). The calculations of total scattering coefficients $\sigma_{mod,org}$ and $\sigma_{mod,adj}$ are performed for ambient size distributions obtained without (dN/d log D_{wet}) and with (dN/d log $D_{wet,adj}$) the RI-based correction, respectively. Recall the RI is obtained at a single wavelength (0.55 μm). Results from previous studies [59,60] suggest that the spectral variability of the real part of RI is quite small within the visible spectral range considered here (0.45–0.7 μm). Therefore, the obtained real RI (0.55 μm) is likely representative for this spectral range and it can be applied to calculate the spectrally-resolved total scattering coefficient of weakly-absorbing aerosol. Note that the values of the imaginary part of RI at 0.55 μm wavelength are quite small (Figure 6b) and thus they represent weakly-absorbing aerosol. For such aerosol, the total scattering is only slightly affected by changes of the imaginary part of RI [59,60]; therefore its spectral dependence can be ignored for the dataset considered here. We perform calculations of the total scattering at three wavelengths (0.45, 0.55, 0.7 μm) using the same complex RI obtained at 0.55 μm wavelength.

Our calculations are based on the Mie code developed by *Barber and Hill* [61] assuming that particles are homogeneous spheres and effective values of the complex RI are size-independent (particles are assumed to be a homogeneous internal mixture; Section 4.2). Note that other computational methods should be applied to calculate optical properties of particles with inhomogeneous internal mixing and aggregate morphology [62] although the influence of the internal mixing and particle geometry on the total scattering is quite small for submicron particles [20,63,64]. We calculate the total scattering coefficients using the original and corrected size distributions with different cut-offs (1- and 2-μm). Large uncertainties (up to 50%) of the measured total scattering associated with coarse mode particles (particle diameter >1 μm) [54], the limited size range of chemical composition data (particle diameter <1.0 μm) (Section 2), and small (<7% on average) relative contribution of coarse mode particles to the scattering coefficient (Appendix B) are the three main factors that led to the selected cut-offs. Note that there is a small difference (~2% on average) between total scattering values calculated for the 1- and 2-μm cut-offs (Appendix B), mainly due to the small fraction of supermicron particles for FLs considered here. Below, we show the ambient scattering coefficients ($\sigma_{mod,org}$ and $\sigma_{mod,adj}$) calculated with the 2-μm cut-off only.

Uncertainties for the calculated scattering coefficient are associated mainly with ambiguities of the required inputs (ambient size distribution and complex RI) and model assumptions (homogeneous internal mixture). According to previous studies with similar model inputs [65], the uncertainties for the calculated dry total scattering are about 20%. Although assumptions associated with the hygroscopic growth factor estimation (Section 4.1) are likely to introduce additional ambiguities associated with the required inputs, we assume that the uncertainties obtained earlier for the dry scattering (20%) are also representative (at least as a lower limit) for the ambient total scattering considered in our study. The influence of the uncertainties mentioned above on agreement between the observed and calculated values at ambient conditions is considered in the next section.

5. Results

This Section outlines the comparison component of our framework for an airborne optical closure experiment (Figure 1; bottom panel) and includes time series and statistics of the observed and calculated total scattering coefficients at ambient conditions. This Section is designed to address questions 1 (*level of agreement for "complete" dataset*) and 2 (*level of agreement for "incomplete" dataset given that the impact of chemical composition data on improved size spectra is ignored*).

Let us start with the time series (Figure 8). The total modeled scattering $\sigma_{mod,org}$ calculated for the original size distribution substantially underestimates the observed scattering σ_{obs} for the majority of FLs considered on 21 July 2012. In contrast to $\sigma_{mod,org}$, the total scattering $\sigma_{mod,adj}$ calculated for the adjusted size distribution matches the observed scattering σ_{obs} reasonably well.

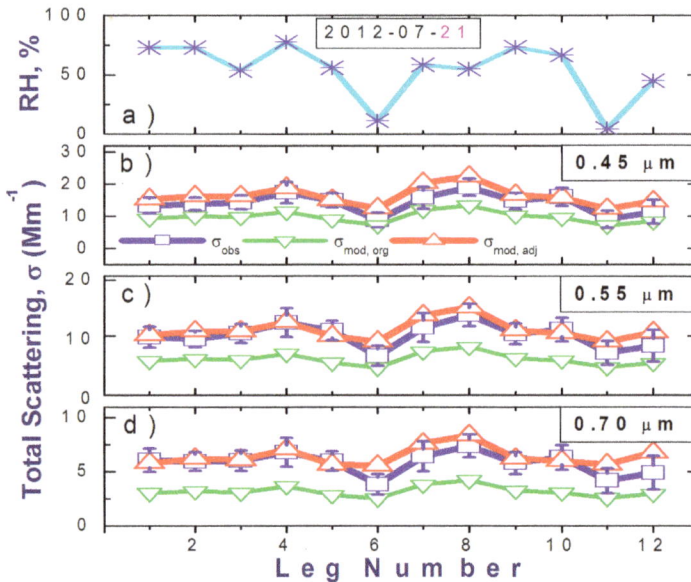

Figure 8. Ambient RH (**a**) and spectral values (**b–d**) of the total scattering coefficient measured (blue) and calculated for the original (green) and RI-based adjusted (red) size distributions for twelve FLs on 21 July 2012 at three wavelengths: (b) 0.45; (c) 0.55 and (d) 0.70 μm. Error bars represent uncertainties of measured scattering coefficient.

The differences in Figure 8 between calculated ($\sigma_{mod,org}$ and $\sigma_{mod,adj}$) and observed (σ_{obs}) scattering illustrate three main points. First, the observed scattering σ_{obs} tends to follow the ambient RH variations. For example, the lowest values of σ_{obs} occurred for dry conditions (FL numbers 6 and 11, where RH < 10%). Second, inclusion of the hygroscopic growth of particles associated with water uptake alone (Section 4.3) is not sufficient for matching the observed ambient scattering ($\sigma_{mod,org}$ versus σ_{obs}). In addition to the hygroscopic growth, application of the RI-based correction to the dry size distributions measured by the OPCs (Section 4.3) is required ($\sigma_{mod,adj}$ versus σ_{obs}). Finally, the calculated scattering $\sigma_{mod,adj}$ reproduces reasonably well the observed scattering σ_{obs} at all three wavelengths considered here (0.45, 0.55 and 0.70 µm), suggesting that our assumption of a spectrally-independent RI seems reasonable for the weakly-absorbing aerosol sampled during the TCAP flights.

The substantial underestimation of $\sigma_{mod,org}$ versus σ_{obs} noticeable in Figure 8 is more obvious in the scatterplot of all considered TCAP FLs shown in Figure 9a. Since both the measured and calculated scattering can exhibit large uncertainties, the bivariate weighted method [66] is used to find the best linear fit. This method owes its popularity [67,68] to its versatility, robustness and ability to use uncertainties of both x and y variables to find the slope and intercept of a best fit straight line to the data [69]; additionally, standard errors of the slope and intercept may be estimated.

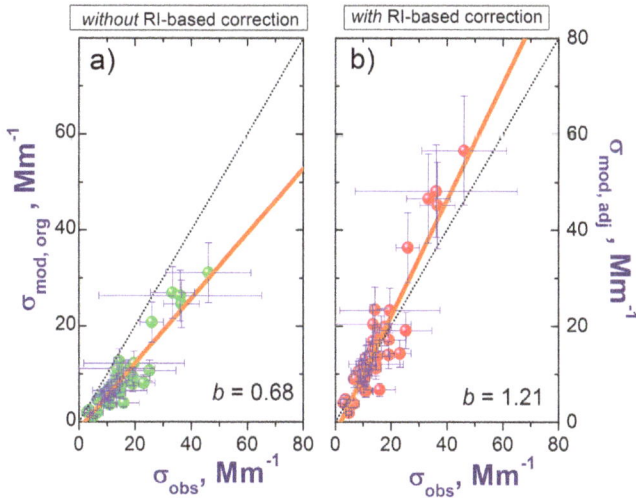

Figure 9. Comparison of the ambient total scattering observed (σ_{obs}) with ambient total scattering calculated (σ_{mod}) for the original (**a**) and adjusted (**b**) size distributions at 0.55 µm wavelength for all TCAP FLs. Here b is the slope of the linear regression fits to the data (straight orange lines). Error bars represent uncertainties of measured (Section 3) and calculated (Section 4.4) scattering coefficients.

The slope of the corresponding linear regression fit is quite small (0.68; Figure 9a). The slope increases from 0.68 to 1.21 when the RI-based correction is applied to the OPC-derived size distributions (Figure 9a *versus* b). There are four points with large observed values of scattering coefficient (σ_{obs} > 30 Mm^{-1}, Figure 9). Close examination reveals that these points represent either large values of RH (>67%) or highly inhomogeneous FLs (standard deviation of σ_{obs}~30 Mm^{-1}). Removal of these points reduces slightly the slope (from 0.68 to 0.65, Figure 9a; and from 1.21 to 1.17, Figure 9b) and increases the intercept (from -1.30 to -1.11, Figure 9a; and from -2.25 to -1.91, Figure 9b). Additionally, such removal reduces the difference between the mean values of σ_{obs} and $\sigma_{mod,adj}$ considerably (roughly from 10% to 0.5%), while having little effect on the discrepancy between the mean values of σ_{obs} and σ_{mod} (about 40%).

Similar changes of parameters (slope and intercept) obtained with the bivariate weighted method can be achieved by increasing the uncertainties of the calculated scattering coefficient but keeping uncertainties of the observed scattering the same (Figure 10). These changes are mostly caused by the influence of four points with large values of the observed/calculated scattering coefficient on the parameters of fitting line. The influence of these points decreases when assumed uncertainties of the calculated scattering are increased. Note that the assumed uncertainties (20%) of the calculated scattering represent dry conditions (Section 4) and thus, they likely are underestimated for FLs with moderate and high RHs (e.g., due to inaccurate treatment of the RH dependence of the HGF_{mix} and its components).

Comparable slopes for linear regression fits ($\sigma_{mod,adj}$ *versus* σ_{obs}) are obtained at the other two wavelengths (0.45 and 0.7 µm) (Table 2). The weak spectral dependence of these slopes suggests that the complex RI obtained at 0.55 µm wavelength could be applied to estimate the spectrally-resolved total scattering within the mid-visible range (0.45–0.70 µm) reasonably well.

On average, the total scattering $\sigma_{mod,org}$ calculated for the original OPC-derived size distribution underestimates the observed scattering σ_{obs} substantially (Table 2). For example, underestimation of the mean value exceeds 40% at 0.55 µm wavelength. However, the total scattering $\sigma_{mod,adj}$ calculated for the RI-adjusted size distributions and the observed scattering σ_{obs} have comparable mean values (Table 2): the relative difference between them is quite small at three given wavelengths (about 13% at 0.45 µm, 7% at 0.55 µm and 12% at 0.70 µm). In comparison with the mean values, the corresponding standard deviations are in moderate agreement (~40%). In addition to the basic statistics of mean and standard deviation, we calculate the corresponding Root Mean Squared Error (RMSE), which is defined as the root mean squared difference between the observed σ_{obs} and the calculated scattering coefficients in question. The corresponding RMSEs are about 7 and 5 Mm^{-1} for

$\sigma_{mod,org}$ and $\sigma_{mod,adj}$ at 0.55 µm wavelength, respectively. In other words, RMSE $(\sigma_{mod,org})$ overestimates RMSE $(\sigma_{mod,adj})$ noticeably (~30%).

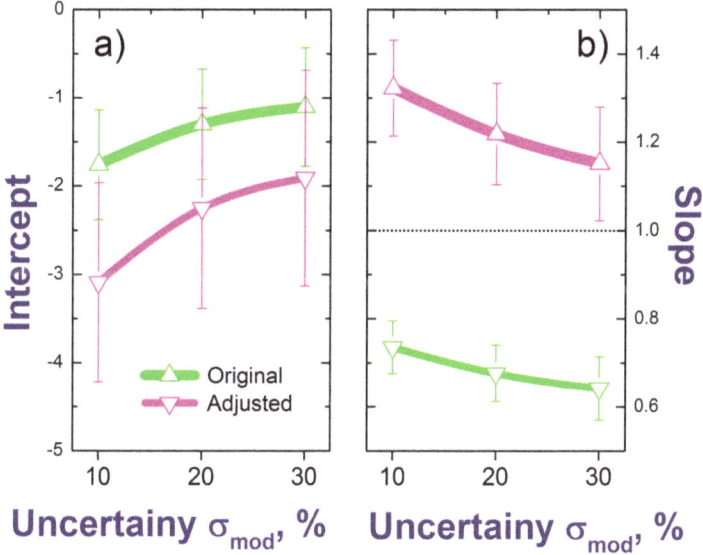

Figure 10. Intercept (**a**) and slope (**b**) of the linear regression fits (e.g., Figure 9) as a function of uncertainty for the ambient total scattering calculated for the original (without RI-based correction) and adjusted (with RI-based correction) size distributions. Error bars represent the standard errors of the displayed parameters (see text for details).

Overall, the quantitative comparisons (Figure 9 and Table 2) demonstrate clearly that application of the RI-based correction improves the agreement between observed and calculated scattering coefficients, most notably in terms of the mean values and RMSE. Thus, our results are in line with findings from previous studies [30,31], which highlighted the importance of such RI-based corrections and suggested corresponding parameterizations for non-absorbing aerosol. These valuable parameterizations were designed assuming that the RI-based correction was a function of the real part of complex RI only. In our approach, both the real and imaginary parts of the complex RI are used as input and thus it can be applied for both non-absorbing and absorbing aerosol. It should be noted that the single-scattering albedo is moderate (0.93 ± 0.03) for the TCAP dataset considered here, therefore this dataset represents slightly-absorbing aerosol. Further studies are needed to examine the feasibility of our approach for improved calculations of total scattering/absorption for strongly-absorbing aerosol.

Table 2. Mean and Standard Deviation (StDv) of the observed (σ_{obs}) and calculated ($\sigma_{mod,org}$ and $\sigma_{mod,adj}$) scattering coefficients obtained for all TCAP FLs at three wavelengths: 0.45 μm (**top** part), 0.55 μm (**middle** part), and 0.70 μm (**bottom** part). The corresponding intercept (*a*), slope (*b*), their standard errors (in parenthesis) and Root Mean Squared Error (RMSE) also are included. The RMSE is defined as the root mean squared difference between the observed and the calculated scattering coefficients.

	Mean	StDv	RMSE	*a*	*b*
			0.45 μm		
σ_{obs}	20.05	12.26	-	-	-
$\sigma_{mod,org}$	13.70	10.42	7.49	−1.58 (0.80)	0.75 (0.06)
$\sigma_{mod,adj}$	22.75	17.09	7.10	−2.54 (1.33)	1.24 (0.10)
			0.55 μm		
σ_{obs}	14.85	8.98	-	-	-
$\sigma_{mod,org}$	8.80	6.78	6.99	−1.30 (0.63)	0.68 (0.06)
$\sigma_{mod,adj}$	15.89	12.17	5.01	−2.25 (1.13)	1.21 (0.11)
			0.70 μm		
σ_{obs}	8.73	5.28	-	-	-
$\sigma_{mod,org}$	4.87	3.78	4.61	−0.73 (0.45)	0.64 (0.07)
$\sigma_{mod,adj}$	9.77	7.57	3.85	−1.40 (0.90)	1.26 (0.15)

6. Sensitivity Study

This section is designed to supplement the previous one and to address question 3 (*level of agreement for "incomplete" dataset given that a non-representative RI is used to correct size spectra*). Note that we are able to use aerosol chemical composition data from the TCAP data set to obtain RI-based corrections for each FL. However, as discussed in Section 1, a more common scenario is one where chemical composition information is available only on a limited basis, or not at all. We wish to investigate the sensitivity of the agreement between calculated and observed scattering to the assumed RI value, particularly if the assumed RI is non-representative of the ambient aerosol. Further motivation for this sensitivity test is the reported lack of agreement (outside the 30% measurement uncertainty) between measured and calculated total scattering coefficients obtained recently for the VOCALS-Rex marine atmosphere campaign [46] where a *universal* refractive index ($n^* = 1.41$ and $k^* = 0$) derived from the entire VOCALS-Rex dataset [29] was used in adjusting observed PCASP size distributions.

Given the strong sensitivity of VOCALS-Rex calculated scattering to size spectra uncertainties [46], it can be hypothesized that the lack of agreement can be associated (at least partially) with the RI specification and its strong impact on the adjusted size distributions. To confirm this hypothesis, we calculate the total scattering from the TCAP data (Section 2) using the procedure previously employed [46] for the VOCALS-Rex dataset. We emphasize that the main difference between their sensitivity-driven procedure and our framework approach (Figure 1) is specification of the RI required for adjusting the OPC-derived size spectra: an assumed universal real RI estimated implicitly from all available aerosol composition measurements (sensitivity-driven procedure) *versus* a variable complex RI estimated explicitly from the complementary chemical composition data (our framework approach, Sections 3–5).

We apply an assumed universal RI ($n^* = 1.41$ and $k^* = 0$), equivalent to that used in VOCALS-Rex analyses, to the entire TCAP dataset and compare the results with those from the variable RI estimated from the TCAP AMS and SP2 data. This assumed RI might reasonably be adopted if TCAP had an "incomplete" data set and we searched the literature for a reasonable universal RI applicable to coastal/marine aerosols sampled aloft. Note that there is a noticeable difference between the assumed RI and the estimated RI (Section 3) for many FLs. Compared with the variable RI, the use of the *universal* RI in the size spectra adjustment increases the relative contribution of particles with moderate diameter (within 0.4–0.8 µm range) (Figure 11), which scatter light in the visible spectral range most effectively. This relative increase in optically important particles, in turn, is responsible for a substantial rise of the calculated scattering coefficient (Figures 12 and 13a). As a result, the corresponding mean value and RMSE (calculations based on variable RI *versus* the universal RI) are enhanced by about 25% and 80%, respectively (Tables 2 and 3). The mean value of the calculated scattering coefficient (based on universal RI) overestimates the mean value of observed scattering by about 35% (Table 3). This substantial overestimation (~35%) of the calculated scattering coefficient: (1) confirms the hypothesis made above regarding the potential strong impact of the RI specification on the calculated scattering and (2) suggests that use of a universal RI specification could be one possible explanation for the lack of agreement noted in the VOCALS-Rex closure study [46].

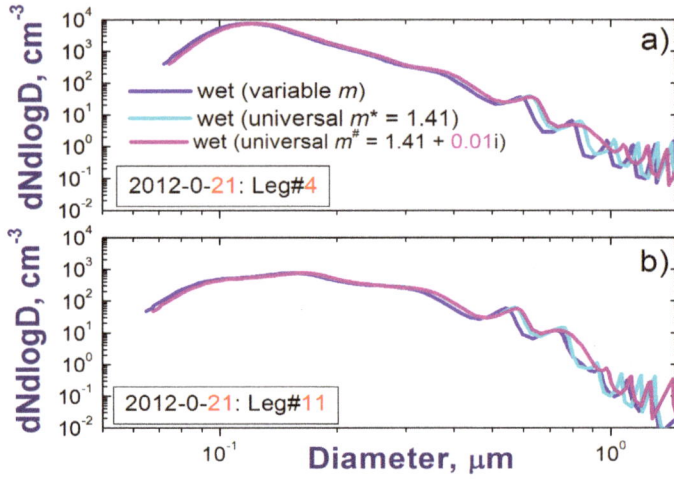

Figure 11. Example of size distributions obtained for two FLs (21 July 2012) and different values of RI. Measured dry size distributions are converted into their wet counterparts with the size-dependent scaling factor calculated for universal (cyan, magenta) and variable (blue) RI. Two values of universal RI are assumed: $m^* = 1.41$ (cyan) for a non-absorbing aerosol and $m^\# = 1.41 + 0.01i$ (magenta) for an absorbing aerosol.

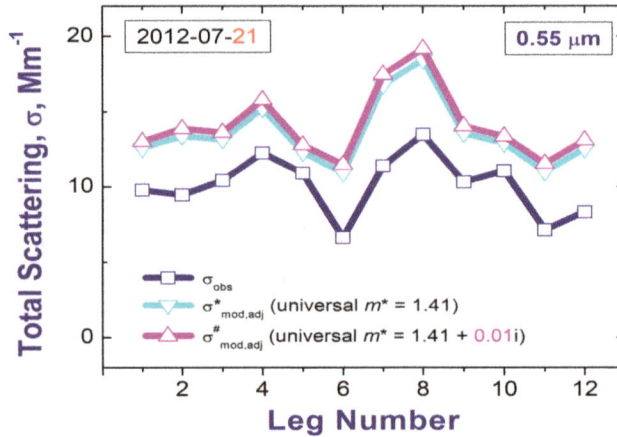

Figure 12. Total scattering coefficient at 0.55 μm wavelength measured (navy blue) and calculated (cyan, magenta) for size distributions measured on 21 July, 2012 and corrected using two assumed universal RIs: $m^* = 1.41$ (cyan) and $m^\# = 1.41 + 0.01i$ (magenta).

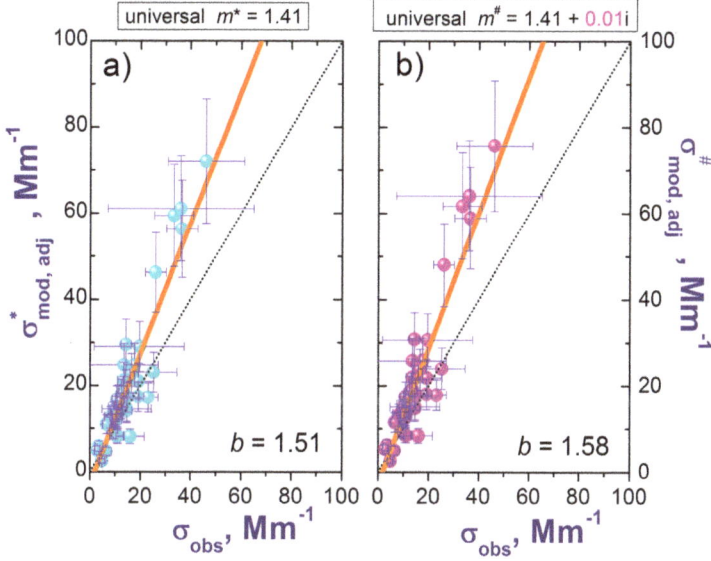

Figure 13. Comparison of the ambient total scattering observed (σ_{obs}) with ambient total scattering calculated (σ_{mod}) for the size distributions adjusted with the universal RI. There are two values of assumed universal RI: $m^* = 1.41$ (**a**) and $m^\# = 1.41 + 0.01i$ (**b**). The observed and calculated values of total scattering are obtained at 0.55 µm wavelength for all TCAP FLs. Here, b is the slope of the linear regression fits to the data (straight orange lines). Error bars represent uncertainties of measured (Section 3) and calculated (Section 4.4) scattering coefficients.

To examine the importance of the imaginary part of the complex RI in changing the adjusted size distribution and thereby in modifying the calculated scattering, we focus on the complex RI for an absorbing aerosol ($m^\#$). We assume that the corresponding universal complex RI has the real ($n^\# = 1.41$) and imaginary ($k^\# = 0.01$) parts. This selected value of the imaginary part ($k^\# = 0.01$) represents roughly the most frequent value observed under dry conditions during the TCAP campaign (Figure 6). Therefore, the imaginary part is the only difference between the complex RIs assumed for non-absorbing ($m^* = 1.41$) and absorbing ($m^\# = 1.41 + 0.01i$) aerosol.

Replacement of the imaginary part of the complex RI ($k^* = 0$ *versus* $k^\# = 0.01$) increases the difference between the corresponding size distributions (Figure 11; cyan curve *versus* magenta curve). Note that this replacement ($k^* = 0$ *versus* $k^\# = 0.01$) makes the adjusted size distributions "look smoother" within the resonance region (Figure 11; cyan curve *versus* magenta curve). The difference between the adjusted size distributions (Figure 11) is responsible for the difference between the corresponding total scattering coefficients (Figure 12, Figure 13 and Table 3). In particular, the replacement of the imaginary part of the complex RI ($k^* = 0$

versus $k^{\#} = 0.01$) increases the RMSE by about 10% (Table 3). Therefore, selection of an inappropriate imaginary part of the assumed complex RI can also increase the discrepancy between the calculated total scattering from the observed total scattering.

Table 3. The same as Table 2 (middle part) except for the scattering coefficients $\sigma^*_{mod,adj}$ (middle row) and $\sigma^{\#}_{mod,adj}$ (bottom row) calculated for size distributions adjusted with two universal RIs $m^* = 1.41$ and $m^{\#} = 1.41 + 0.01i$, respectively.

	Mean	StDv	RMSE	a	b
σ_{obs}	14.85	8.98	-	-	-
$\sigma^*_{mod,adj}$	19.82	15.46	9.17	−2.85 (1.42)	1.51 (0.14)
$\sigma^{\#}_{mod,adj}$	20.67	16.19	10.20	−2.96 (1.48)	1.58 (0.15)

The results presented in this Section illustrate that selecting inappropriate values of the real part of RI when adjusting OPC-derived size distributions in an optical closure study can cause substantial *overestimation* of the calculated scattering coefficient. Moreover, selecting inappropriate values of the imaginary part of RI can increase this overestimation noticeably, even for weakly absorbing aerosol. Therefore, one should not expect closure studies based on such "incomplete" data sets to be as exact as closure studies using "complete" data sets because of all the assumptions that must be made and the high probability that the RI estimates based on these assumptions potentially will be inappropriate. The error in RI estimates likely will be larger for aircraft sampling regimes because of the strong temporal and spatial variability of aerosol sampled by research aircraft. When faced with an "incomplete" data set, another possible approach is to use conventional iterative or optimization schemes, which apply a set of assumed representative RI values for minimizing differences between the measured and calculated aerosol properties of interest [34,40,70]. Such iterative or optimization schemes may possibly improve the RI estimation relative to the approach of assuming a *universal* RI.

7. Summary

We extend methods for calculating total aerosol scattering at ambient RH, originally developed for ground-based measurements [17,21,41] to the challenging situation of airborne measurements. The importance of such extension is now widely recognized [6,45]. Our extended framework is suitable for conducting optical closure studies using "complete" aircraft data sets, where "complete" means that collocated and concurrent information on particle chemical composition is available. Our approach takes advantage of the existing information on aerosol chemical constituents and explicitly uses it to obtain improved ambient size spectra derived from complementary Optical Particle Counter (OPC) data, and therefore to obtain improved estimates of the total scattering under ambient conditions with low-to-moderate values of relative humidity (RH < 80%).

To illustrate the performance of our approach, we use "complete" aerosol data collected by the DOE G-1 aircraft during the recent Two-Column Aerosol Project (TCAP; http://campaign.arm.gov/tcap/) over the North Atlantic Ocean and US coastal region (Cape Cod, Massachusetts). The integrated data set collected by the G-1 aircraft includes: (1) size distributions measured by three OPCs: an Ultra-High Sensitivity Aerosol Spectrometer (UHSAS; 0.06–1 µm), a Passive Cavity Aerosol Spectrometer (PCASP; 0.1–3 µm) and a Cloud and Aerosol Spectrometer (CAS; 0.6– > 10 µm), (2) chemical composition data measured by an Aerosol Mass Spectrometer (AMS; 0.06–0.6 µm) and a Single Particle Soot Photometer (SP2; 0.06–0.6 µm) and (3) the dry total scattering coefficient measured by TSI integrating nephelometer at three wavelengths (0.45, 0.55, 0.7 µm) and f(RH) measured with a humidification system at three RHs (near 45%, 65% and 90%) at a single wavelength (0.525 µm). To illustrate the importance of the chemical composition data in the scattering closure study, we also utilize "incomplete" aerosol data, where "incomplete" means that information on particle chemical composition is not used to obtain the corrected ambient size spectra. The main conclusions are organized along the three main questions we posed at the start of our study:

- Analysis based on using the "complete" data set addresses our first question, namely: *What level of agreement between the in-flight measured and calculated values of total scattering coefficient can be achieved at ambient RH?* We demonstrate that despite the well-known limitations of airborne measurements and the assumptions required by our approach, we can obtain good agreement between the observed and calculated scattering at three wavelengths (about 13% at 0.45 µm, 7% at 0.55 µm, and 12% at 0.7 µm on average) using the RI-based correction for OPC-derived size spectra and the best available chemical composition data for the RI estimation. We calculate the total scattering coefficient from the combined size spectra (UHSAS, PCASP and CAS data) and aerosol composition (AMS and SP2 data) at ambient conditions with a wide range of relative humidity values (from 5% to 80%). These calculations involve several assumptions, such as the homogeneous internal mixture assumption for estimating the hygroscopic growth factor and complex refractive index (RI) at ambient conditions, and simplified specification of particle geometry (homogeneous spheres) for Mie calculations.
- Analysis based on using an "incomplete" dataset addresses our second question, namely: *What is the effect of ignoring the influence of chemical composition data on this agreement?* We illustrate that ignoring the RI-based correction in the TCAP data can cause a substantial *underestimation* (about 40% on average) of the ambient

calculated scattering when noticeable discrepancies between the actual RIs and those used for the OPC calibration have occurred. Our findings are in harmony with previous studies, which have highlighted the importance of the RI-based correction and have suggested its parameterization for non-absorbing aerosol assuming that the RI-based correction is a function of real RI only [30,31]. In comparison with these important parameterizations, our approach is more flexible in terms of available inputs (complex RI is estimated explicitly from the complementary chemical composition data), and therefore in terms of the expected applications (both non-absorbing and absorbing aerosol sampled by ground-based and airborne instruments).

- Analysis based on using an "incomplete" dataset also addresses our third question, namely: *How sensitive is this agreement to the assumed RI value, particularly if the assumed RI is non-representative of the ambient aerosol?* We illustrate in a sensitivity study that using a non-representative universal RI instead of the actual RI can result in a large *overestimation* (about 35% on average) of the calculated total scattering at ambient RH, and this overestimation is sensitive to specification of the imaginary part of the complex RI, even for weakly-absorbing aerosol. This sensitivity study suggests that the usefulness of assumptions required for universal RI estimation could be marginal, particularly when applied to the strong temporal and spatial variability of aerosol sampled by research aircraft. As a result, calculations of aerosol optical properties based on these assumptions should be used with caution and other possible approaches should be considered to improve the RI estimation. These possibilities include application of conventional iterative or optimization schemes where a set of assumed representative RI values is used to minimize differences between the measured and calculated aerosol properties of interest [34,40,70].

To our knowledge, this work represents the first optical closure study that uses explicitly airborne chemical composition measurements of both non-absorbing and absorbing aerosol components in improving the OPC-derived size spectra. These measurements are employed to extend the capability of well-established methods originally developed for use with comprehensive ground-based measurements. Given the extended flexibility of these methods and the increasing availability of aerosol composition data collected from aircraft platforms, we expect that our approach can be successfully applied for improved understanding of a wide range of sophisticated processes and phenomena related to aerosols, including the time evolution of aerosol properties and dynamical aerosol-cloud interactions [71,72]. We further expect that closer agreement between measured and calculated aerosol properties, indicating the consistency of the observational data set, will improve confidence in, and use of, such observational data sets in global and regional climate model evaluations. For example, appropriate adjustments to optical particle counter data are needed to better understand and evaluate predictions of cloud-aerosol interactions, since cloud condensation nuclei (CCN) calculations are dependent on the aerosol size distribution.

Acknowledgments: The ARM Aerial Facility team is gratefully acknowledged for collecting the aircraft data during TCAP which was supported by the Department of Energy (DOE) Office of Science Atmospheric Radiation Measurement (ARM) and Atmospheric System Research (ASR) Programs. The ARM Aerial Facility is an integral part of the DOE ARM Program. This research was supported by the ARM and ASR Programs. The Pacific Northwest National Laboratory is operated by Battelle Memorial Institute under contract DE-AC06-76RLO 1830. We appreciate valuable discussions with Elaine Chapman (PNNL) and thoughtful comments from two anonymous reviewers that helped improve our paper.

Author Contributions: Abstract (E.K., L.B.), Introduction (E.K., L.B.), Approach (E.K., L.B., M.P., C.F., J.B.), Data (M.P., D.C., J.T., J.S., A.S., B.S.), Model and adjustments (E.K., L.B., M.P., M.O., J.S.), Results (all), Sensitivity study (E.K., L.B., M.P., M.O.), Summary (all). E.K. led the research and manuscript preparation. All authors contributed to data analysis. J.F. supervised research and manuscript writing.

Conflicts of Interest: The authors declare no conflict of interest.

Appendix A. Merging of Size Distributions

This appendix explains how data from overlapping measurements from three instruments (UHSAS, PCASP, and CAS) are merged into the combined size distributions (Figure 2). The best estimate aerosol size distribution $\vec{N}$ can be recovered from the merged raw counts $\vec{C}$ if a Kernel function (**R**) can first be quantified:

$$\vec{N}\,\mathbf{R} = \vec{C} \tag{A1a}$$

or

$$\begin{bmatrix} n_1 \\ n_2 \\ \vdots \\ n_{j-1} \\ n_j \end{bmatrix} \begin{bmatrix} R_{UHSAS11} & R_{UHSAS12} & & & \\ R_{UHSAS21} & R_{UHSAS22} & & & \\ & & \ddots & & \\ & & & R_{CASx-1\,j-1} & R_{CASx-1\,j} \\ & & & R_{CASx\,j-1} & R_{CASxx\,j} \end{bmatrix} = \begin{bmatrix} C_{UHSAS_1} \\ C_{UHSAS_2} \\ \vdots \\ C_{CAS_{x-1}} \\ C_{CAS_x} \end{bmatrix} \tag{A1b}$$

The raw counts are measured by the UHSAS, PCASP, and CAS. Furthermore, **R** can be defined as:

$$R_{UHSAS_{i,j}} = Q_{UHSAS}\,t\,e_{UHSAS_{i,j}}, \quad R_{PCASP_{i,j}} = Q_{PCASP}\,t\,e_{PCASP_{i,j}}, \quad R_{CAS_{i,j}} = V_{TAS}\,A\,t\,e_{CAS_{i,j}} \tag{A2}$$

where Q, V_{TAS}, $e_{i,j}$, A and t are the flow, airspeed, collection efficiency for each probe, CAS laser sample area and the data collection integration time, respectively. From calibrations we know the collection efficiency ($e_{i,j}$) of each probe within a specified bin size, and the CAS laser sample area (A). The other variables Q (flow) and V_{TAS} (airspeed) are measured by the probes themselves during operation.

To start the recovery process, a rough estimate of $\vec{N}$ has to be provided. The size distribution values dN/d log D_p for UHSAS, CAS, PCASP are first interpolated to the same bin space. The concentration values from the UHSAS, CAS, and or PCASP that overlap in the same bin are averaged together then converted back to dN by multiplying by the d log D_p value for that bin. Using the estimation for $\vec{N}$, Twomey smoothing is started. The initial number distribution is smoothed until the roughness of the solution has decreased to a set value, normally around 0.96. Roughness of the solution is measured by the average value of the second derivative. This initial trial solution is then ingested into a loop which continues until either the roughness of the solution has decreased below a set limit (typically 0.98), the goodness of the fit decreases, or the maximum number of iterations has been reached. After the loop successfully exits the resulting number distribution is considered the best estimate aerosol size distribution.

Appendix B. Contributions from Particles of Different Sizes to Scattering

To quantify contributions from particles of different sizes to the total scattering, we calculate the ambient scattering coefficient $\sigma_{mod,adj} = \int_{D_{min}}^{D^*} C_{sca}(m, D)\, dN/dlog\, D\, dlog\, D$ at single wavelength (0.55 µm), where C_{sca} is the scattering cross section, which is a function of particle diameter D and *size-independent* complex refractive index (m; Equation (5)), dN/d log D is the ambient and corrected number size distribution (Equation (9b)), $D_{min} \sim 0.07$ µm and D^* is an assumed cut-off (between 0.1 to 5 µm). Then we calculate normalized scattering coefficient $\rho(D^*) = \sigma_{mod,adj}(D^*)/\sigma_{mod,adj}(D^* = 5$ µm), which represents the relative contribution of particles with $D < D^*$ to the total scattering from all particles smaller than 5 µm. Figure B1a shows the mean and standard deviation for $\rho(D^*)$ computed for all 45 TCAP flight legs considered in this study (Section 3). Sub-micron particles clearly dominate the total scattering in this case with $\rho(D^*)$ reaching about 0.93 and 0.96 for $D^* = 1$ µm and $D^* = 2$ µm, respectively (Figure B1a). The obtained small (~7%, on average) contribution by the super-micron particles to the ambient scattering coefficient confirms the applicability of the simplified approach (Appendix C) to the TCAP airborne data. For events where super-micron particles dominate, for example, aerosol plumes resulting from dust storms [73] or volcanic eruptions [34] application of the strict approach (Appendix C) would be more relevant.

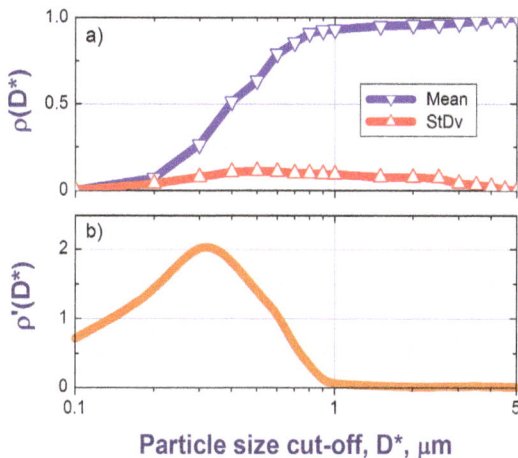

Figure B1. The normalized scattering coefficient (**a**) and its derivative (**b**) as a function of particle size cut-off. The blue line shows the mean for all TCAP flight legs and the red line represents the corresponding standard deviations.

To further illustrate the relative contribution to $\sigma_{mod,adj}$ from different particle size ranges (for a given complex refractive index), we note that the above relation for the ambient scattering coefficient can be rearranged as $\sigma_{mod,adj} = \int_{D_{min}}^{D^*} \left(d\sigma_{mod,adj}(D)/dD \right) dD$. Using this rearranged relation, we calculate the derivative of the normalized scattering coefficient as $\rho'(D^*) = \sigma'_{mod,adj}(D^*)/\sigma_{mod,adj}(D^* = 5 \,\mu m)$, where $\sigma'_{mod,adj}(D) = d\sigma_{mod,adj}(D)/dD$. Figure B1b shows that on average particles in the size range between 0.3 and 0.4 μm contribute most strongly to the ambient scattering coefficient, while the corresponding contribution of particles in the "bumpy" region (particle sizes larger than 0.5 μm) is smaller.

Appendix C. Correction of OPC-derived Size Spectra

This appendix includes details related to the calculations of the size-dependent theoretical response curves. The latter are required to obtain an adjusted dry diameter of particles (Equation (6)) and the corresponding dry (Equation (7)) and ambient (Equation (9b)) size distributions using the OPC-derived size spectra. Recall that during the OPC measurements individual particles are illuminated by a laser beam and then the light scattered by the particles is collected over a large solid angle [32–34,74]. The wavelength of the illuminating light and solid angle limits are known properties of a given detector [12]. For example, the UHSAS and PCASP collect side-scattered light (~35–135 degrees) at 0.6328 μm and 1.054 μm wavelength, respectively; while the CAS collects forward-scattered (~ 4–13 degrees) and back-scattered (~5–14 degrees) light at 0.685 µm wavelength. Therefore, strictly speaking, the scaling factors should be calculated for each instrument independently using the specified properties, and then the individual *corrected* size spectra (UHSAS, PCASP and CAS) should be used to obtain the corresponding merged size distributions. We shall call this the "strict approach." Another potential approach could include the development of empirical relationships between the scattering measured over the limited range of angles and the total scattering for given compositions and ambient conditions. Such an "empirical approach" would not involve Mie calculations and therefore would relax assumptions and data requirements for the RI-based corrections.

For our study, the individual *uncorrected* size spectra (UHSAS, PCASP and CAS) are combined to obtain the corresponding merged size distributions (Section 3) and then these merged distributions are corrected. We shall call this the "simplified approach." The basis of this simplified approach is the assumption that the properties of the PCASP (both the wavelength of the illuminating light and the solid angle limits) are representative of all merged size distributions. Such a simplification is motivated by the fact that the PCASP-derived size distributions represent particles in the 0.13–3 µm size range (Section 2), which scatter light most effectively in the visible spectral range considered here (about 0.2–0.7 µm). It can therefore be expected that the PCASP-derived size distributions contribute most of the total scattering measured by this airborne nephelometer.

We apply the simplified approach to calculate the theoretical response curves (Figure C1) and adjust particle diameters (Figure C2). Let us start with Figure C1. The first curve represents an assumed "experimental" one, which would be obtained from laboratory calibrations for particles with a known RI. Given that uncertainties of the experimental calibration are unknown, the "experimental" curve is the smoothed version of the corresponding theoretical response curve. A polynomial (curvilinear) regression model ([75]; pages 342–347) is applied to generate this smoothed version. Similar smoothed versions have been used previously for different OPCs [34]. The second and third curves represent low and high values of the imaginary part of RI obtained for two FLs (Figure 6b) and illustrate two important points. First, the corresponding curves are comparable and "smooth" for particles with diameter smaller than 0.75 µm; (Figure C1; upper zoom-in panel). In contrast, the curves are quite different and "bumpy" for large particles with diameter between 1.5 µm and 3.0 µm (Figure C1; lower zoom-in panel). Second, the observed "bumpy" behavior is responsible for the well-known multivalued issue, when particles with different diameter can produce the same response (Figure C1; lower zoom-in panel, green curve). In the presented calculations, we assume that the corresponding response is due to a particle with minimal diameter (Figure C1; lower zoom-in panel, red open circle on green curve). The observed differences between the second and third curves

(Figure C1) are responsible for the corresponding differences between the adjusted diameters (Figure C2).

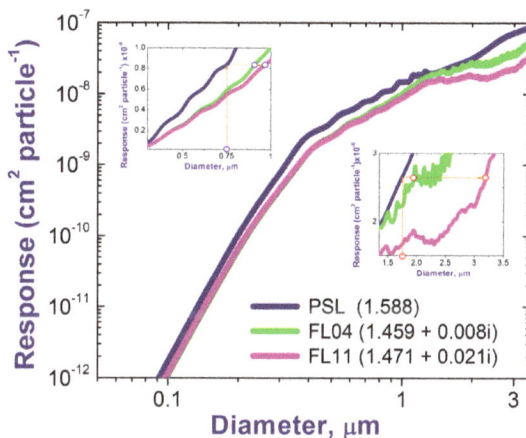

Figure C1. Example of theoretical response curves as a function of dry diameter (D_{dry}) calculated for spherical particles and different values of complex RI. These values represent polystyrene latex spheres (PSL; navy blue) and those calculated for two FLs (FL = 4, green; FL = 11, magenta; 21 July 2012) from chemical composition measurements under dry conditions. Two complementary zoom-in versions of the calculated response curves are also included to illustrate that these curves may exhibit "smooth" and "bumpy" behavior for different ranges of diameter (see text for details).

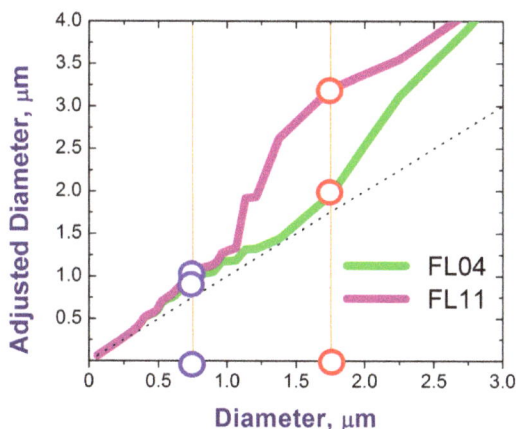

Figure C2. Adjusted dry diameter ($D_{dry,adj}$) as a function of original dry diameter (D_{dry}) obtained for two FLs (FL = 4, green; FL = 11, magenta; 21 July 2012) from the calculated response curves (Figure C1).

75

References

1. Wang, X.; Heald, C.L.; Ridley, D.A.; Schwarz, J.P.; Spackman, J.R.; Perring, A.E.; Coe, H.; Liu, D.; Clarke, A.D. Exploiting simultaneous observational constraints on mass and absorption to estimate the global direct radiative forcing of black carbon and brown carbon. *Atmos. Chem. Phys.* **2014**, *14*, 10989–11010.

2. Sundström, A.-M.; Arola, A.; Kolmonen, P.; Xue, Y.; de Leeuw, G.; Kulmala, M. On the use of a satellite remote-sensing-based approach for determining aerosol direct radiative effect over land: A case study over China. *Atmos. Chem. Phys.* **2015**, *15*, 505–518.

3. Adachi, K.; Chung, S.H.; Buseck, P.R. Shapes of soot aerosol particles and implications for their effects on climate. *J. Geophys. Res.* **2010**, *115*, D15206.

4. Bond, T.C.; Doherty, S.J.; Fahey, D.W.; Forster, P.M.; Berntsen, T.; DeAngelo, B.J.; Flanner, M.G.; Ghan, S.; Kärcher, B.; Koch, D.; *et al.* Bounding the role of black carbon in the climate system: A scientific assessment. *J. Geophys. Res. Atmos.* **2013**, *118*, 5380–5552.

5. China, S.; Mazzoleni, C.; Gorkowski, K.; Aiken, A.C.; Dubey, M.K. Morphology and mixing state of individual freshly emitted wildfire carbonaceous particles. *Nat. Comm.* **2013**, *4*, 2122.

6. Fast, J.D.; Allan, J.; Bahreini, R.; Craven, J.; Emmons, L.; Ferrare, R.; Hayes, P.L.; Hodzic, A.; Holloway, J.; Hostetler, C.; *et al.* Modeling regional aerosol and aerosol precursor variability over California and its sensitivity to emissions and long-range transport during the 2010 CalNex and CARES campaigns. *Atmos. Chem. Phys.* **2014**, *14*, 10013–10060.

7. Washenfelder, R.A.; Attwood, A.R.; Brock, C.A.; Guo, H.; Xu, L.; Weber, R.J.; Ng, N.L.; Allen, H.M.; Ayres, B.R.; Baumann, K.; *et al.* Biomass burning dominates brown carbon absorption in the rural southeastern United States. *Geophys. Res. Lett.* **2015**, *42*, 653–664.

8. Müller, D.; Hostetler, C.A.; Ferrare, R.A.; Burton, S.P.; Chemyakin, E.; Kolgotin, A.; Hair, J.W.; Cook, A.L.; Harper, D.B.; Rogers, R.R.; *et al.* Airborne Multiwavelength High Spectral Resolution Lidar (HSRL-2) observations during TCAP 2012: Vertical profiles of optical and microphysical properties of a smoke/urban haze plume over the northeastern coast of the US. *Atmos. Meas. Tech.* **2014**, *7*, 3487–3496.

9. Russell, P.B.; Kacenelenbogen, M.; Livingston, J.M.; Hasekamp, O.P.; Burton, S.P.; Schuster, G.L.; Johnson, M.S.; Knobelspiesse, K.D.; Redemann, J.; Ramachandran, S.; Holben, B. A multiparameter aerosol classification method and its application to retrievals from spaceborne polarimetry. *J. Geophys. Res. Atmos.* **2014**, *119*, 9838–9863.

10. Chaikovskaya, L.; Dubovik, O.; Litvinov, P.; Grudo, J.; Lopatsin, A.; Chaikovsky, A.; Denisov, S. Analytical algorithm for modeling polarized solar radiation transfer through the atmosphere for application in processing complex lidar and radiometer measurements. *J. Quant. Spectrosc. Radiat. Trans.* **2015**, *151*, 275–286.

11. Kokhanovsky, A.A.; Davis, A.B.; Cairns, B.; Dubovik, O.; Hasekamp, O.P.; Sano, I.; Mukai, S.; Rozanov, V.V.; Litvinov, P.; Lapyonok, T.; *et al.* Space-based remote sensing of atmospheric aerosols: The multi-angle spectro-polarimetric frontier. *Earth-Sci. Rev.* **2015**.

12. Baumgardner, D.; Brenguier, J.L.; Bucholtz, A.; Coe, H.; DeMott, P.; Garrett, T.J.; Gayet, J.F.; Hermann, M.; Heymsfield, A.; Korolev, A.; *et al.* Airborne instruments to measure atmospheric aerosol particles, clouds and radiation: A cook's tour of mature and emerging technology. *Atmos. Res.* **2011**, *102*, 10–29.

13. McFarquhar, G.; Schmid, B.; Korolev, A.; Ogren, J.A.; Russell, P.B.; Tomlinson, J.; Turner, D.D.; Wiscombe, W. Airborne instrumentation needs for climate and atmospheric research. *Bull. Amer. Meteor. Soc.* **2011**, *92*, 1193–1196.

14. Konwar, M.; Panicker, A.S.; Axisa, D.; Prabha, T.V. Near-cloud aerosols in monsoon environment and its impact on radiative forcing. *J. Geophys. Res. Atmos.* **2015**, *120*.

15. Russell, P.B.; Kinne, S.A.; Bergstrom, R.W. Aerosol climate effects: Local radiative forcing and column closure experiments. *J. Geophys. Res.* **1997**, *102*, 9397–9407.

16. Schmid, B.; Livingston, J.M.; Russell, P.B.; Durkee, P.A.; Jonsson, H.H.; Collins, D.R.; Flagan, R.C.; Seinfeld, J.H.; Gassó, S.; Hegg, D.A.; *et al.* Clear-sky closure studies of lower tropospheric aerosol and water vapor during ACE-2 using airborne sunphotometer, airborne in-situ, space-borne, and ground-based measurements. *Tellus B* **2000**, *52*, 568–593.

17. Malm, W.C.; Day, D.E.; Carrico, C.; Kreidenweis, S.M.; Collett, J.L., Jr.; McMeeking, G.; Lee, T.; Carrillo, J.; Schichtel, B. Intercomparison and closure calculations using measurements of aerosol species and optical properties during the Yosemite Aerosol Characterization Study. *J. Geophys. Res.* **2005**, *110*, D14302.

18. Mack, L.A.; Levin, E.J.T.; Kreidenweis, S.M.; Obrist, D.; Moosmüller, H.; Lewis, K.A.; Arnott, W.P.; McMeeking, G.R.; Sullivan, A.P.; Wold, C.E.; *et al.* Optical closure experiments for biomass smoke aerosols. *Atmos. Chem. Phys.* **2010**, *10*, 9017–9026.

19. Marshall, J.; Lohmann, U.; Leaitch, W.R.; Lehr, P.; Hayden, K. Aerosol scattering as a function of altitude in a coastal environment. *J. Geophys. Res.* **2007**, *112*, D14203.

20. Ma, N.; Birmili, W.; Müller, T.; Tuch, T.; Cheng, Y.F.; Xu, W.Y.; Zhao, C.S.; Wiedensohler, A. Tropospheric aerosol scattering and absorption over central Europe: A closure study for the dry particle state. *Atmos. Chem. Phys.* **2014**, *14*, 6241–6259.

21. Zieger, P.; Fierz-Schmidhauser, R.; Poulain, L.; Müller, T.; Birmili, W.; Spindler, G.; Wiedensohler, A.; Baltensperger, U.; Weingartner, E. Influence of water uptake on the aerosol particle light scattering coefficients of the Central European aerosol. *Tellus B* **2014**, *66*, 22716.

22. Wang, J.; Flagan, R.C.; Seinfeld, J.H.; Jonsson, H.H.; Collins, D.R.; Russell, P.B.; Schmid, B.; Redemann, J.; Livingston, J.M.; Gao, S.; *et al.* Clear-column radiative closure during ACE-Asia: Comparison of multiwavelength extinction derived from particle size and composition with results from sunphotometry. *J. Geophys. Res.* **2002**, *107*, 4688.

23. Parworth, C.; Fast, J.; Mei, F.; Shippert, T.; Sivaraman, C.; Tilp, A.; Watson, T.; Zhang, Q. Long-term measurements of submicrometer aerosol chemistry at the Southern Great Plains (SGP) using an Aerosol Chemical Speciation Monitor (ACSM). *Atmos. Environ.* **2015**, *106*, 43–55.

24. Schmid, B.; Tomlinson, J.M.; Hubbe, J.M.; Comstock, J.M.; Mei, F.; Chand, D.; Pekour, M.S.; Kluzek, C.D.; Andrews, E.; Biraud, S.C.; *et al.* The DOE ARM Aerial Facility. *Bull. Am. Meteor. Soc.* **2014**, *95*, 723–742.

25. Kassianov, E.; Flynn, C.; Redemann, J.; Schmid, B.; Russell, P.B.; Sinyuk, A. Initial assessment of the Spectrometer for Sky-Scanning, Sun-Tracking Atmospheric Research (4STAR)-based aerosol retrieval: Sensitivity study. *Atmosphere* **2012**, *3*, 495–521.

26. Dunagan, S.; Johnson, R.; Zavaleta, J.; Russell, P.; Schmid, B.; Flynn, C.; Redemann, J.; Shinozuka, Y.; Livingston, J.; Segal-Rosenhaimer, M. Spectrometer for Sky-Scanning Sun-Tracking Atmospheric Research (4STAR): Instrument technology. *Remote Sens.* **2013**, *5*, 3872–3895.

27. Segal-Rosenheimer, M.; Russell, P.B.; Schmid, B.; Redemann, J.; Livingston, J.M.; Flynn, C.J.; Johnson, R.R.; Dunagan, S.E.; Shinozuka, Y.; Herman, J.; *et al.* Tracking elevated pollution layers with a newly developed hyperspectral Sun/Sky spectrometer (4STAR): Results from the TCAP 2012 and 2013 campaigns. *J. Geophys. Res. Atmos.* **2014**, *119*, 2611–2628.

28. Allen, G.; Coe, H.; Clarke, A.; Bretherton, C.; Wood, R.; Abel, S.J.; Barrett, P.; Brown, P.; George, R.; Freitag, S.; *et al.* South East Pacific atmospheric composition and variability sampled along 20° S during VOCALS-Rex. *Atmos. Chem. Phys.* **2011**, *11*, 5237–5262.

29. Kleinman, L.I.; Daum, P.H.; Lee, Y.-N.; Lewis, E.R.; Sedlacek, A.J., III; Senum, G.I.; Springston, S.R.; Wang, J.; Hubbe, J.; Jayne, J.; *et al.* Aerosol concentration and size distribution measured below, in, and above cloud from the DOE G-1 during VOCALS-REx. *Atmos. Chem. Phys.* **2012**, *11*, 207–223.

30. Liu, Y.; Daum, P.H. The effect of refractive index on size distributions and light scattering coefficients derived from optical particle counters. *J. Aerosol. Sci.* **2000**, *31*, 945–957.

31. Ames, R.B.; Hand, J.L.; Kreidenweis, S.M.; Day, D.E.; Malm, W.C. Optical measurements of aerosol size distributions in Great Smokey Mountains National Park: Dry aerosol characterization. *J. Air Waste Manag. Assoc.* **2000**, *50*, 665–676.

32. Garvey, D.M.; Pinnick, R.G. Response characteristics of the particle measuring systems active scattering aerosol spectrometer probe (ASASP-X). *Aerosol Sci. Technol.* **1983**, *2*, 477–488.

33. Kim, Y.J.; Boatman, J.F. Size calibration corrections for the Forward Scattering Spectrometer Probe (FSSP) for measurement of atmospheric aerosols of different refractive indices. *J. Atmos. Oceanic Technol.* **1990**, *7*, 681–688.

34. Bukowiecki, N.; Zieger, P.; Weingartner, E.; Juranyi, Z.; Gysel, M.; Neininger, B.; Schneider, B.; Hueglin, C.; Ulrich, A.; Wichser, A.; *et al.* Ground-based and airborne *in-situ* measurements of the Eyjafjallajökull volcanic aerosol plume in Switzerland in spring 2010. *Atmos. Chem. Phys.* **2011**, *11*, 10011–10030.

35. Kondo, Y. Effects of black carbon on climate: Advances in measurement and modeling. *Monogr. Environ. Earth Planets* **2015**, *3*, 1–85.

36. Kreidenweis, S.M.; Asa-Awuku, A. Aerosol Hygroscopicity: Particle water content and its role in atmospheric processes. *Treat. Geochem.: Second Ed.* **2013**, *5*, 331–361.

37. Jensen, T.L.; Kreidenweis, S.M.; Kim, Y.; Sievering, H.; Pszenny, A. Aerosol distributions measured in the North Atlantic marine boundary layer during ASTEX/MAGE. *J. Geophys. Res.* **1996**, *101*, 4455–4467.

38. Swietlicki, E.; Zhou, J.; Berg, O.H.; Martinsson, B.G.; Frank, G.; Cederfelt, S.-I.; Dusek, U.; Berner, A.; Birmili, W.; Wiedensohler, A.; *et al.* A closure study of sub-micrometer aerosol particle hygroscopic behavior. *Atmos. Res.* **1999**, *50*, 205–240.

39. Dusek, U.; Frank, G.P.; Massling, A.; Zeromskiene, K.; Iinuma, Y.; Schmid, O.; Helas, G.; Hennig, T.; Wiedensohler, A.; Andreae, M.O. Water uptake by biomass burning aerosol at sub- and supersaturated conditions: Closure studies and implications for the role of organics. *Atmos. Chem. Phys.* **2011**, *11*, 9519–9532.

40. Kassianov, E.; Barnard, J.; Pekour, M.; Berg, L.K.; Shilling, J.; Flynn, C.; Mei, F.; Jefferson, A. Simultaneous retrieval of effective refractive index and density from size distribution and light-scattering data: Weakly absorbing aerosol. *Atmos. Meas. Tech.* **2014**, *7*, 3247–3261.

41. Titos, G.; Jefferson, A.; Sheridan, P.J.; Andrews, E.; Lyamani, H.; Alados-Arboledas, L.; Ogren, J.A. Aerosol light-scattering enhancement due to water uptake during the TCAP campaign. *Atmos. Chem. Phys.* **2014**, *14*, 7031–7043.

42. Collins, D.R.; Flagan, R.C.; Seinfeld, J.H. Improved inversion of scanning DMA data. *Aerosol Sci. Technol.* **2002**, *36*, 1–9.

43. Schmid, B.; Ferrare, R.; Flynn, C.; Elleman, R.; Covert, D.; Strawa, A.; Welton, E.; Turner, D.; Jonsson, H.; Redemann, J.; *et al.* How well do state-of-the-art techniques measuring the vertical profile of tropospheric aerosol extinction compare? *J. Geophys. Res.* **2006**, *111*, D05S07.

44. Zelenyuk, A.; Imre, D.; Wilson, J.; Zhang, Z.; Wang, J.; Mueller, K. Airborne Single Particle Mass Spectrometers (SPLAT II& miniSPLAT) and new software for data visualization and analysis in a geo-spatial context. *J. Am. Soc. Mass Spectrom.* **2015**, *26*, 257–270.

45. Berg, L.; Fast, J.D.; Barnard, J.C.; Burton, S.P.; Cairns, B.; Chand, D.; Comstock, J.M.; Dunagan, S.; Ferrare, R.A.; Flynn, C.J.; *et al.* The Two-Column Aerosol Project: Phase I overview and impact of elevated aerosol layers on aerosol optical depth. *J. Geophys. Res. Atmos.* **2015**. under review.

46. Esteve, A.R.; Highwood, E.J.; Morgan, W.T.; Allen, G.; Coe, H.; Grainger, R.G.; Brown, P.; Szpek, K. A study on the sensitivities of simulated aerosol optical properties to composition and size distribution using airborne measurements. *Atmos. Environ.* **2014**, *89*, 517–524.

47. Twomey, S. Comparison of constrained linear inversion and an iterative nonlinear algorithm applied to indirect estimation of particle-size distributions. *J. Comput. Phys.* **1975**, *18*, 188–200.

48. Markowski, G.R. Improving Twomey's Algorithm for Inversion of Aerosol Measurement Data. *Aerosol Sci. Technol.* **1987**, *7*, 127–141.

49. Moteki, N.; Kondo, Y. Effects of mixing state on black carbon measurements by laser-induced incandescence. *Aerosol Sci. Technol.* **2007**, *41*, 398–417.

50. Sedlacek, A.J., III; Lewis, E.R.; Kleinman, L.; Xu, J.; Zhang, Q. Determination of and evidence for non-core-shell structure of particles containing black carbon using the Single-Particle Soot Photometer (SP2). *Geophys. Res. Lett.* **2012**, *39*.

51. Pekour, M.S.; Schmid, B.; Chand, D.; Hubbe, J.M.; Kluzek, C.D.; Nelson, D.A.; Tomlinson, J.M.; Cziczo, D.J. Development of a new airborne humidigraph system. *Aerosol Sci. Technol.* **2013**, *47*, 201–207.

52. Shinozuka, Y.; Johnson, R.R.; Flynn, C.J.; Russell, P.B.; Schmid, B.; Redemann, J.; Dunagan, S.E.; Kluzek, C.D.; Hubbe, J.M.; Segal-Rosenheimer, M.; *et al.* Hyperspectral aerosol optical depths from TCAP flights. *J. Geophys. Res. Atmos.* **2013**, *118*, 12180–12194.

53. Anderson, T.L.; Ogren, J.A. Determining aerosol radiative properties using the TSI 3563 Integrating Nephelometer. *Aerosol Sci. Technol.* **1998**, *29*, 57–69.

54. Hallar, A.G.; Strawa, A.W.; Schmid, B.; Andrews, E.; Ogren, J.; Sheridan, P.; Ferrare, R.; Covert, D.; Elleman, R.; Jonsson, H.; *et al.* Atmospheric Radiation Measurements Aerosol Intensive Operating Period: Comparison of aerosol scattering during coordinated flights. *J. Geophys. Res.* **2006**, *111*, D05S09.

55. Barnard, J.C.; Fast, J.D.; Paredes-Miranda, G.; Arnott, W.P.; Laskin, A. Technical note: Evaluation of the WRF-Chem "Aerosol chemical to aerosol optical properties" module using data from the MILAGRO campaign. *Atmos. Chem. Phys.* **2010**, *10*, 7325–7340.

56. Hu, D.; Chen, J.; Ye, X.; Li, L.; Yang, X. Hygroscopicity and evaporation of ammonium chloride and ammonium nitrate: Relative humidity and size effects on the growth factor. *Atmos. Environ.* **2011**, *45*, 2349–2355.

57. Healy, R.M.; Evans, G.J.; Murphy, M.; Jurányi, Z.; Tritscher, T.; Laborde, M.; Weingartner, E.; Gysel, M.; Poulain, L.; Kamilli, K.A.; *et al.* Predicting hygroscopic growth using single particle chemical composition estimates. *J. Geophys. Res. Atmos.* **2014**, *119*, 9567–9577.

58. Pilinis, C.; Charalampidis, P.E.; Mihalopoulos, N.; Pandis, S.N. Contribution of particulate water to the measured aerosol optical properties of aged aerosol. *Atmos. Environ.* **2014**, *82*, 144–153.

59. Dubovik, O.; Holben, B.; Eck, T.F.; Smirnov, A.; Kaufman, Y.J.; King, M.D.; Tanré, D.; Slutsker, I. Variability of absorption and optical properties of key aerosol types observed in worldwide locations. *J. Atmos. Sci.* **2002**, *59*, 590–608.

60. Kokhanovsky, A.A. *Aerosol Optics: Light Absorption and Scattering by Particles in the Atmosphere*; Springer-Berlin: Heidelberg, Germany, 2008; p. 148.

61. Barber, P.W.; Hill, S.C. *Light Scattering by Particles: Computational Methods*; World Scientific Publishing: Singapore, 1990.

62. Scarnato, B.V.; Vahidinia, S.; Richard, D.T.; Kirchstetter, T.W. Effects of internal mixing and aggregate morphology on optical properties of black carbon using a discrete dipole approximation model. *Atmos. Chem. Phys.* **2013**, *13*, 5089–5101.

63. Lesins, G.; Chylek, P.; Lohmann, U. A study of internal and external mixing scenarios and its effect on aerosol optical properties and direct radiative forcing. *J. Geophys. Res. Atmos.* **2002**, *107*, 4094–4106.

64. Freney, E.J.; Adachi, K.; Buseck, P.R. Internally mixed atmospheric aerosol particles: Hygroscopic growth and light scattering. *J. Geophys. Res. Atmos.* **2010**, *115*, D19210.

65. Wex, H.; Neususs, C.; Wendisch, M.; Stratmann, F.; Koziar, C.; Keil, A.; Wiedensohler, A.; Ebert, M. Particle scattering, backscattering, and absorption coefficients: An *in situ* closure and sensitivity study. *J. Geophys. Res. Atmos.* **2012**, *107*, 8122.

66. York, D.; Evensen, N.M.; Lopez Martinez, M.; de Basabe Delgado, J. Unified equations for the slope, intercept, and standard errors of the best straight line. *Am. J. Phys.* **2004**, *72*, 367–375.

67. Petäjä, T.; Mauldin, R.L., III; Kosciuch, E.; McGrath, J.; Nieminen, T.; Paasonen, P.; Boy, M.; Adamov, A.; Kotiaho, T.; Kulmala, M. Sulfuric acid and OH concentrations in a boreal forest site. *Atmos. Chem. Phys.* **2009**, *9*, 7435–7448.

68. Zieger, P.; Weingartner, E.; Henzing, J.; Moerman, M.; de Leeuw, G.; Mikkilä, J.; Ehn, M.; Petäjä, T.; Clémer, K.; van Roozendael, M.; *et al.* Comparison of ambient aerosol extinction coefficients obtained from in-situ, MAX-DOAS and LIDAR measurements at Cabauw. *Atmos. Chem. Phys.* **2011**, *11*, 2603–2624.

69. Cantrell, C.A. Technical Note: Review of methods for linear least-squares fitting of data and application to atmospheric chemistry problems. *Atmos. Chem. Phys.* **2008**, *8*, 5477–5487.

70. Hand, J.L.; Kreidenweis, S.M. A new method for retrieving particle refractive index and effective density from aerosol size distribution data. *Aerosol Sci. Tech.* **2002**, *36*, 1012–1026.

71. McComiskey, A.; Feingold, G.; Frisch, A.S.; Turner, D.D.; Miller, M.A.; Chiu, J.C.; Min, Q.; Ogren, J.A. An assessment of aerosol-cloud interactions in marine stratus clouds based on surface remote sensing. *J. Geophys. Res.* **2009**, *114*, D09203.

72. Modini, R.L.; Frossard, A.A.; Ahlm, L.; Russell, L.M.; Corrigan, C.E.; Roberts, G.C.; Hawkins, L.N.; Schroder, J.C.; Bertram, A.K.; Zhao, R.; *et al.* Primary marine aerosol-cloud interactions off the coast of California. *J. Geophys. Res. Atmos.* **2015**, *120*.

73. Yu, H.; Chin, M.; Bian, H.; Yuan, T.; Prospero, J.M.; Omar, A.H.; Remer, L.A.; Winker, D.M.; Yang, Y.; Zhang, Y.; Zhang, Z.; *et al.* Quantification of trans-Atlantic dust transport from seven-year (2007–2013) record of CALIPSO lidar measurements. *Remote Sens. Environ.* **2015**, *159*, 232–249.

74. Barnard, J.C.; Harrison, L.C. Monotonic responses from monochromatic optical particle counters. *Appl. Opt.* **1988**, *3*, 584–592.

75. Kennedy, W.J.; Gentle, J.E. *Statistical Computing*; Marcel Dekker: New York, NY, USA, 1980.

The Impact of Selected Parameters on Visibility: First Results from a Long-Term Campaign in Warsaw, Poland

Grzegorz Majewski, Wioletta Rogula-Kozłowska, Piotr O. Czechowski, Artur Badyda and Andrzej Brandyk

Abstract: The aim of this study was to investigate how atmospheric air pollutants and meteorological conditions affected atmospheric visibility in the largest Polish agglomeration. The correlation analysis, principal component analysis (PCA) and generalized regression models (GRMs) were used to accomplish this objective. The meteorological parameters (temperature, relative humidity, precipitation, wind speed and insolation) and concentrations of the air pollutants (PM_{10}, SO_2, NO_2, CO and O_3) were recorded in 2004–2013. The data came from the Ursynów-SGGW, MzWarszUrsynów and Okęcie monitoring stations, located in the south of Warsaw (Poland). It was shown that the PM_{10} concentration was the most important parameter affecting the visibility in Warsaw. The concentration, and indirectly the visibility, was mainly affected by the pollutant emission from the flat/building heating (combustion of various fuels). It changed intensively during the research period. There were also periods in which this emission type did not have a great influence on the pollutant concentrations (mainly PM_{10}) and visibility. In such seasons, the research revealed the influence of the traffic emission and secondary aerosol formation processes on the visibility.

Reprinted from *Atmosphere*. Cite as: Majewski, G.; Rogula-Kozłowska, W.; Czechowski, P.O.; Badyda, A.; Brandyk, A. The Impact of Selected Parameters on Visibility: First Results from a Long-Term Campaign in Warsaw, Poland. *Atmosphere* **2015**, *6*, 1154–1174.

1. Introduction

The visibility deterioration caused by atmospheric pollution is a global problem. It occurs in many densely populated areas that have experienced population growth and industrialization. However, visibility is a complex issue. On one hand, it is directly affected by the anthropogenic air pollution. On the other hand, it is influenced by the meteorological conditions [1].

The anthropogenic air pollution effect on human health and visibility has been examined for decades. Many studies were conducted not only to assess the benefits for human health resulting from air pollutant emission reduction but also to understand how air pollutants negatively affect visibility. Generally, visibility makes a good index for the air pollution extent. It can also be used as a surrogate for assessing the human health effects [2,3].

The visibility impairment is mainly attributed to the scattering and absorption of the visible light caused by suspended particles and gaseous pollutants in the atmosphere [4,5]. The visibility impairment in the urban atmosphere is closely related to the air pollution from anthropogenic sources, such as car exhaust fumes, fuel combustion, solid waste incineration, and industrial emissions [6–10].

The visibility impairment is mainly influenced by the airborne particulate matter (PM), particularly its fine particles with aerodynamic diameters smaller than 2.5 μm ($PM_{2.5}$). In urban areas, the major $PM_{2.5}$ components, such as ammonium, sulphates, nitrates, organic matter and elemental carbon [11], are the main factors contributing to the light absorption and scattering. Therefore, their presence effectively reduces visibility [12,13]. The specific content of $PM_{2.5}$ is the most important aspect when analysing the $PM_{2.5}$ effect on visibility. The size and chemical composition of each component particle affects its ability to refract, scatter, and absorb light [14]. There is a strong correlation between the presence of $PM_{2.5}$ and PM_{10} (particles with aerodynamic diameters smaller than 10 μm), to the extent that a targeted reduction in PM_{10} is likely to lead to an increase in the atmospheric visibility [15,16]. In addition to the air pollutants, the meteorological parameters (*i.e.*, wind speed and direction, relative air humidity, air temperature, atmospheric pressure and precipitation) can also directly or indirectly affect atmospheric visibility as they influence the local and regional air quality in urban areas [17–22].

Air quality monitoring has already been performed in Warsaw for about 20 years. Nevertheless, the measurement standards were adjusted to comply with the European Union (EU) regulations and requirements in 2004. Since then, the air quality has been successfully recorded in order to warn the community of high pollutant levels. The system also contributes to the research on air pollutant influence on human health [23,24]. The collected data also enables investigations into the air pollution impact on visibility. Taking into consideration the necessity to improve the visibility in urban areas, the underlying mechanisms must be well understood, particularly when it comes to aspects such as the main contributing air pollutants and their origin. This research field has a worldwide significance. Nonetheless, it needs further development in Poland. Among the key statistical methods applied in this study, the researchers found the correlation analysis and the related Principal Component Analysis (PCA) and Generalized Regression Models (GRMs) particularly useful. All the models served to identify the air pollutants and meteorological parameters influencing visibility in an urban area.

2. Materials and Methods

2.1. Research Area

Warsaw is the biggest city in Poland and the ninth biggest city in Europe (517.24 km^2). It is under the influence of the warm transitional temperate climate. The mean yearly air temperature is 8.6 °C. The mean precipitation sum is 550–650 mm. In Warsaw, the visible influence of a large urban agglomeration on the climate is seen as the so-called *urban heat island* [25]. The mean air temperatures and precipitation sums are higher whereas the wind speed is lower in the city center. As the air in the city is highly polluted, the cloud cover is bigger and the air transparency deteriorates. Consequently, the direct solar radiation decreases, whereas the diffuse sky radiation increases.

Even though the air quality has improved tremendously in Poland over the last 30 years [26–28], it is still unsatisfactory in Warsaw. What is definitely positive is the fact that the air pollutant levels (especially CO and SO$_2$) have been gradually lowered over the last few years. Nonetheless, the permissible levels of PM$_{10}$ are still exceeded in the capital of Poland [29,30]. PM comes from many sources there, including, among others, the energy production sector and vehicular transport. The area of Warsaw is additionally threatened by the air inflowing from heavily polluted southern Poland [31]. PM composition in large urban areas of Poland differs significantly from the compositions observed in other urban areas in Europe [32–34]. There are higher contents of elemental (soot) and organic carbon and lower percentage of the secondary inorganic matter in PM. It seems that the differences must have a considerable impact on the visibility in the research area.

2.2. Air Quality Data and Visibility Observation

The study was based on the measurement results obtained from the MzWarszUrsynów monitoring station of the atmospheric air quality ($\lambda E = 21°02'$; $\varphi N = 52°09'$), located in the south of Warsaw (Figure 1). The researchers used the data on the mean hourly concentrations of the following air pollutants, measured by proper type of analyzers: sulphur dioxide (SO$_2$)—MLU 100A, carbon monoxide (CO)—MLU 300, ozone (O$_3$)—MLU 400, nitrogen oxides (NO$_2$)—MLU 200A, and PM$_{10}$—TEOM1400a. They were monitored with pulsed fluorescence, infrared absorption, ultraviolet light absorption, chemiluminescence, and a β-gauge automated particle sampler, respectively. The meteorological data came from the Ursynów-SGGW meteorological station (λE 21°02'; φN 52°09'). The following information was investigated: mean hourly air temperature values (T), insolation intensity (Rad), relative air humidity (RH), precipitation intensity (P) and wind speed (Ws). The measurements taken at the station were performed according to the instruction for network of stations belonging to the Institute of Meteorology and Water Management (IMGW). The data on the visibility were obtained from the only station taking such measurements, *i.e.*, the Okęcie station (λ_E 20°59′; φ_N 52°09′). The distance between the stations was approximately 6 km. The visibility measurements were carried out with a visibility meter equipped with an atmospheric phenomenon detector—Vaisala FS11 (wavelength 875 nm). It performed the functions of a visibility meter using light dispersion measurements and an atmospheric phenomenon detector. The horizontal visibility measurements were performed in the range of 10 m–50 km. The data (1-h values) were shared by the IMGW. The information used in the study came from 2004–2013. For the whole research period, the following numbers of data (n) were obtained: visibility $n = 87,634$; SO_2, $n = 85,416$; PM_{10} $n = 83,016$; NO_2 $n = 85,985$; O_3 $n = 85,148$; T $n = 85,936$; RH $n = 86,205$; Rad $n = 85,384$.

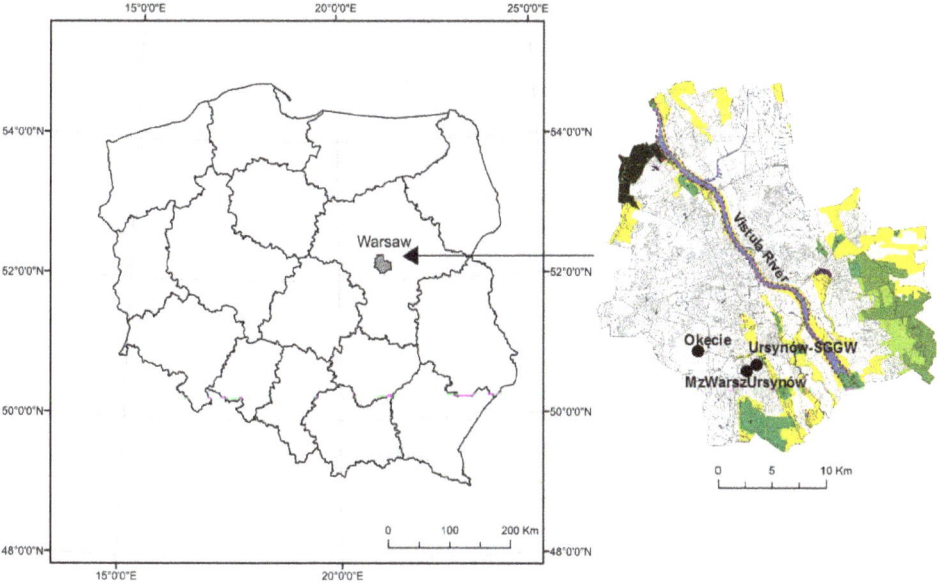

Figure 1. Location of the measurement stations in Warsaw (Poland).

2.3. Statistical Method

The correlation analysis, principal component analysis (PCA) and generalized regression models (GRMs) were applied in this research. The analyses helped to identify factors affecting visibility.

Generally, the PCA is a data reduction exercise. It is achieved through finding linear combinations (principal components) of the original variables, which account for as much of the original total variance as possible [16,17,19,35,36]. PCA is used to identify factors and their synergies in strong measurements scales: interval and ratio. Here, a scale or a level of measurement is the classification that describes the nature of information within the numbers assigned to variables. The interval type allows for the degree of difference between items, but not the ratio between them. The ratio type takes its name from the fact that measurement is the estimation of the ratio between a magnitude of a continuous quantity and a unit magnitude of the same kind.

GRM's models were used to identify factors in weak measurements scales: nominal and ordinal (ex. seasonality), and additionally, jointly with variables in strong scales. Nominal scales refer to the construction of classification of items, while ordinal ones allow for rank order by which data can be sorted. GRM is considered to be a path, embracing more than a model. It enabled to find the "best" model, describing the analysed phenomenon, out of an available range of models, and to replace many classical ones (e.g., ANOVA, MANOVA). Such an approach is more efficient for replication and cross-validation studies, less costly to put into practice in predicting and controlling the outcome in the future. Finally, it allows use of a wider range of fit statistics and diagnostic calculations, than using single models separately [21,37,38].

3. Results and Discussion

3.1. General Description of Visibility, Meteorology and Air Pollution

The yearly course of the air temperature (average in the period 2004–2013) was typical for the temperate and transitional climate in Poland. July was the warmest month (mean air temperature = 20.5 °C). January was the coldest month (mean air temperature = −2.1 °C). The lowest wind speed values were measured in August and September (approx. 2.1 m/s), whereas the highest ones were observed in March (3.1 m/s)—Figure 2.

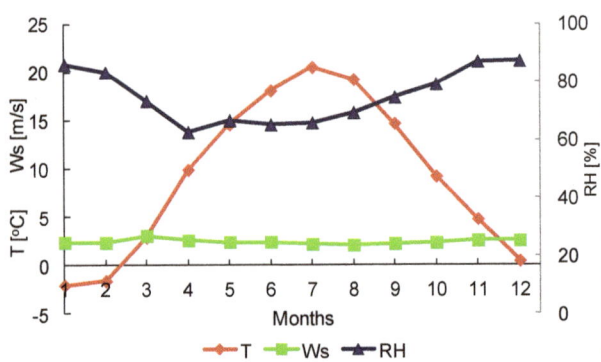

Figure 2. Monthly variations of meteorological parameters over Warsaw in 2004–2013.

Figure 3 presents the diurnal patterns of visibility for the period 2004–2013. Visibility shows an obvious diurnal variation in each season of the year. In spring (March, April, May) and summer (June, July, August), a valley appears in early morning, from 04:00 to 06:00, while in autumn (September, October, November) and spring it is observed slightly later, between 06:00 and 07:00. The peak appears generally in the afternoon, from 13:00 to 15:00, except for winter (December, January, February), for which the peak is slightly later, at about 18:00. Visibility shows stronger diurnal cycles in summer and autumn, while it exhibits a much weaker variation in winter. Apart from this, the diurnal patterns during different seasons are desynchronized, which is attributed to the difference in weather patterns and stability of atmospheric boundary layer. The diurnal variation of visibility, characteristic for the city of Warsaw, is similar to that over several cities in China, however, hourly visibility values are found to be three times higher than those recorded in the Chinese cities [19].

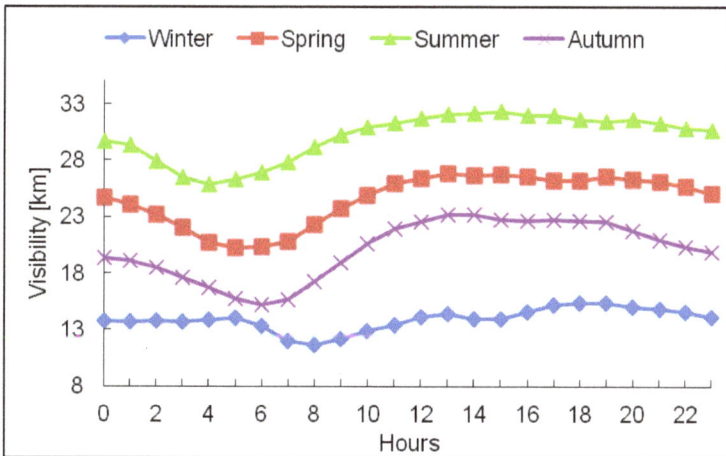

Figure 3. Diurnal variations of visibility over Warsaw for winter (December, January, February), spring (March, April, May), summer (June, July, August) and autumn (September, October, November) in 2004–2013.

For the whole research period (2004–2013) monthly visibility was in a wide range of 6.7–34.1 km (Figure 4). Within the researched period, noticeable seasonal changes in visibility were found. Visibility was generally higher in summer and lower in late autumn and winter. The average seasonal visibilities were 24.5, 30.1, 20.1, and 13.9 km in spring, summer, autumn, and winter, respectively (Figure 4).

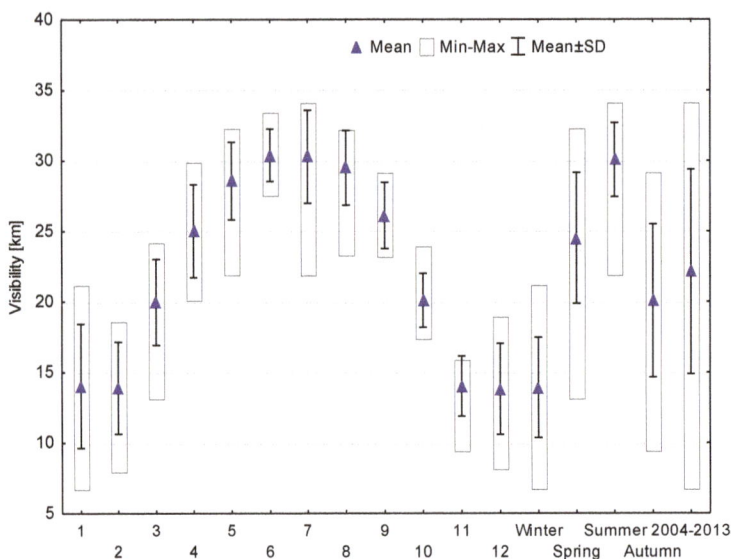

Figure 4. Descriptive statistics of monthly and seasonal visibility over Warsaw in 2004–2013.

Monthly visibility in Warsaw exhibited, in general, a considerable winter period variation (January–March, October–November) from 2004–2013. The summer period was characterized by much lower variation.

The mean yearly visibilities were in the range of 19.8–23.7 km (Figure 5). This proves that it did not show significant variability year by year. However, monthly values exhibit an increasing trend throughout the research period, but are statistically insignificant; $p < 0.05$ (Figure 6). Average visibility for the whole period 2004–2013 was equal to 22.1 km and is from 5.3 km to over 13.1 km higher than the one over large, highly urbanized cities of China [1,19].

From 2004–2013 yearly air pollutants concentration didn't show much variation, alike visibility. In the analysed years, the mean yearly concentrations of NO_2 were 21.4–28.0 $\mu g \cdot m^{-3}$, which made 53.6%–70% of the permissible value (40 $\mu g \cdot m^{-3}$)—Figure 5. The mean yearly concentrations of SO_2 were 5.5–11.5 $\mu g \cdot m^{-3}$ (the permissible value; 20 $\mu g \cdot m^{-3}$), and the mean yearly concentrations of CO were 365.7–549.5 $\mu g \cdot m^{-3}$. The mean yearly concentrations of PM_{10} did not exceed the permissible value as well (40 $\mu g \cdot m^{-3}$) and were in the range of 28.0–37.2 $\mu g \cdot m^{-3}$. The only pollutant, that, according to the Polish applicable laws, exceeded the permissible limit, was the ozone O_3. The mean yearly concentrations of O_3 were 43.5–50.4 $\mu g \cdot m^{-3}$. The permissible level of the 8-h O_3 concentration is 120 $\mu g \cdot m^{-3}$ and can be exceeded about 25 days in each year. The biggest number of days with the exceeded value was observed in 2005. In the research area, there was no steady trend in the changes for O_3, which is a secondary pollutant. The changes in its concentration mainly resulted from the changes in the weather conditions (insolation intensity, air temperature) and the participation of the O_3 precursors (e.g., nitrogen oxides, hydrocarbons and other pollutants participation in the O_3 formation) in the atmospheric air [30,39].

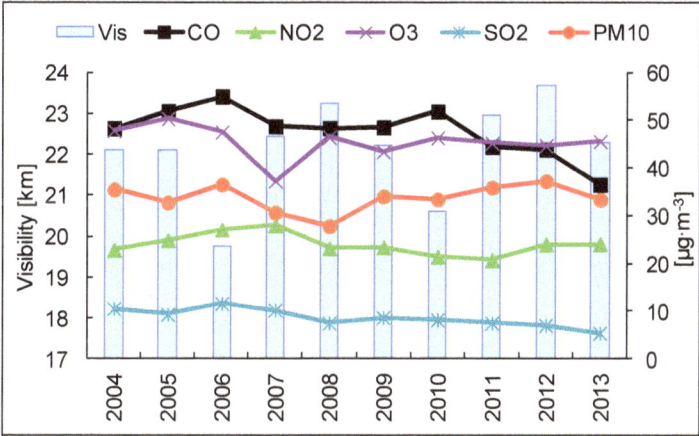

Figure 5. Yearly variations of visibility and ambient concentrations of NO_2, O_3, SO_2, PM_{10} and CO (divided by 10 in the figure) over Warsaw in 2004–2013.

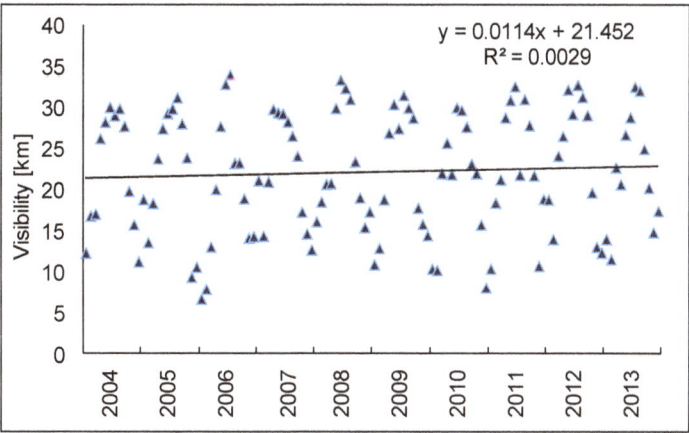

Figure 6. Visibility time series of monthly mean and its' linear regression trend over Warsaw in 2004–2013.

Since the visibility is strongly affected by air pollutants [1,40,41], the presence of its' weaker variation over Warsaw from 2004–2013 is found. In Poland, considerable changes in air pollution were observed from 1980–2000 [42]. In that period, political and economical transformation was related with a sudden decrease of industrial emission due to large factories closure and limited production in remaining ones [43,44]. On the other hand, such transformation contributed to the knowledge on negative consequences of air pollution, and for this reason in 1980s industrial emissions were largely restricted in Poland. Unfortunately, no reliable air pollution measurements were then performed within the research area.

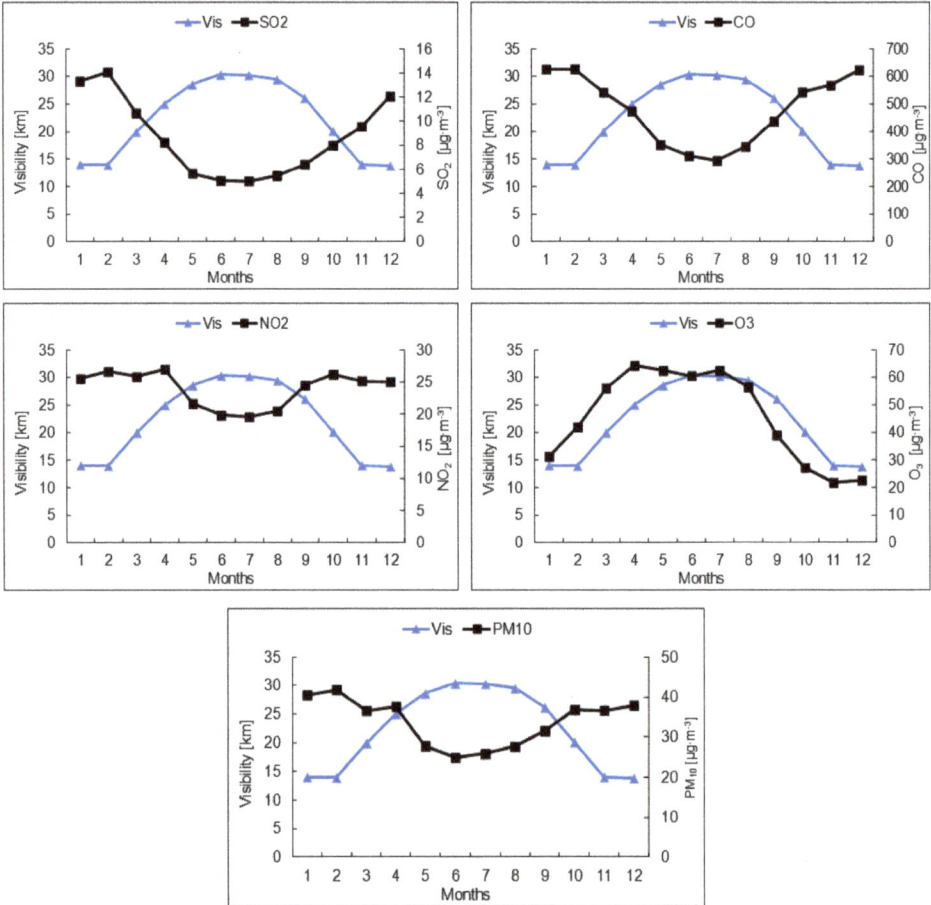

Figure 7. Monthly variations of visibility and ambient concentrations of air pollutants over Warsaw in 2004–2013.

Monthly visibilities and monthly ambient concentrations of air pollutants (averages for the whole period 2004–2013) are shown in Figure 7. Comparing the maximum and minimum values for the monthly concentrations shows that O_3, CO and SO_2 have the largest variability; while much lower variability was shown by PM_{10} and NO_2. Visibility values were inversely proportional to all analysed pollutants, except for O_3. Ambient concentrations of PM_{10}, SO_2, NO_2 and CO tend to be strongly affected by the emission from fuel combustion for heating purposes [31,45]. It is obvious that their concentration is higher in winter months, while the visibility becomes lower (Figure 7). Except for the higher winter emission in Poland, the increase of air pollution is also attributed to meteorological conditions (lowering of air mixing layer, stagnant air) that deteriorate ventilation and ability of self-cleaning [46]. The fact that concentrations of CO, NO_2 and SO_2 are higher in the cold season (Figure 7) is attributed to increased emissions—however, the atmospheric lifetime of these compounds is also generally longer in the winter and this is likely to explain some of the increase as well.

The mean NO_2 concentration was slightly higher in the cold season than in the warm one. Such a situation was related to the low variation of the yearly NO_2 emission. Typically, the data indicated two daily peaks in the ambient concentrations of both NO_x forms (NO and NO_2), which pointed to the traffic-related pollution [30,39]. Emission from the traffic contributes also to ambient concentrations of volatile organic compounds (VOCs). VOCs and NO react with O_3 giving a loss of O_3 near to pollution sources. On the other hand, higher O_3 in summer is probably because of higher solar insolation, and visibility is lower because of the lack of cold season emissions and boundary layer effects. For those reasons, the course of monthly ozone concentrations is different from NO_2 and other pollutants, and resembles the course of visibility.

3.2. Visibility during Periods Differing in the Air Pollution with PM

Table 1 presents the mean values of visibility, PM_{10}, gaseous pollutants and meteorological parameters (calculated on the basis of 1-h values from the entire research period) grouped within three categories: clear days (PM_{10} concentration did not exceed 50 $\mu g \cdot m^{-3}$), moderate days (PM_{10} concentration was 50–200 $\mu g \cdot m^{-3}$) and episode days (PM_{10} concentration exceeded 200 $\mu g \cdot m^{-3}$).

The NO_2 concentration was 3.7 times higher during the episodes than on the clear days (20.4 $\mu g \cdot m^{-3}$), while the CO concentration was approximately six times higher. A similar situation was observed for SO_2. For O_3, an inverse correlation was observed. Its lowest concentrations were found for the episode days. The episodes were characterized by low mean visibility (6.0 km), low air temperature, low wind speed, and higher relative air humidity. On the other hand, the clear days were characterized by the highest mean visibilities (24.3 km), higher wind speeds and lower relative air humidity. Episode days occurred mostly in winter, while clear days were found in warm periods, mostly in summer. Unfortunately, the researchers did not have data on the mixed layer height and atmospheric pressure. As different studies show, high atmospheric pressure leads to the lower mixed layer height and low wind speed, which causes increased pollutant concentrations close to the pollutant sources and visibility deterioration. Low atmospheric pressure results in the higher mixed layer height and high wind speed, which provides effective ventilation for cities and good dispersion of pollutants [6]. The research conducted by Majewski *et al.* [29] into the atmospheric pressure influence on the PM_{10} concentration in Warsaw showed that the increase in the PM concentration was significantly statistically related to the increase in the atmospheric pressure.

Table 1. Values of visibility and other parameters calculated as arithmetic means on the basis of the hourly values from 2004–2013 for three periods differing in the air pollution with PM.

Parameter	Air Quality		
	$PM_{10} < 50$ [a] $\mu g \cdot m^{-3}$ Clear Days	$PM_{10} > 50$ [a] $\mu g \cdot m^{-3}$ Moderate Days	$PM_{10} > 200$ [b] $\mu g \cdot m^{-3}$ Episode
Occurrence number (hours)	68,652	14,285	146
PM_{10} ($\mu g \cdot m^{-3}$)	25.7	75.1	236.1
SO_2 ($\mu g \cdot m^{-3}$)	6.8	12.6	28.1
CO ($\mu g \cdot m^{-3}$)	378.8	822.1	2342.4
NO_2 ($\mu g \cdot m^{-3}$)	20.4	40.5	76.0
O_3 ($\mu g \cdot m^{-3}$)	48.5	28.9	14.0
Visibility (km)	24.3	13.6	6.0
Temperature (°C)	10.5	4.7	−6.9
Wind speed ($m \cdot s^{-1}$)	2.7	1.7	1.1
Relative humidity (%)	73.4	77.7	78.9

[a] Permissible level for the 24-h PM_{10} concentration due to the human health protection in Poland. [b] Threshold value for informing the inhabitants about the risk of exceeding the alert level for PM_{10} in Poland.

3.3. Weekend/Weekday Differences in Visibility and Air Pollution

When referring to the visibility and air quality studies, there is a phenomenon known as the *weekend effect*, Studies on this phenomenon can help to better understand the emission characteristics of air pollutants in urban areas and the weekend effect has been reported in America since the 1970s [47,48]. Contemporary research is especially focusing on visibility variations and the effect of air quality on visibility on weekdays compared with weekend days. In order to investigate potential weekend effect over Warsaw, mean weekend and weekday levels of visibility as well as air pollutants were calculated and subjected to Fisher–Snedecore test. Visibilities on weekends were slightly better than on weekdays, and the differences were statistically significant. Mean hourly visibility was equal to 22.61 ± 12.57 km at weekends and slightly over 22.02 ± 12.71 on weekdays (Table 2). For PM_{10}, slightly higher and statistically significant weekday concentrations were observed, as well as for SO_2, which was slightly lower during weekends. We also found CO concentrations to become lower on weekends (statistically significant differences) and, moreover, mean weekend and weekday NO_2 concentrations over Warsaw showed statistically significant differences, with NO_2 concentration higher on weekdays. This is probably due to less vehicular emission on weekends.

All concentrations of airborne pollutants were higher during weekdays, (statistically significant differences), except ozone. O_3 concentration was higher at weekends even though the O_3 pollutant precursor concentrations (such as NO_x and volatile organic compounds) are lower on weekends [47,48]. The weekend effect in the O_3 concentrations is the most likely to be attributable to decreased O_3 destruction by NO_x, as there are lower emissions from its' main source—communication and vehicular transportation [48,49]. The result of this study, concerning lower concentrations of airborne pollutants on weekends, excluding ozone, is similar to that obtained by Tsai [17].

Table 2. Visibility and air pollutant concentrations calculated as arithmetic means on the basis of the hourly values from 2004–2013 for two different periods—weekends and week days.

	Vis	PM_{10} ($\mu g \cdot m^{-3}$)	SO_2 ($\mu g \cdot m^{-3}$)	CO ($\mu g \cdot m^{-3}$)	NO_2 ($\mu g \cdot m^{-3}$)	O_3 ($\mu g \cdot m^{-3}$)
Weekend	22.61 ± 12.57	32.03 ± 24.20	8.08 ± 6.95	456.45 ± 327.25	20.20 ± 15.74	48.94 ± 29.46
	(n = 25,044)	(n = 23,652)	(n = 24,485)	(n = 23,782)	(n = 24,585)	(n = 24,444)
Weekday	22.02 ± 12.71	34.66 ± 25.38	8.75 ± 8.15	485.49 ± 340.51	25.50 ± 17.60	44.51 ± 29.94
	(n = 62,590)	(n = 59,364)	(n = 60,931)	(n = 59,495)	(n = 61,400)	(n = 60,704)
Fisher test	62,590	59,364	60,931	59,495	61,400	60,704
p-value	0.0000	0.0000	0.0000	0.0000	0.0000	0.0000

3.4. Correlations between Pollutants, Meteorological Variables and Visibility

Total correlations found between the results of the visibility measurements and other measurements (mean hourly values) in the whole research period are shown in Table 3. The visibility measurement results were negatively correlated with CO, PM_{10}, SO_2 and NO_2. Therefore, the increase in CO and SO_2 concentrations corresponds with visibility decrease, and PM_{10} and NO_2 are those species, that can directly contribute to visual range limitations.

Visibility was positively correlated with the O_3 concentrations. The atmospheric O_3 is a secondary air pollutant formed in photochemical reactions. Hence, summer is the period of the most intense O_3 formation. It resulted from the high insolation intensity during this season. The observed correlation was caused by the fact that both parameters (visibility and O_3 concentrations) increased and decreased in the same periods (Figure 7).

Table 3. Correlation matrix for total parameters measured for the entire period of 2004–2013 (correlations calculated for hourly values) and arithmetic means and standard deviations of the hourly value sets for the measured parameters.

	A	SD	PM$_{10}$	SO$_2$	CO	NO$_2$	O$_3$	T	Ws	RH	Rad	P
Vis (km)	22.19	12.67	−0.33	−0.27	−0.37	−0.21	0.47	0.52	0.14	−0.60	0.37	−0.10
PM$_{10}$ (μg·m^{-3})	33.91	25.07	1.00	0.37	0.66	0.56	−0.26	−0.23	−0.25	0.05	−0.13	−0.04
SO$_2$ (μg·m^{-3})	8.56	7.83		1.00	0.34	0.25	−0.15	−0.38	−0.10	0.10	−0.07	−0.03
CO (μg·m^{-3})	477.20	337.03			1.00	0.62	−0.41	−0.38	−0.28	0.21	−0.25	−0.03
NO$_2$ (μg·m^{-3})	23.98	17.26				1.00	−0.52	−0.19	−0.36	0.15	−0.31	−0.03
O$_3$ (μg·m^{-3})	45.78	29.87					1.00	0.47	0.22	−0.71	0.56	0.02
T (°C)	9.33	9.33						1.00	0.03	−0.52	0.48	0.05
Ws (m/s)	2.42	1.68							1.00	−0.08	0.16	0.04
RH (%)	74.97	18.82								1.00	−0.58	0.06
Rad (W/m^2)	117.77	200.98									1.00	−0.03
P (mm)	0.07	0.52										1.00

Notes: A: Mean value; SD: Standard deviation; Vis: Visibility; T: Temperature; RH: Relative air humidity; Ws: Wind speed; Rad: Insolation intensity; P: Precipitation.

In addition to their impact on visibility via changing the concentration of pollutants, meteorological variables can affect the visibility more directly - as humidity increases, hygroscopic aerosols increase in size and thus the scattering of light by them increases. At some point, aerosols activate and become fog [50,51]. This transition is very complex and has a very large impact on visibility [50,51].

Hourly visibility exhibited low, negative correlation with precipitation (Table 3). Generally, the precipitation lowered the air pollutant concentrations through the precipitation scavenging. Thereby, visibility increased [52]. However, the air purification effect and related visibility improvement appears with a delay after rainy days. During heavy precipitation visibility will be reduced. However, after the precipitation has stopped, the aerosols concentration are likely to be lower, thus giving two opposite impacts with a visibility increase following rain. In fact, the visibility reduction is more likely caused by the scattering of light by the hydrometeors. There is a negative correlation between precipitation and the primary pollutants, but this is weak and may relate to improved boundary layer ventilation when there is precipitation rather than scavenging. This is suggested by the fact that the relatively insoluble CO and NO$_2$ have a negative correlation with precipitation, which is the same size as that of the more soluble SO$_2$.

Visibility was positively correlated with the three remaining meteorological parameters, air temperature, insolation intensity, and wind speed. Most likely under clear sky conditions, the temperatures increase, the relative humidity falls and so the aerosols shrink, thus increasing the visibility.

3.5. Principal Component Analysis (PCA)

PCA served to extract four principal components (new variables: PC1, PC2, PC3 and PC4) with eigenvalues >1.0, which accounted for 73% of the total variance (Table 4). PC1, PC2, PC3 and PC4 can be interpreted as major factors that control visibility [1,6,17,40]. Visibility (Vis), O_3 concentration (O_3), NO_2 concentration (NO_2), CO concentration (CO), air temperature (T), relative air humidity (RH), PM_{10} concentration (PM_{10}) and insolation intensity (Rad) were most strongly correlated with PC1. Knowing the research area and the influence of various emission sources in this region, it can be assumed that PC1 could be related to fuel combustion for heating (mainly hard coal, wood/biomass and heating/crude oil) [31,44]. The emission increased over the year along with the T drop and Rad decrease. At the same time, the pollutant concentrations (*i.e.*, PM_{10}, NO_2, CO, or SO_2) in the air increased. Simultaneously, the photochemical pollutant concentrations (represented by O_3) decreased. Visibility was reduced in periods of high air pollution, related with heating (opposite signs for Vis/PC1 correlation and PM_{10}, NO_2, CO, SO_2/PC1 correlation).

The PCA performed only for the cold season data (Table 5) confirmed the same correlations between PC1 and the air pollution with NO_2, SO_2 and CO, and the inverse (opposite signs) correlation between PC1 and visibility, O_3 concentration, air temperature and insolation. For the warm season, the PCA revealed a very strong correlation between PC1 and NO_2 and an inverse strong correlation (opposite signs) between PC1 and visibility, O_3 concentration and insolation. Most likely, in those periods when there is less pollution from heating, and temperature rises, air pollution is mainly shaped by traffic emissions [31]. Then, the increase of NO_2 is observed, which reacts with O_3 and in consequence the concentration of O_3 in the air decreases. The reaction intensity becomes higher when the insolation intensity is stronger (O_3 concentrations and insolation were correlated with PC1 in the same way). Under such conditions, visibility may be improved (contrary signs of Vis/PC1 and NO_2/PC1). Nonetheless, it is not possible to discuss the cause-and-effect correlation between the traffic emission (and the related photochemical reactions) and the visibility increase or decrease. Most probably, visibility increased because the air pollution with PM was lower and the temperature and insolation were higher in summer.

PC2 was most strongly but differently (opposite signs) correlated with the PM_{10} concentrations and relative air humidity (Table 4). Visibility, pollutant concentrations, temperature and insolation were correlated with PC2 in the same way as PM_{10} but to a lesser extent. Thus, PC2 reflected the situation when the concentrations of all the observed pollutants and visibility decreased whereas the relative air humidity and precipitation increased. While humidity increases, hygroscopic aerosols increase in size and thus the scattering of light by them increases, so visibility drops. Air humidity concentration in the air was high due to precipitation at high temperature. During precipitation, concentrations of all the pollutants decreased due to leaching. The situation concerned PM_{10} to the largest extent.

Table 4. Correlations between the principal components (PC1, PC2, PC3, PC4) and measured parameter values. The PCA was performed for the hourly data collected in 2004–2013.

Component	PC1	PC2	PC3	PC4
Vis	**0.7**	0.3	0.2	0.2
PM_{10}	**−0.6**	**0.6**	−0.1	−0.1
SO_2	−0.4	0.3	**−0.6**	0.0
CO	**−0.7**	0.5	0.0	−0.1
NO_2	**−0.7**	0.5	0.3	0.0
O_3	**0.8**	0.3	−0.3	−0.1
T	**0.7**	0.3	0.4	−0.2
Ws	0.4	−0.3	−0.5	0.0
RH	**−0.7**	**−0.6**	0.1	0.0
Rad	**0.6**	0.4	−0.3	−0.1
P	0.0	−0.2	0.0	**−0.9**
eigenvalue	4.06	1.79	1.12	1.02
% total variance	37	16	10	9
Cumul. % variance	37	53	63	73

Table 5. Correlations between the principal components (PC1, PC2, PC3) and measured parameter values. The PCA was performed for the hourly data collected in 2004–2013, separately for the cold (January–March and October–December) and warm (April–September) seasons.

Component	Season	PC1	PC2	PC3	Season	PC1	PC2	PC3
Vis.		**−0.7**	0.3	−0.1		−0.5	−0.3	−0.5
PM_{10}		**0.7**	0.5	−0.1		0.4	**−0.7**	0.2
SO_2		0.4	0.4	0.4		0.1	−0.4	0.5
CO		**0.8**	0.4	−0.1		**0.6**	−0.5	0.1
NO_2		**0.7**	0.3	−0.3		**0.7**	−0.5	−0.1
O_3		**−0.7**	0.4	0.4		**−0.8**	−0.2	0.2
T	cold	−0.4	−0.1	**−0.8**	warm	**−0.6**	−0.4	0.0
Ws		**−0.6**	0.0	0.0		−0.4	0.3	0.4
RH		0.5	**−0.8**	0.1		**0.7**	**0.6**	0.0
Rad.		−0.4	**0.6**	−0.1		**−0.7**	−0.2	0.2
P		0.0	−0.3	0.0		0.0	0.2	**0.6**
Eigen value		3.59	1.84	1.12		3.46	1.99	1.14
% total variance		33	17	10		31%	18%	10%

3.6. Generalized Regression Model (GRM)

The GRM identification was performed to finally confirm the influence of the analysed factors on visibility. It concerned the observations of the measured parameters and other defined factors, such as the influence of a season or specific year or combination of these factors. It was assumed that a given factor would be introduced into the model, if the value F (F—Fischer-Snecedor distribution) characterizing the significance of the factor contribution into the dependable variable forecasting (visibility) was higher than $F1$. The factor was removed if its F was lower than $F2$ (Table 6).

Table 6. Generalized Regression Model (GRM) summary: Variables introduced into the model due to estimation.

Variable	Model Steps	Degrees of Freedom	F2 for out	p2 for out	F1 for in	P1 for in	Effect
RH	16	1	1594.25	0.00000			In model
lnPM$_{10}$		1	389.95	0.00000			In model
Season		1	64.60	0.00000			In model
Precipitation Y$|$N		1	233.19	0.00000			In model
O$_3$		1	66.46	0.00000			In model
lnCO		1	92.79	0.00000			In model
T		1	24.39	0.00000			In model
Rad		1	41.06	0.00000			In model
Ws		1	36.40	0.00000			In model
YEAR		9	6.42	0.00000			In model
Year * Season		9	5.78	0.00000			In model
lnSO$_2$		1	21.37	0.00000			In model
Year * Prec.Y$|$N		9	3.68	0.00013			In model
Season * Prec.Y$|$N		1	13.53	0.00023			In model
lnNO$_2$		1	10.67	0.00109			In model
Year * Season * Prec.Y$|$N		9			0.629	0.773	Out of model

Before the identification, the variables underwent necessary analyses and transformations. The variables that were at least in the interval scales were submitted to the quality assessment and logarithming (Box-Cox transformation with the Lambda parameter = 0.5). The precipitation variable was taken to the nominal scale, where 0 and 1 meant the hours without and with precipitation, respectively (variable: Prec.Y$|$N).

Table 6 presents the results of the stepwise estimation for the GRM. The variables, seasonal factors and their interactions marked "in model" turned out to be significantly affecting visibility. Hence, they were introduced into the model. Table 7 presents the adjustment of the model that was finally selected as the best one with consideration for the maximum adjustment criterion and minimum number of the independent variables.

Table 7. Assessment of the GRM adjustment.

	Multipl.—R	Multipl.—R²	Correct.—R²	SS—Model	df—Model	MS—Model	SS—Rest	Df—Rest	MS—Rest	F	p
Vis	0.75	0.56	0.56	6499948	39	166665.3	5033049	71764	70.133	2376.4	0.00

To picture the significance of the variables, they were ordered following the values of the Student's t-distribution for the assessment of the model parameter significance (Figure 8—Pareto chart). The higher the t-value was, the more important the significance of the factor was for explaining its influence on visibility. Relative air humidity and PM_{10} were two most important factors affecting visibility on the basis of the identified model. The influence exerted by SO_2 and NO_2 on visibility was much lower from the impact that other analysed pollutants had. The results also show the influence of the seasonality, atmospheric precipitation (its presence or lack; variable: Prec.Y|N) and interactions between the periods (e.g., season * month) on visibility.

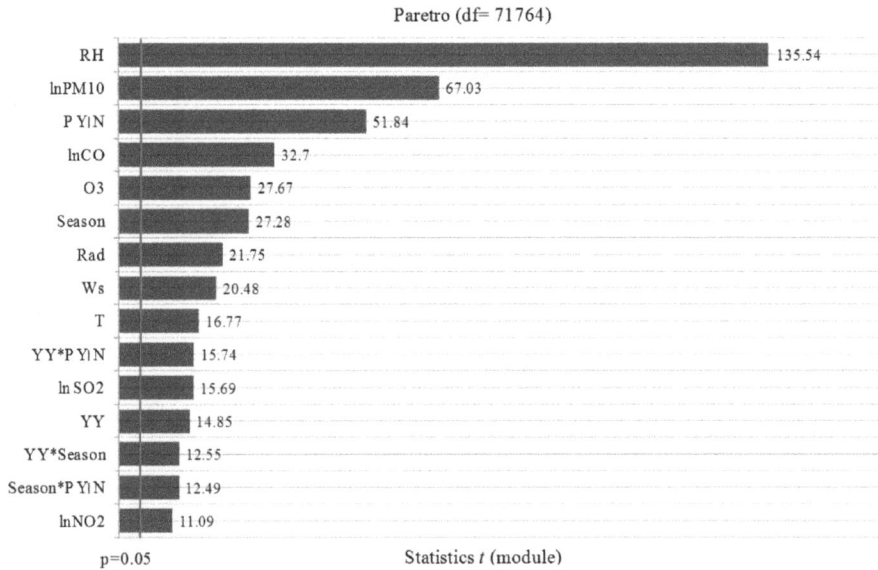

Paretro (df= 71764)

RH — 135.54
lnPM10 — 67.03
P Y|N — 51.84
lnCO — 32.7
O3 — 27.67
Season — 27.28
Rad — 21.75
Ws — 20.48
T — 16.77
YY*P Y|N — 15.74
ln SO2 — 15.69
YY — 14.85
YY*Season — 12.55
Season*P Y|N — 12.49
lnNO2 — 11.09

p=0.05   Statistics t (module)

Figure 8. Pareto chart for the significance of factors affecting visibility in the GRM.

Table 8 presents estimated parameter values for the selected GRM (Tables 6 and 7), together with estimations for the parameter errors, t-distribution and p-value indicating the variable significance and coincidence with visibility. In a statistical sense, it was shown that the relative air humidity and wind speed variations corresponded with visibility more than the temperature. It was also revealed that visibility was most likely sensitive to the changes in the PM_{10} and CO concentrations. Specifically, a 50% drop in the PM_{10} concentration could be associated with visibility improvement by 2.9 km. On the other hand, a drop in the CO concentration by 50% corresponded with the visibility being increased by 2.2 km, however, there was no evidence for a direct cause-and-effect relationship. The 50% decrease in the O_3 concentration might only slightly affect the visibility increase, as suggested by the performed regression analyses.

Table 8. Estimated parameter values for the selected GRM with estimations of parameter errors, t-distribution and p-value indicating the variable significance.

Regression Coefficien	Level/Effect	Estimated Regression Coefficient	Estimated Standard Deviation	t	p Value
Intercept		82.615	0.633	130.468	0.000
$\ln SO_2$		−0.902	0.057	−15.692	0.000
$\ln NO_2$		0.827	0.075	11.089	0.000
$\ln CO$		−3.122	0.095	−32.699	0.000
O_3		−0.054	0.002	−27.673	0.000
T		0.097	0.006	16.766	0.000
Ws		0.450	0.022	20.479	0.000
RH		−0.395	0.003	−135.537	0.000
Rad		−0.005	0.000	−21.751	0.000
$\ln PM_{10}$		−4.220	0.063	−67.032	0.000
YY	2004	0.727	0.163	4.467	0.000
YY	2005	−0.736	0.164	−4.495	0.000
YY	2006	−1.829	0.168	−10.867	0.000
YY	2007	0.880	0.175	5.024	0.000
YY	2008	0.913	0.165	5.540	0.000
YY	2009	2.028	0.137	14.847	0.000
YY	2010	−1.619	0.162	−9.997	0.000
YY	2011	1.174	0.115	10.182	0.000
YY	2012	−0.287	0.142	−2.016	0.044
Season	warm	1.695	0.062	27.284	0.000
Prec.Y|N	Non-raining	2.698	0.052	51.837	0.000
Year * Season	1	0.197	0.096	2.048	0.041
Year * Season	2	−0.857	0.095	−9.028	0.000
Year * Season	3	−1.132	0.093	−12.119	0.000
Year * Season	4	−0.189	0.104	−1.815	0.070
Year * Season	5	−1.052	0.105	−10.023	0.000
Year * Season	6	−0.276	0.092	−3.010	0.003
Year * Season	7	0.814	0.101	8.036	0.000
Year * Season	8	0.531	0.093	5.730	0.000
Year * Season	9	1.121	0.089	12.555	0.000
Year * Prec.Y|N	1	0.433	0.162	2.675	0.007
Year * Prec.Y|N	2	0.714	0.163	4.386	0.000
Year * Prec.Y|N	3	−0.081	0.167	−0.487	0.626
Year * Prec.Y|N	4	0.368	0.173	2.127	0.033

Table 8. *Cont.*

Regression Coefficien	Level/Effect	Estimated Regression Coefficient	Estimated Standard Deviation	t	p Value
Year * Prec.Y|N	5	−0.136	0.161	−0.843	0.399
Year * Prec.Y|N	6	0.132	0.137	0.964	0.335
Year * Prec.Y|N	7	1.426	0.161	8.880	0.000
Year * Prec.Y|N	8	−1.756	0.112	−15.736	0.000
Year * Prec.Y|N	9	−0.045	0.140	−0.319	0.750
Season * Prec.Y|N	1	0.624	0.050	12.487	0.000

The results of the GRM analysis, performed herein, correspond with previous, abundant research, pointing that visibility degradation is due to particulate matter, as well as relative humidity, that can greatly enhance degradation in the presence of hygroscopic aerosols [6,17,53].

4. Conclusions

In Warsaw, the changes in the visibility exhibited a seasonal character. The visibility increased in summer and decreased in late autumn and winter. Mean seasonal visibilities were 24.5, 30.1, 20.1, and 13.9 km in spring, summer, autumn, and winter, respectively. The mean yearly visibility values were in the range of 19.8–23.7 km.

When the meteorological conditions were unfavourable for dispersion and transportation, the visibility was adversely affected by high pollutant concentrations (PM_{10}—236.1 $\mu g \cdot m^{-3}$; SO_2—28.1 $\mu g \cdot m^{-3}$; CO—2342.4 $\mu g \cdot m^{-3}$; and NO_2—76 $\mu g \cdot m^{-3}$). Consequently, the visibility was low (6.0 km). The unfavourable meteorological conditions involved low wind speed (approx. 1.1 m/s) and low air temperature.

On clear days, air quality was found to be good. Mean pollutant concentrations were: PM_{10}—25.7 $\mu g \cdot m^{-3}$; SO_2—6.8 $\mu g \cdot m^{-3}$; CO—378.8 $\mu g \cdot m^{-3}$; and NO_2—20.4 $\mu g \cdot m^{-3}$. Mean wind speed was 2.7 m/s. Generally, the mean visibility value was 24.3 km for the good air quality days in Warsaw.

PCA helped to find that the biggest changes in the visibility in Warsaw were observed with changes in air temperature, concentrations of PM_{10}, CO, NO_2 and O_3, and insolation. Generally, in the cold season, a fall in temperature corresponded to air pollution increase and visibility deterioration. An underlying cause for such a situation is the increase of emissions, related with fuel combustion for heating purposes.

There might also be an indirect correlation between the visibility and traffic emission in the warm (non-heating) season in Warsaw. Traffic emission influence on air quality in the warm season manifested itself with the increasing NO_2 concentrations in the air without simultaneous increase of other pollutants' concentration. At the same time, there existed a decrease in concentrations of ozone, reacting photochemically under strong insolation. In such conditions the visibility increased.

Those conclusions correspond well with the GRM analyses, which demonstrated that the visibility in Warsaw was clearly affected by the measurement season and the factors-variables containing combinations of variables constructed from different measurement periods (e.g., season * month).

It was unequivocally proven that the PM_{10} concentration was the most important parameter affecting the visibility in Warsaw. The GRM results demonstrated that the reduction in the PM_{10} concentrations by 50% (with all the remaining parameters unchanged) contributed to the increase in the visibility by 2.9 km.

Acknowledgments: The work was carried out within the project No 2012/07/D/ST10/02895 (ID 202319) financed by the National Science Centre Poland (NCN). This research was supported by The Faculty of Civil and Environmental Engineering basic (statutory) research projects.

Author Contributions: Conceived and designed the experiments: (GM). All authors (GM, WR-K, PC, AB, AB) contributed to the data analysis and manuscript writing.

Conflicts of Interest: The authors declare that there is no conflict of interests regarding the publication of this paper.

References

1. Deng, J.; Xing, Z.; Zhuang, B.; Du, K. Comparative study on long-term visibility trend and its affecting factors on both sides of the Taiwan Strait. *Atmos. Res.* **2014**, *143*, 266–278.
2. Huang, W.; Tan, J.; Kan, H.; Zhao, N.; Song, W.; Song, G.; Chen, G.; Jiang, L.; Jiang, C.; Chen, R.; et al. Visibility, air quality and daily mortality in Shanghai, China. *Sci. Total Environ.* **2009**, *407*, 3295–3300.
3. Thach, T.Q.; Wong, C.M.; Chan, K.P.; Chau, Y.K.; Chung, Y.N.; Ou, C.Q.; Yang, L.; Hedley, A.J. Daily visibility and mortality: Assessment of health benefits from improved visibility in Hong Kong. *Environ. Res.* **2010**, *110*, 617–623.
4. Latha, K.M.; Badarinath, K.V.S. Black carbon aerosols over tropical urban environment—A case study. *Atmos. Res.* **2003**, *69*, 125–133.
5. Lee, J.Y.; Jo, W.K.; Chun, H.H. Long-Term Trends in Visibility and Its Relationship with Mortality, Air-Quality Index, and Meteorological Factors in Selected Areas of Korea. *Aerosol Air Qual. Res.* **2015**.
6. Tsai, Y.I.; Kuo, S.C.; Lee, W.; Chen, C.L.; Chen, P.T. Long-term visibility trends in one highly urbanized, one highly industrialized, and two rural areas of Taiwan. *Sci. Total Environ.* **2007**, *382*, 324–341.
7. Deng, X.; Tie, X.; Wu, D.; Zhou, X.; Bi, X.; Tan, H.; Li, F.; Jiang, C. Long-term trend of visibility and its characterizations in the Pearl River Delta (PRD) Region, China. *Atmos. Environ.* **2008**, *42*, 1424–1435.

8. Pusheng, Z.; Zhang, X.; Xu, X.; Zhao, X. Long-Term Visibility trends and characteristics in the region of Beijing, Tianjin, and Hebei, China. *Atmos. Res.* **2011**, *101*, 711–718.

9. Fajardo, O.A.; Jiang, J.; Hao, J. Assessing young people's preferences in urban visibility in Beijing. *Aerosol Air Qual. Res.* **2013**, *13*, 1536–1543.

10. Zhuang, X.; Wang, Y.; Zhu, Z.; Querol, X.; Alastuey, A.; Rodríguez, S.; Wei, H.; Xu, S.; Lu, W.; Viana, M.; *et al.* Origin of PM10 pollution episodes in an industrialized mega-city in central China. *Aerosol Air Qual. Res.* **2014**, *14*, 338–346.

11. Rogula-Kozłowska, W.; Klejnowski, K.; Rogula-Kopiec, P.; Ośródka, L.; Krajny, E.; Błaszczak, B.; Mathews, B. Spatial and seasonal variability of the mass concentration and chemical composition of PM2.5 in Poland. *Air Qual. Atmos. Health* **2014**, *7*, 41–58.

12. Lee, C.G.; Yuan, C.S.; Chang, J.C.; Yuan, C. Effects of aerosol species on atmospheric visibility in Kaohsiung city, Taiwan. *J. Air Waste Manag. Assoc.* **2005**, *55*, 1031–1041.

13. Jung, J.; Lee, H.; Kim, Y.J.; Liu, X.; Zhang, Y.; Gu, J.; Fan, S. Aerosol chemistry and the effect of aerosol water content on visibility impairment and radiative forcing in Guangzhou during the 2006 Pearl River Delta campaign. *J. Environ. Manag.* **2009**, *90*, 3231–3244.

14. Appel, B.R.; Tokiwa, Y.; Hsu, J.; Kothny, E.L; Hahn, E. Visibility as related to atmospheric aerosol constituents. *Atmos. Environ. (1967)* **1985**, *19*, 1525–1534.

15. Malm, W.C.; Day, D.E. Estimates of aerosol species scattering characteristics as a function of relative humidity. *Atmos. Environ.* **2001**, *35*, 2845–2860.

16. Tsai, Y.I.; Lin, Y.H.; Lee, S.Z. Visibility variation with air qualities in the metropolitan area in Southern Taiwan. *Water Air Soil Pollut.* **2003**, *144*, 19–40.

17. Tsai, Y.I. Atmospheric visibility trends in an urban area in Taiwan 1961–2003. *Atmos. Environ.* **2005**, *39*, 5555–5567.

18. Wen, C.C.; Yeh, H.H. Comparative influences of airborne pollutants and meteorological parameters on atmospheric visibility and turbidity. *Atmos. Res.* **2010**, *96*, 496–509.

19. Deng, J.; Wang, T.; Jiang, Z.; Xie, M.; Zhang, R.; Huang, X.; Zhu, J. Characterization of visibility and its affecting factors over Nanjing, China. *Atmos. Res.* **2011**, *101*, 681–691.

20. Du, K.; Mu, C.; Deng, J.; Yuan, F. Study on atmospheric visibility variations and the impacts of meteorological parameters using high temporal resolution data: An application of Environmental Internet of Things in China. *Int. J. Sustain. Dev. World Ecol.* **2013**, *20*, 238–247.

21. Majewski, G.; Czechowski, P.O.; Badyda, A.; Brandyk, A. Effect of air pollution on visibility in urban conditions. Warsaw Case Study. *Environ. Prot. Eng.* **2014**, *40*, 47–64.

22. Chen, J.; Qiu, S.; Shang, J.; Wilfrid, O.M.F.; Liu, X.; Tian, H.; Boman, J. Impact of relative humidity and water soluble constituents of PM2.5 on visibility impairment in Beijing, China. *Aerosol Air Qual. Res.* **2014**, *14*, 260–268.

23. Badyda, A.; Dąbrowiecki, P.; Lubiński, W.; Czechowski, P.O.; Majewski, G.; Chciałowski, A.; Kraszewski, A. Influence of traffic-related air pollutants on lung function. *Adv. Exp. Med. Biol.* **2013**, *788*, 229–235.

24. Badyda, J.; Dąbrowiecki, P.; Czechowski, P.O.; Majewski, G.; Doboszyńska, A. Traffic-Related Air Pollution and Respiratory Tract Efficiency. *Adv. Exp. Med. Biol.* **2015**.

25. Majewski, G.; Kleniewska, M.; Przewoźniczuk, W. The effect of urban conurbation on the modification of human thermal perception, as illustrated by the example of Warsaw (Poland). *Theor. Appl. Climatol.* **2014**, *116*, 147–154.

26. Pastuszka, J.S.; Wawroś, A.; Talik, E.; Paw U, K.T. Optical and chemical characteristics of the atmospheric aerosol in four towns in southern Poland. *Sci. Total Environ.* **2003**, *309*, 237–251.

27. Czarnecka, M.; Nidzgorska-Lencewicz, L. Impact of weather conditions on winter and summer air quality. *Int. Agrophysics* **2011**, *25*, 7–12.

28. Rogula-Kozłowska, W.; Kozielska, B.; Błaszczak, B.; Klejnowski, K. The mass distribution of particle-bound PAH among aerosol fractions: A case-study of an urban area in Poland. In *Organic Pollutants Ten Years after the Stockholm Convention—Environmental and Analytical Update*; Puzyn, T., Mostrag-Szlichtyng, A., Eds.; InTech: Rijeka, Croatia, 2012; pp. 163–190.

29. Majewski, G.; Kleniewska, M.; Brandyk, A. Seasonal variation of particulate matter mass concentration and content of metals. *Pol. J. Environ. Stud.* **2011**, *20*, 417–427.

30. Rozbicka, K.; Majewski, G.; Rozbicki, T. Seasonal variation of air pollution in Warsaw conurbation. *Meteorol. Z.* **2014**, *23*, 175–179.

31. Majewski, G.; Rogula-Kozłowska, W. The elemental composition and origin of fine ambient particles in the largest Polish conurbation: First results from the short-term winter campaign. *Theor. Appl. Climatol.* **2015**.

32. Rogula-Kozłowska, W.; Klejnowski, K.; Rogula-Kopiec, P.; Mathews, B.; Szopa, S. A study on the seasonal mass closure of ambient fine and coarse dusts in Zabrze, Poland. *Bull. Environ. Contam. Toxicol.* **2012**, *88*, 722–729.

33. Rogula-Kozłowska, W.; Rogula-Kupiec, P.; Mathews, B.; Klejnowski, K. Effects of road traffic on the ambient concentrations of three PM fractions and their main components in a large Upper Silesian city. *Ann. Wars. Univ. Life Sci.* **2013**, *45*, 243–253.

34. Rogula-Kozłowska, W. Traffic-Generated Changes in the Chemical Characteristics of Size-Segregated Urban Aerosols. *Bull. Environ. Contam. Toxicol.* **2014**, *93*, 493–502.

35. Harrison, R.M.; Deacon, A.R.; Jones, M.R. Sources and processes affecting concentrations of PM_{10} and $PM_{2.5}$ particulate matter in Birmingham (UK). *Atmos. Environ.* **1997**, *31*, 4103–4117.

36. Statheropoulos, M.; Vassiliadis, N.; Pappa, A. Principal component and canonical correlation analysis for examining air pollution and meteorological data. *Atmos. Environ.* **1998**, *32*, 1087–1095.

37. Kleinbaum, D.G.; Kupper, L.L.; Muller, K.E.; Nizam, A. *Applied Regression Analysis and Other Multivariable Methods*; Duxbury Press: Pacific Grove, CA, USA, 1998; p. 798.

38. Izenman, A.J. *Modern Multivariate Statistical Techniques: Regression. Classification, and Manifold Learning*; Springer Texts in Statistics: New York, NY, USA, 2008.

39. Rozbicka, K.; Rozbicki, T. Spatiotemporal variations of tropospheric ozone concentrations in the Warsaw Agglomeration (Poland). *Ann. Wars. Univ. Life Sci.* **2014**, *46*, 247–261.

40. Xue, D.; Li, C.; Liu, Q. Visibility characteristics and the impacts of air pollutants and meteorological conditions over Shanghai, China. *Environ. Monit. Assess.* **2015**, *187*, 363.

41. Cao, J.J.; Wang, Q.Y.; Chow, J.C.; Watson, J.G.; Tie, X.X.; Shen, Z.X.; Wang, P.; An, Z.S. Impacts of aerosol compositions on visibility impairment in Xi'an, China. *Atmos. Environ.* **2012**, *59*, 559–566.

42. Rogula-Kozłowska, W.; Błaszczak, B.; Kozielska, B.; Klejnowski, K. *The Mass Distribution of Particle-Bound PAH among Aerosol Fractions: A Case-Study of an Urban Area in Poland*; InTECH Open Access Publisher: Rijeka, Croatia, 2012.

43. Houthuijs, D.; Breugelmans, O.; Hoek, G.; Vaskövi, É.; Miháliková, E.; Pastuszka, J.S.; Jirik, V.; Sachelarescu, S.; Lolova, D.; Meliefste, K.; *et al.* PM$_{10}$ and PM$_{2.5}$ concentrations in Central and Eastern Europe:: Results from the Cesar study. *Atmos. Environ.* **2001**, *35*, 2757–2771.

44. Pastuszka, J.S.; Okada, K. Features of atmospheric aerosol particles in Katowice, Poland. *Sci. Total Environ.* **1995**, *175*, 179–188.

45. Reizer, M.; Juda-Rezler, K. Explaining the high PM$_{10}$ concentrations observed in Polish urban areas. *Air Qual. Atmos. Health* **2015**.

46. Pastuszka, J.S.; Rogula-Kozłowska, W.; Zajusz-Zubek, E. Characterization of PM$_{10}$ and PM$_{2.5}$ and associated heavy metals at the crossroads and urban background site in Zabrze, Upper Silesia, Poland, during the smog episodes. *Environ. Monit. Assess.* **2010**, *168*, 613–627.

47. Cleveland, W.S.; Graedel, T.E.; Kleiner, B.; Warner, J.L. Sunday and workday variations in photochemical air pollutants in New Jersey and New York. *Science* **1974**, *186*, 1037–1038.

48. Qin, Y.; Tonnesen, G.S.; Wang, Z. Weekend/weekday differences of ozone, NOx, Co, VOCs, PM$_{10}$ and the light scatter during ozone season in southern California. *Atmos. Environ.* **2004**, *38*, 3069–3087.

49. Marr, L.C.; Harley, R.A. Spectral analysis of weekday-weekend differences in ambient ozone, nitrogen oxide, and non-methane hydrocarbon time series in California. *Atmos. Environ.* **2002**, *36*, 2327–2335.

50. Stull, R.B. *Meteorology for Scientists and Engineers*; Brooks/Cole: Pacific Grove, CA, USA, 2000.

51. Elias, T.; Haeffelin, M.; Drobinski, P.; Gomes, L.; Rangognio, J.; Bergot, T.; Chazette, P.; Raut, J.C.; Colomb, M. Particulate contribution to extinction of visible radiation: Pollution, haze, and fog. *Atmos. Res.* **2009**, *92*, 443–454.

52. Geertsema, G.T.; Schreur, B.G.J. The effect of improved nowcasting of precipitation on air quality modeling. *Atmos. Environ.* **2009**, *43*, 4924–4934.

53. Pitchford, M.; Malm, W.; Schichtel, B.; Kumar, N.; Lowenthal, D.; Hand, J. Revised algorithm for estimating light extinction from IMPROVE particle speciation data. *J. Air Waste Manag.* **2007**, *57*, 1326–1336.

Study of Black Sand Particles from Sand Dunes in Badr, Saudi Arabia Using Electron Microscopy

Haider Abbas Khwaja, Omar Siraj Aburizaiza, Daniel L. Hershey, Azhar Siddique, David A. Guerrieri P. E., Jahan Zeb, Mohammad Abbass, Donald R. Blake, Mirza Mozammel Hussain, Abdullah Jameel Aburiziza, Malissa A. Kramer and Isobel J. Simpson

Abstract: Particulate air pollution is a health concern. This study determines the microscopic make-up of different varieties of sand particles collected at a sand dune site in Badr, Saudi Arabia in 2012. Three categories of sand were studied: black sand, white sand, and volcanic sand. The study used multiple high resolution electron microscopies to study the morphologies, emission source types, size, and elemental composition of the particles, and to evaluate the presence of surface "coatings or contaminants" deposited or transported by the black sand particles. White sand was comprised of natural coarse particles linked to wind-blown releases from crustal surfaces, weathering of igneous/metamorphic rock sources, and volcanic activities. Black sand particles exhibited different morphologies and microstructures (surface roughness) compared with the white sand and volcanic sand. Morphological Scanning Electron Microscopy (SEM) and Laser Scanning Microscopy (LSM) analyses revealed that the black sand contained fine and ultrafine particles (50 to 500 nm ranges) and was strongly magnetic, indicating the mineral magnetite or elemental iron. Aqueous extracts of black sands were acidic (pH = 5.0). Fe, C, O, Ti, Si, V, and S dominated the composition of black sand. Results suggest that carbon and other contaminant fine particles were produced by fossil-fuel combustion and industrial emissions in heavily industrialized areas of Haifa and Yanbu, and transported as cloud condensation nuclei to Douf Mountain. The suite of techniques used in this study has yielded an in-depth characterization of sand particles. Such information will be needed in future environmental, toxicological, epidemiological, and source apportionment studies.

Reprinted from *Atmosphere*. Cite as: Khwaja, H.A.; Aburizaiza, O.S.; Hershey, D.L.; Siddique, A.; Guerrieri P. E., D.A.; Zeb, J.; Abbass, M.; Blake, D.R.; Hussain, M.M.; Aburiziza, A.J.; Kramer, M.A.; Simpson, I.J. Study of Black Sand Particles from Sand Dunes in Badr, Saudi Arabia Using Electron Microscopy. *Atmosphere* **2015**, *6*, 1175–1194.

1. Introduction

Particulate air pollution significantly impacts public health in both developed and developing countries. Major sources of particulate matter (PM) include large industries, such as power plants, petrochemical plants, cement plants, vehicular traffic, windblown dust, crustal erosion, volcanic eruptions, wildfires, sea spray, and combustion processes [1]. Amongst all of these, combustion is the main source for pollutant emissions. PM varies in composition, origin, and size, ranging from submicron (sub-μm), 1–30 μm, and up to 50 μm or more. Large particles (>10 μm) have a relatively low residence time in the air and tend to settle quickly through gravitational subsidence. Fine particles with aerodynamic diameters less than 2.5 μm ($PM_{2.5}$) and "coarse" particles with aerodynamic diameters less than 10 μm (PM_{10}) remain airborne for a longer period of time. Ultrafine particles are particulate matter of nanoscale size (less than 100 nm in diameter). Fine particles in the 0.05–2 μm range can travel distances up to approximately 1000 km [2].

Exposure to PM has been associated with numerous effects on human health, including increased morbidity and mortality, respiratory problems, cardiovascular diseases, lung cancer, renal and brain damage, and human metal poisonings [3–6]. Both fine and coarse particles are readily inhaled into the human respiratory tract. It is suggested that fine particles are more strongly implicated in cardiovascular effects than coarse particles, while both impact respiratory end points [7]. In epidemiological studies, positive correlations have been established between elevated levels of inhaled airborne PM, especially $PM_{2.5}$, and acute adverse health effects [8–10]. Aside from the amount of PM inhaled, particle-related parameters, such as size, composition, and solubility have also been linked to health effects in toxicology studies. Size affects the site deposition in the human respiratory tract and the consequent degree of toxicity that may result. Therefore, ultrafine particles are more toxic than coarse particles and are linked to pulmonary diseases in human and animals [11–14]. Particle sizes also reveal the origin and the formation of airborne particles: larger sized particles are of crustal origin whereas fine particles originate from combustion processes or gas-to-particle conversion in the atmosphere [15].

Particulate air pollution is a health concern among the residents of Badr, Saudi Arabia. Deposition of black sand particles has recently become common on Douf Mountain (Figure 1), located to the west of Badr. The soil surrounding Medina consists mostly of basalt, while the hills to the southeast and northeast are volcanic sand that dates to the Paleozoic era. The geology of Badr is represented by granophyre and alkali-feldspar granite. Dominant lithologies are andesitic and dacitic to rhyolitic tuffs, lavas, and volcanic rocks [16]. The town of Badr, in Medina province (23°47′N 38°47′E; 123 m altitude), is located about 150 km southwest of the holy city of Medina, towards the western outlet of Wadi-e-Safra (Figure 2a). The total area of Badr is 8226 km^2 with a population of 61,600. Average annual rainfall is

80 mm. June, July, August, and September are the hottest months in Badr with an average temperature of 40 °C, while the coldest are January and February at 14 °C. Historically, Badr was a market, situated on the road which connects Sham (Syria, Lebanon and Palestine now) with Mecca (Figure 2b).

(a)

(b)

Figure 1. (a) Photograph of black sand on Douf Mountain, Badr, Saudi Arabia; (b) Photograph of black sand collected from Douf Mountain on February 2012.

Despite concerns about the health and environmental significance of the black sand particles, not much is known about the size, morphology, and specific chemical constituents of these sand particles. It is generally known that certain chemical forms are more toxic than others: in general, soluble forms are potentially more toxic than those tied up as insoluble oxides or forming glassy structures [17]. Knowledge of the morphology and chemical characterization of sand particles can aid in control strategies and help to interpret and predict chemical interactions in the atmosphere, downwind fallout rates, potential damage to vegetation, deterioration of materials and structures, and, in particular, impact on human health. Among microanalytical techniques for characterizing black sand particles, Scanning Electron Microscopy with an attached Energy Dispersive X-ray Spectrometer (SEM/EDS) is the most versatile technique, capable of simultaneously obtaining information on particle size, morphology, elemental composition, and microstructure [18,19]. Here the microscopic make-up of different varieties of sand particles collected from a dune near Badr is presented.

(a)

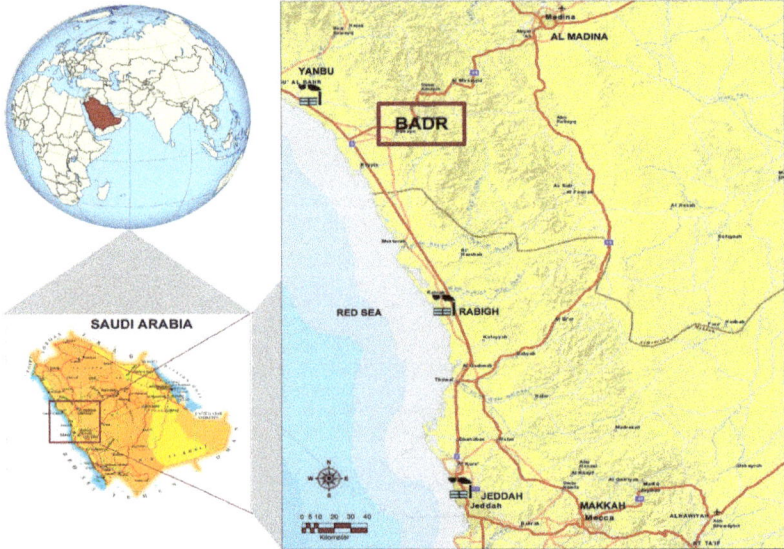

(b)

Figure 2. (**a**) Map of Saudi Arabia; (**b**) Map of western Saudi Arabia showing location of sampling site, Badr.

108

2. Experimental

2.1. Sample Collection

Representative random composite sand samples about 4–6 inches thick were collected in 2012 from various locations in the sand dunes (Figure 1b) using hand towels and gloves. The randomly collected samples were composited and thoroughly mixed, and a representative sub-sample was taken for analysis. The sample collected was very black in color, and it was then separated into three groups in the laboratory: magnetic black sand, non-magnetic black sand, and white sand. Of the black sand sampled, there were two distinct types: magnetic and non-magnetic. Solutions of the black sands, immersed in deionized water gave a low pH of 5.0, indicating acidity. A volcanic sand sample was also collected from the area. For this study, the two types of black sand were compared to "normal" white sand and to volcanic sand. Four samples were selected for microscopic examination and morphology, size, and elemental composition were determined. Several hundred individual particles of each type (magnetic black sand, non-magnetic black sand, white sand, and volcanic sand) were viewed for comparison purposes.

2.2. Equipment and Analytical Methodology

Sand particles were characterized at the New York State Department of Environmental Conservation (NYS DEC) microscopy laboratory in Rensselaer NY, using an Olympus SZX12 optical stereo microscope (OSM), SEM/EDS (model JEOL 6490LV SEM equipped with a Sahara EDS), and an Olympus LEXT laser scanning microscope (LSM, model OLS 3100 LSM). This suite of techniques provided images showing morphology of particles, size, elemental composition, and specific source types. The stereo microscope provided a colored representation of the differences in the particles. The SEM provided much more detail at higher magnifications: high resolution of up to $300,000\times$ magnification can be achieved. For this study, $12,000\times$ microscope magnification was used. In addition, elemental composition was determined using an energy dispersive X-ray spectrometer (EDS) attachment to the SEM. An Olympus-LEXT instrument was used to locate, determine, and characterize fine particles adhered to the sand particles. An LSM or confocal scanning microscope (CLSM) was used to create high resolution optical images that show depth, creating a 3-dimensional image of the object being scanned, which can show how the particles are adhered to the particle surface. This technique allows the surface contours to be imaged for profile viewing. The LSM offers better detail on surface morphology and layering of the surface and produces a surface structure in submicron units (minimum resolution of 0.1 µm).

The analysis of the samples was performed by a SEM/EDS to determine particle morphology and elemental composition. Electron microscopic particle identifications were based on visual comparisons to in-house reference standards and to data in McCrone's Atlas [20]. The EDS system can collect data from X-rays equivalent to the depth of the secondary electron formation. This is dependent on the sample density and keV. The approximate secondary electron collection depth is between 0.5 and 1 micron. For this study, both secondary electron imaging and backscattered electron imaging were used. Operating conditions for the SEM analysis were: accelerating voltage = 15 keV; filament current = 112 µA; working distance = 19 mm; analytical time = 200 s; dead time = varied from sample to sample. The X-ray spectra were generated from specified areas on sand grains placed on aluminum SEM stubs and collected for 90 s with probe current of 50 nA. The weight-percent of the elemental analysis was determined by using ZAF correction factors for each element.

Attached to the JEOL SEM was a Bruker-energy dispersive X-ray spectrometer (EDS; "Spirit" system), that was able to quantify elements with atomic number $Z \geq 6$. The EDS detector was controlled by Bruker "Spirit" system. In combination, the two provided detailed particle analysis information. A second EDS system, equipped with TEAM software, owned and operated by EDAX Inc., was used for back up and confirmation purposes. Intensities of the EDS spectrum lines were converted to corresponding weight-percent concentrations based on their excitation energy (Kv). Calibration was referenced to the excitation energy of Al and Cu. Weight-percent concentrations of elements were determined semi-quantitatively using the TEAM software. Forty-one X-ray regions of interest were used to detect the presence of C, O, Na, Mg, Al, Si, P, S, Cl, K, Ca, Ti, V, Cr, Mn, Fe, Co, Ni, Cu, Zn, Ga, As, Se, Br, Sr, Y, Zr, Mo, Pd, Ag, Cd, In, Sn, Sb, Ba, La, Au, Hg, Tl, Pb, and U. An element was considered to be present if its peak counts were at least three times the square root of the background counts in the region of interest. Uncertainty in the reported percentages is within $\pm 3\%$–5% for each element studied.

Source contributions for individual samples were determined by comparing elemental ratios, morphological identifiers [20] and verified using a variety of microscopy techniques. Most of these techniques include visual observations, comparison to reference standards, comparisons to the source samples, and digital mapping. When setting up the individual source classes, miscellaneous and unidentified particles were placed in the biological/miscellaneous category. Particle size and elemental ratios were determined with the aid of multiple digital maps. The particle size was determined using the Feret diameter. In addition to the above, the identification of questionable particles was compared to reference standards and to McCrone's Atlas [20]. SEM/EDS elements were calibrated at different energies using computer generated standards, in-house standards, and a copper grid (ASTM) prior to sample analysis.

The Dionex ICS 2500 ion chromatography system used for the determination of anions and cations was configured with an auto sampler, a pump, and a conductivity detector. Anions were eluted from an analytical column (AS14 4 mm × 250 mm) and a guard column (AG14 4 mm) using 3.5 mM Na_2CO_3/1.0 mM $NaHCO_3$ eluent. Cations were eluted from an analytical column (CS14 4 mm × 250 mm) and a guard column (CG14 4 mm) using 10 mM methylsulfonic acid eluent. For quantification, a linear calibration curve (r > 0.995) was established for each analyte and an independently made quality control sample was analyzed to validate the calibration curve. Sand samples were extracted in water (1 g dry weight/5 mL of water), filtered through a pre-cleaned 0.2 µM PTFE filter, and collected in a scintillation vial.

2.3. Sample Preparation

The sample was placed on a 12.5 mm diameter aluminum stub in order to directly observe the particles in an undisturbed condition. Each sample was affixed to the stub with double-sided carbon tape. To obtain representative elemental composition, and to avoid artificial damage and distortion of particles, the samples were not coated with carbon or gold.

3. Results and Discussion

3.1. Morphology of Sand Samples

OSM, SEM, LSM Analysis

Figure 3 shows four stereo microscope images (63×) of non-magnetic black sand, magnetic black sand, white sand, and volcanic sand. White sand appears as a clean crystalline structure (quartz—SiO_2), whereas the volcanic sand and the magnetic black sand both have reddish tints indicative of iron oxide. Even at relatively low SEM magnification (220×), significant structural and elemental features were evident. The volcanic sand illustrates a very angular structure while ovoid structures were evident in the remaining three sands. The iron oxide particles were either irregular or spherical.

The surface differences became more pronounced at higher SEM magnification (450×; Figure 4a). Black sand particles, both magnetic and non-magnetic were more spherical than the oval white sand particles and angular volcanic sand.

111

Figure 3. Stereo micrograph of non-magnetic black, magnetic black, white, and volcanic sand (63× magnification).

SEM images at a magnification of 2000× (Figure 4b) revealed different surface structures in all four samples. At this magnification the fine particles began to appear, attached to the surfaces of the sand grains, especially the black sand particles. There are distinct differences between the images of the black sand particles and the images of the white sand and volcanic sand particles. While white sand and the volcanic sand are crustal in nature (Figure 4b), the white sand particles have been polished over time due to weathering; however, they otherwise appear similar to the volcanic sand. At 7000× magnification, fine particles attached to the surface of the white and black sand grains were round in shape, whereas particles on the volcanic sand were jagged and irregular. Examination at 7000× clearly indicated the predominance of fine and ultrafine particles (50 to 500 nm range) in the black sand sample. Morphology of these small round particles suggests industrial contributions and/or byproducts of fossil-fuel combustion. At higher magnification (12,000×), differences in the fine particles embedded on the grains became more evident (Figure 4c). Since they are complex aggregates, their total surface area is enormous in contrast to coarser PM. Due to the longer residence times and better mobility of fine particles, their dispersion in the atmosphere is more efficient than that of coarse particles, and they are very readily transported long distances by air streams.

112

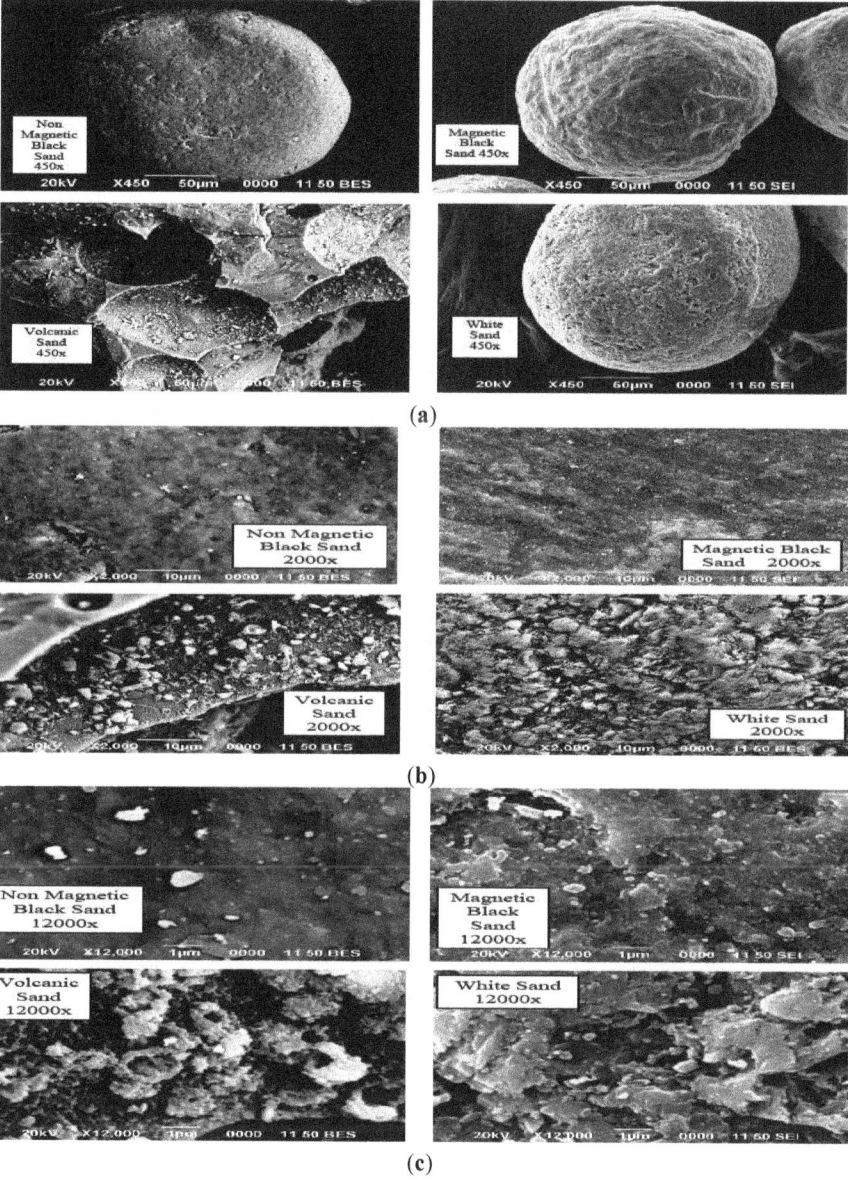

Figure 4. Scanning electron micrographs of sand at (**a**) 450× magnification; (**b**) (2000× magnification showing ultrafine particles; and (**c**) 12,000× magnification.

The first series of LSM images, taken at 10× microscope magnification (mm optical, Figure 5a), showed characteristics of an oil-like product on the surface of the black sand samples with additional fine particles attached to the surface. The other three sand samples appeared similar in size, shape, and surface textures. The coloration of the three sand samples is somewhat similar; however, the magnetic black sand was less reflective because of the oil contribution. The LSM showed that the volcanic sand sample had a different appearance. It was less smooth and appeared more crystalline, and the surface appeared to be comprised of many smaller particles stuck together with many different color variations. At higher magnification ($20\times$), all four sand samples did show differences in shape, as well as color variations. Black sand samples appeared more spherical whereas the white sand sample appeared more oval, as seen previously by SEM. The surface of the magnetic black sand had some roughness, similar to the white sand, and there were variations in coloration that were not as apparent at lower magnification. The volcanic sand looked to be very rough and composed of material of many colors. The black sand appeared to be mostly gold-colored at this magnification, with similar surface roughness to that of the magnetic black sand.

The first four LSM images (Figure 5a) were created using the optical settings of the microscope. The next two images (Figure 5b,c) were created using the confocal imaging capabilities of the LSM. In Figure 5b, a $100\times$ color image of a magnetic black sand particle, individual fine particles are seen coating the surface. Figure 5c depicts a $100\times$ color LSM image of a white sand particle, where fine black particles are seen coating the surface. They may be the beginning of a black coating, but most likely are anthropogenic. A cross section of a black sand particle ($90\times$ magnification; Supplementary Figure S1), shows a white interior with a black coating on the outside. In this image, taken with the stereo microscope, the particle on the left shows where the surface coating was removed and the white core exists. The magnetic black sand clearly depicts the reddish iron oxide particles.

One of the black sand grains was fractured and divided into quarters, similar to an interior slice of an orange. In Figure 6a, an SEM image ($370\times$), the left side shows the interior of the sand grain with no deposition or evidence of fine and ultrafine particles. The roundish outside clearly shows a "fingernail" effect. There is a thin surface layer (1 μm) embedding a large number of particles from the atmospheric deposition, and the coating can be seen as having a rough texture. A similar observation was seen using the LSM in the confocal mode ($100\times$), in which the rough textured outside coating is apparent with individual submicron particles attached to the surface, and the inside is smooth with no coating (Figure 6b). The 1 μm deep surface coating is also apparent. It is known that emissions from combustion of fossil-fuel contain carbon soot and sulfur that is easily absorbed by soil particles during long-range transportation [21,22].

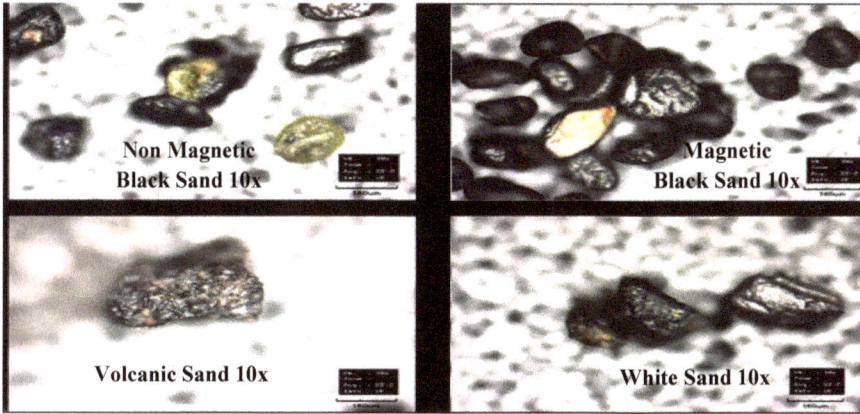

(a)

(b)

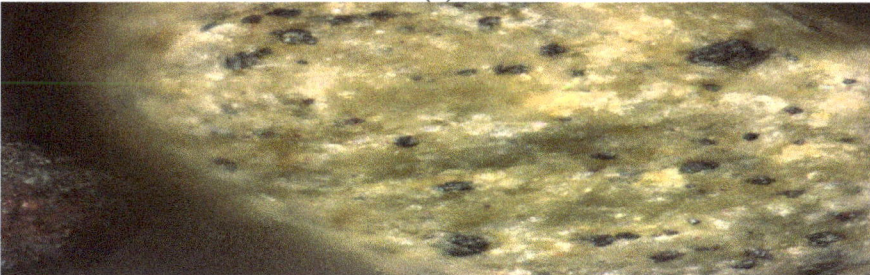

(c)

Figure 5. LSM images (**a**) sand (10× magnification); (**b**) magnetic black sand (100× magnification); and (**c**) white sand (100× magnification).

The coating of the sand grains with the oil-like material and the fine and ultrafine particles were identified independently and are believed to be correct. It does not appear likely that the black sands originated from a natural oil seep (leaking oil field) as opposed to having anthropogenic origin. Oil fields are located in eastern Saudi Arabia, and not in western Saudi Arabia where the study took place. In addition, the predominant wind direction is from the northwest, in the direction of a

115

highly industrialized area (Yanbu, Figure 2). Indeed, three of the country's largest oil refineries operate 83 km upwind of the sampling site. Together these results point to anthropogenic origin of the black sands, rather than impact from natural oil seeps.

(a)

(b)

Figure 6. (a) Scanning electron micrograph of magnetic black sand (370× magnification); **(b)** LSM image of magnetic black sand (100× magnification).

3.2. Elemental Composition (SEM/EDS)

Elemental determinations were performed to indicate the possible chemical species and quantities present. In an energy range of up to 7.5 keV, the following elements were detected: C, O, Na, Mg, Al, Si, P, S, K, Ca, Ti, V, Cr, Mn, and Fe. Carbon, O, Na, Mg, Al, Si, K, Ca, Ti, and Fe peaks were observed at 0.28, 0.53, 1.04, 1.25, 1.50, 1.74, 3.31, 3.69, 4.51, and 6.40 keV, respectively. These elements were the most common, found in varying intensity in all samples (Table 1). A clear P peak at 2.01 keV appeared in the magnetic black sand and volcanic sand samples. Only the magnetic black sand sample exhibited a peak at 2.31 keV, which clearly indicates S. Well-defined V and Mn peaks at energies of 4.95 and 5.89 keV, respectively, were detected in the non-magnetic black and volcanic sand sample. A Cr peak at 5.41 keV in the volcanic sand was clearly pronounced. The striking difference in elemental analysis between the black and white sand confirms that the two distinct entities were being separately resolved. The analysis of black sand and the comparison with white and volcanic sand samples that were performed in this study enables one to attribute the observed elements to anthropogenic emissions.

Table 1. Elemental Comparison of Four Sand Grains (Weight %).

Element	Magnetic Black Sand	Non-Magnetic Black Sand	White Sand	Volcanic Sand
C	19.0	8.8	5.4	6.4
O	13.2	37.6	51.9	42.3
Na	0.3	0.5	3.1	2.5
Mg	0.3	0.7	3.5	3.1
Al	1.1	1.8	8.3	6.4
Si	2.0	3.4	20.0	16.8
P	0.1	0.0	0.0	0.4
S	0.3	0.0	0.0	0.0
K	1.0	0.1	1.9	0.7
Ca	1.0	0.4	1.6	6.2
Ti	6.7	18.0	0.8	1.5
V	0.0	0.7	0.0	0.3
Cr	0.0	0.0	0.0	0.3
Mn	0.0	0.4	0.0	0.5
Fe	55.0	27.6	3.5	12.6

The high sample to sample variability of the Fe concentration represents the most striking result shown on Table 1: the Fe weight contribution represents 55.0% of the magnetic black sand sample, compared to 27.6%, 12.6%, and 3.5% in the non-magnetic black sand, volcanic sand, and white sand samples, respectively. The higher O peak in the energy dispersive X-ray spectrum of the magnetic black and non-magnetic black sand indicates that the Fe may be present as iron oxides specifically hematite (Fe_2O_3), and magnetite (Fe_3O_4). The black color of the sand is a result of the large quantity of C-bearing particles with a nm size. The magnetic black sand sample had the highest C content (19.0 wt %), in contrast with the volcanic sand (6.4 wt %) and white sand (5.4 wt %). Elemental carbon is predominantly a product of the combustion process and is a good tracer for combustion-generated particles. Of the heavy metals, only Ti was present in levels approaching 6.7 wt %. Potassium, Ca, Al, and Si were present at 1–2 wt %. The presence of a low intensity but discernible

117

S peak in addition to C and O peaks in magnetic black sand suggests soot particles derived from fossil-fuel combustion. Similar results were observed in other electron microscopy studies of ambient particles [23–25]. A study by Chen *et al.* [25] reported that soot particles exhibiting a sulfur EDS peak, in addition to carbon and oxygen peaks, are prominently derived from combustion of coal, residual oil, gasoline, and diesel oil. Generally, the S content of soot particles is much less than that of char particles [23]. Since soot particles have smaller density compared to soil particles, their lifetime in the atmosphere is longer and they are more easily transported far away from their source areas [26]. A study carried out by Ali-Mohamed and Matter [27] revealed that soot from black smoke of the Kuwaiti oil-well fires settled 400 km downwind in Bahrain. Particles of combustion source origin included C, Al, S, Ti, and Fe-rich particles from metal industries and iron deposits. Many studies have reported the prevalence of C-rich particles in the fine fraction of PM_{10} in several urban environments [28–31]. By analyzing precipitation samples collected at Kanazawa, Japan, Tazaki *et al.* [32] reported that the high concentration of nm sized C-bearing particles and powdery dusts were produced by oil field combustion and a sandstorm in Iraq, and the particles were transported as cloud nuclei from Iraq to Japan by westerly winds.

Non-magnetic black sand is characterized by C, O, Na, Mg, Al, Si, P, K, Ca, Ti, V, Mn and Fe (Table 1) and is highly enriched in Ti (18.0 wt %). Titanium oxide particles are common in ambient PM samples [33,34]. Carbon and Fe were observed in lesser amounts (8.8 and 27.6 wt %) as compared to the magnetic black sand sample (19.0 and 55.0 wt %). Levels of Mn (0.40 wt %) in non-magnetic black sand were comparable to the volcanic sand sample (0.50 wt %). The presence of transition metals on sand particles is consistent with the anthropogenic process where volatilized metal, H_2SO_4 vapor, and SO_2 condense on submicron ash particles, forming a complex aerosol mixture that is eventually emitted into the ambient air [35]. Vanadium products can indicate a portion of fly ash from the combustion of heavy fuel oil [17]. The principal metals generally contained in fuel oils include C (1.5%–69%), Na (0.2%–3.91%), Mg (2.26%–18.4%), Al (0.01%–1.42%), Si (0.05%–0.31%), K (0.1%–0.13%), Ca (0.1%–1.0%), V (1.1%–12.85%), Fe (0.40%–0.61%), and Ni (0.35%–2.28%). Possible compound compositions of oil-fired fly ash are: C as C; Na as Na_2SO_4; Mg as MgO, Mg as $MgSO_4.H_2O$; Al as Al_2O_3, Al as $Al_2 (SO_4)_3$; Si as SiO_2; K as K_2SO_4; Ca as CaO, and Ca as $CaSO_4$; V as V_2O_5, V as $VOSO_4.3H_2O$; and Fe as Fe_2O_3, Fe as $FeSO_4$ [17].

White sand is rich in C, O, Na, Mg, Al, Si, K, Ca, Ti, and Fe (Table 1). Oxygen (51.9 wt %), Si (20 wt %), Al (8.3 wt %), Mg (3.5 wt %), Na (3.1 wt %), and K (1.9 wt %) had the highest values of the four sand grain samples. The white sand had the lowest carbon content (5.4 wt %), which is likely associated with Ca (as $CaCO_3$—calcite, the main component of limestone) and other alkaline earth metals e.g., Mg (as CaMg (CO_3)—dolomite), due to the corresponding presence of O [18]. It should be noted that the sample contained 2.4 times more Al than Fe. Both the white sand and the volcanic sand exhibit similar ratios of O to Si (2.59 and 2.52, respectively) and Si to Al (2.41 and 2.63, respectively). The silicates present in the white sand and volcanic sand are mainly silicon oxides and alumino-silicate particles originating from both natural (e.g., earth's crustal matter) and anthropogenic (e.g., combustion of fossil-fuels) sources [25]. Silica-rich particles have been reported as significant constituents of the fine fractions of aerosols collected in arid desert areas of southern Utah [36], the northern Sahara [37], Cairo, Egypt [38], and Phoenix, Arizona [30].

Table 1 reveals that predominant elements in volcanic sand are Ca, Al, C, Fe, Si, and O: the weight concentration of these elements varies from 6.2% to 42.3%. Particles are rich in Ca, likely as $CaCO_3$, and contain minor amounts of V, Cr, P, Mn, K, Ti, Na, and Mg (0.30% to 3.1%). Phosphorous had the highest value (0.40 wt %) of the four sand grain samples. Phosphorous–bearing particles normally occur as calcium phosphates [25]. Silicates present in the sample, within which Ca-rich silicates are prevalent, come from the soil [39].

3.3. Enrichment Factors (EFs)

Enrichment factors (EFs) were calculated to identify the role of potential sources of crustal or anthropogenic pollution:

$$EF = (C_x/C_{Al})_{sand}/(C_x/C_{Al})_{crust} \qquad (1)$$

where C_x and C_{Al} represent the concentrations of an element and the abundance of aluminum, respectively. The main advantage of the EF is that the enrichment of all elements can be promptly compared. Aluminum was selected as the reference element [40]. Usually, an EF value of <10 indicates an element of crustal origin, while an EF $\geq$ 10 is ascribed to an element of anthropogenic source [41]. The EFs of Na, Mg, Si, K, and Ca in magnetic black sand ranged from 0.5 to 3, demonstrating that these elements have a significant crustal origin. EFs of Fe, S, and Ti were 74, 84, and 88, respectively, revealing a mainly anthropogenic origin. For the non-magnetic black sand, a strong enrichment was observed for Mn (EF = 20), Fe (EF = 23), Ti (EF = 147), and V (EF = 237), while for all other elements EF values were less than 1. The EFs of V and Cr in volcanic sand were 26 and 37, respectively, which suggests an anthropogenic component.

3.4. Extractable Ion Analysis by IC

In order to quantify anthropogenic contributions and further assess SEM/EDS results, ion chromatography analysis of magnetic black sand was also performed, and it confirmed the presences of the following water-soluble anions and cations: sulfate (SO_4^{2-}), nitrate (NO_3^-), chloride (Cl^-), sodium (Na^+), potassium (K^+), magnesium (Mg^{2+}), and calcium (Ca^{2+}). Among the anions, SO_4^{2-} had the highest concentration (59 ppm), followed by Cl^- (17 ppm) and NO_3^- (5.2 ppm). Sulfate and NO_3^- are related to anthropogenic emissions. Generally, S-bearing particles exist in the forms of SO_4^{2-}, HSO_4^-, and/or H_2SO_4 with or without NH_4^+ [26]. In fact, studies have concluded that atmospheric SO_4^{2-} can either be emitted directly as primary particles or result from gas-to-particle conversion reactions in the atmosphere [15,42]. Sulfur dioxide from anthropogenic sources is oxidized into H_2SO_4, which in turn accumulates to form condensation nuclei [43]. Sulfur-bearing particles are often observed in airborne PM samples [25,44–46]. Sulfate has been found as the dominant chemical species in aerosol particles in urban atmospheres [47–49]. For example, a study by Zhang *et al.* [47] indicates the dominance of SO_4^{2-} (>70%) in fine particles from the urban area of Beijing during non-dust-storm periods. Bassett and Seinfeld [50] reported that NO_3^- is hardly formed on fine particles owing to its volatility. Therefore, NO_3^- may be formed on soot particles through heterogeneous conversions of nitrogen oxides. Our observation is supported by an aerosol particle study by Ganor *et al.* [51] at Haifa Bay, which has revealed that SO_4^{2-} and NO_3^- concentrations were 5–10 times higher in the land breeze than in sea breeze, as a consequence of the emissions from local industries or the long-range transportation of pollutants. The high abundance of Cl^- is likely to be linked to Fe, Na, Ca, K, and Mg, and may have been transported long-range by way of sea [52].

As evidenced earlier, the black sand particles have very different morphologies and microstructures compared to the white sand and volcanic sand. An important finding in our study is the predominance of fine and ultrafine particles (50 to 500 nm ranges). Anthropogenic and natural sources have been implicated for the observed particles. It should be pointed out that Badr is situated 83 km southeast of Yanbu (Figure 2), which was established in 1975 as one of the country's two industrial centers, and has several oil-fueled power plants, three oil refineries, heavy petrochemical industries, a large cement factory, desalination plants, mineral and metal industries, and several other small industries and workshops. There are naturally occurring lava fields and iron deposits located upwind, near and around Badr. Also, Badr is about 800 km southeast of Haifa's (Figure 2) extensive industrial zone [52]. In fact, Badr is at a crossroads where PM from different sources may converge: long-range transported polluted air masses from Yanbu and Bay of Haifa, sea spray from the Red Sea itself, and mineral components from the deserts of North Africa (Sahara) and the Arabian Peninsula. Occasionally, Saudi Arabia and nearby countries are hit by sandstorms. The opaque mass of dust, debris, and fine particles in suspension is usually transported by strong winds to high mountains [32]. Backward-in-time trajectories using NOAA HYSPLIT model [53] suggest that the dominant air mass movement is northwesterly from the highly industrialized areas (Yanbu and Haifa) towards the area of concern (Badr). Thus, the morphological analysis, composition, proximity to the site, and air mass movements all suggest that emissions by these industries have contributed to particulate air pollution. They were formed as cloud condensation nuclei were readily long-range transported by northwesterly air currents, and finally settled on Douf Mountain.

4. Conclusions

High resolution electron microscopy, viz. optical stereo microscope (OSM), scanning electron microscopy with energy dispersive X-ray spectrometry (SEM/EDS), and laser scanning microscopy (LSM), was a powerful tool for in-depth analysis and identification of particles by revealing details of the microstructure, morphology, emission source types, size, and elemental composition of each type of sand particles. The physical and chemical characterization of sand samples from Badr, Saudi Arabia by different complimentary techniques has contributed to our knowledge and understanding of the sand particles. These particles could have long-term health consequences to the residents of Badr, but a dedicated health analysis would be needed to investigate this possibility.

White sand contained silicates (alumino-silicates), quartz (SiO_2, clean crystalline structure, milky, rose), calcite, olivine, feldspar, and magnetite. Compared to white sand and volcanic sand, the black sand particles exhibited very different morphologies and microstructures (surface roughness). Morphological SEM and LSM analyses showed that the black sand contained fine and ultrafine particles (50 to 500 nm ranges) and there was a thin surface layer (1 μm) embedding a large number of particles from the atmospheric deposition. Ion chromatography analyses confirmed the presence of the following water-soluble anions and cations in variable concentrations in magnetic black sand: SO_4^{2-}, NO_3^-, Cl^-, Na^+, K^+, Mg^{2+}, Ca^{2+}.

X-ray energy dispersive microanalyses of sand samples revealed varying weight concentrations of C, O, Na, Mg, Al, Si, S, Ca, Ti, V, and Fe, ranging from 0.1% to 55% in all four samples. The predominant elements in black sand were Fe, C, O, Ti, Si, Al, V, and S. Enrichment factor (EF) analysis results demonstrated that Na, Mg, Al, Si, K, and Ca in black sand originated from natural sources, while Fe, C, Ti, V, and S were linked to anthropogenic sources. Fossil-fuel combustion and industrial emissions in Haifa and Yanbu can be significant sources of carbon and other contaminant particles, which can be easily adhered to soil particles during long-range transport.

This study presents the first in-depth characterization of black sand particles from sand dunes of Saudi Arabia. Reliable baseline data on particulate air pollution are needed for setting standards and objectives of air pollution controls, and to investigate the role of local and long-range transported anthropogenic emissions. Examining the types, dimensions, and the amount of carbon and other contaminant particles deposited in the coming years will indicate effectiveness of the pollution control devices which have been installed on industries producing energy from fossil-fuels. A comprehensive study should be considered to cover the whole area, trace wind pathways from source regions, and outreach to local people to locate other areas where black sand has been deposited. Future research into the issues such as the real size of the ultrafine particles, mechanisms of long-range transport in the region and mechanisms of adhesion of the ultrafine particles to the larger sand grains would help explain these issues more fully. Further research into health outcomes (respiratory diseases, cardiovascular) in the general population and susceptible groups in Badr and other major cities of Saudi Arabia should also be performed.

Acknowledgments: The authors express their appreciation to Abdul Rahman Al Youbi, Vice President for Academic Affairs, King Abdulaziz University and Abdullah Ahmad A. Ghamdi, Dean Research and Consulting Institute, King Abdulaziz University for their support. We gratefully acknowledge the support by the Wadsworth Center, New York State Department of Health. We also thank Curtis Boynton, a summer intern with the NYS DEC, for his help. The authors wish to thank to Tara Nylese of EDAX Inc. for providing the results for confirmation purposes. The authors also wish to thank Brian Frank, Katherine Alben, Simeen Tabatabai, and Jessica Stark for valuable comments and editing the manuscript.

Author Contributions: The study was completed with the cooperation between all the authors. Haider A. Khwaja, Omar S. Aburizaiza, Azhar Siddique, Abdullah J. Aburiziza, and Donald Blake conceived and designed the experiment. The corresponding authors Jahan Zeb, Mohammad Abbass, and Azhar Siddique were in-charge of sampling and collected the relevant data. Daniel Hershey was instrumental in the design of the study, sample analyses by SEM and LSM, and interpretation of the images and X-ray diffraction spectra. David Guerrieri performed all the LSM analyses and data interpretation. Malissa Kramer analyzed the sample by SEM, checked and reported the results. Mirza M. Hussain was in-charge of inorganic anions and cations analyses by Ion Chromatography. Isobel Simpson interpreted the results and critical revision. Paper was written by Haider A. Khwaja with a significant contribution by Omar Aburizaiza, Daniel Hershey and David Guerrieri.

Conflicts of Interest: The authors declare no conflict of interest.

References

1. D'Almeida, G.A.; Koepke, P.; Shettle, E.P. *Atmospheric Aerosols: Global Climatology and Radiative Characteristics*; A Deepak Publishing: Hampton, VA, USA, 1991.
2. Jaenicke, R. Physical properties of atmospheric particulate sulfur compounds. *Atmos. Environ.* **1978**, *12*, 161–169.
3. Goyer, R. *Issue Paper on the Human Health Effects of Metals*; U.S. Environmental Protection Agency: Lexington, MA, USA, 2004.
4. Jarup, L.; Akesson, A. Current status of cadmium as an environmental health problem. *Toxicol. Appl. Pharmacol.* **2009**, *238*, 201–208.
5. Mamtani, R.; Stren, P.; Dawood, I.; Cheema, S. Metals and Disease: A Global Primary Health Care Perspective. *J. Toxicol.* **2011**, *2011*.
6. Yaman, M. Comprehensive comparison of trace metal concentrations in cancerous and non-cancerous human tissues. *Curr. Med. Chem.* **2006**, *13*, 2513–2525.

7.   U.S. Environmental Protection Agency (US EPA). *Third External Review Draft of Air Quality Criteria for Particulate Matter (April 2002)*; U.S. Environmental Protection Agency: Research Triangle Park, NC, USA, 2002; Volume 1.

8.   Dockery, D.W.; Pope, C.A. Acute respiratory effects of particulate air pollution. *Ann. Rev. Public Health* **1994**, *15*, 107–132.

9.   Donaldson, K.; MacNee, W. The mechanism of lung injury caused by PM_{10}. In *Air Pollution and Health, Issues in Environmental Science and Technology*; Hester, R.E., Harrison, R.M., Eds.; Royal Society of Chemistry: Cambridge, UK, 1998; Volume 10, pp. 21–32.

10. Lipmann, M.; Ito, K.; Nadas, A.; Burnett, R.T. *Association of Particulate Matter Components with Daily Mortality and Morbidity in Urban Populations*; Research Report 95; Health Effects Institute: Cambridge, MA, USA, 2000.

11. Environmental Protection Agency (EPA). *Air Quality Criteria for Particulate Matter*; EPA 600/P-651001CF; National Center for Environmental Assessment: Research Triangle Park, NC, USA, 1996; Volume III.

12. HEI Perspectives. In *Understanding the Health Effects of Components of the Particulate Matter Mix: Progress and Next Steps*; Health Effects Institute: Cambridge, MA, USA, 2002.

13. Aust, A.; Smith, K.R.; Veranth, J.M.; Hu, A.; Lightly, J.S.; Ball, J.C.; Stracci, A.M.; Young, W.C. *Particle Characteristics Responsible for Effects on Human Lung Epithelial Cells*; Health Effects Institute: Cambridge, MA, USA, 2002.

14. Oberdorster, G.; Finkelstein, J.N.; Johnston, C.; Gelein, R.; Cox, C.; Baggs, R.; Elder, A.C.P. *Acute Pulmonary Effects of Ultrafine Particles in Rats and Mice*; Research Report 96; Health Effects Institute: Cambridge, MA, USA, 2000.

15. Post, J.E.; Buseck, P.R. Characterization of individual particles in the Phoenix urban aerosol using electron-beam instruments. *Environ. Sci. Technol.* **1984**, *18*, 35–42.

16. Clark, M.D. *Geological Map of the Al Hamra Quadrangle, Sheet 23 C, Kingdom of Saudi Arabia*; Directorate General of Mineral Resources: Jeddah, Saudi Arabia, 1981.

17. Henry, W.M.; Knapp, K.T. Compound forms of fossil fuel fly ash emissions. *Environ. Sci. Technol.* **1980**, *14*, 450–456.

18. Conny, J.M.; Norris, G.A. Scanning electron microanalysis and analytical challenges of mapping elements in urban atmospheric particles. *Environ. Sci. Technol.* **2011**, *45*, 7380–7386.

19. Perrone, M.R.; Turnone, A.; Buccolieri, A.; Buccolieri, G. Particulate matter characterization at a coastal site in south-eastern Italy. *J. Environ. Monit.* **2006**, *8*, 183–190.

20. McCrone, W.C. *The Particle Atlas. Edition Two*; Ann Arbor Science Publishers Inc.: Ann Arbor, MI, USA, 1980.

21. Friedlander, S.K. A review of the dynamics of sulfate containing aerosols. *Atmos. Environ.* **1978**, *12*, 187–195.

22. Kleefeld, S.; Hoffer, A.; Krivacsy, Z.; Jennings, S.G. Importance of organic and black carbon in atmospheric aerosols at Mace Head, on the West Coast of Ireland (53°19′ N, 9°54′ W). *Atmos. Environ.* **2002**, *36*, 4479–4490.

23. Chen, Y.; Shah, N.; Huggins, F.E.; Huffman, G.P. Investigation of the microcharacteristics of $PM_{2.5}$ in residual oil fly ash (ROFA) by analytical transmission electron microscopy. *Environ. Sci. Technol.* **2004**, *38*, 6553–6560.

24. Chen, Y.; Shah, N.; Huggins, F.E.; Huffman, G.P. Transmission electron microscopy investigation of ultrafine coal fly ash particles. *Environ. Sci. Technol.* **2005**, *39*, 1144–1151.

25. Chen, Y.; Shah, N.; Huggins, F.E.; Huffman, G.P. Microanalysis of ambient particles from Lexington, KY by electron microscopy. *Atmos. Environ.* **2006**, *40*, 651–663.

26. Zhang, D.; Iwasaka, Y.; Shi, G. Soot particles and their impacts on the mass cycle in the Tibetan atmosphere. *Atmos. Environ.* **2001**, *35*, 5883–5894.

27. Ali-Mohamed, A.Y.; Matter, H.A. Determination of inorganic particulates: (Cationic, anionic and heavy metals) in the atmosphere of some areas in Bahrain during the Gulf crisis in 1991. *Atmos. Environ.* **1996**, *30*, 3497–3503.

28. He, K.; Yang, F.; Ma, Y.; Zhang, Q.; Yao, X.; Chan, C.K.; Cadle, S.; Chan, T.; Mulawa, P. The characteristics of $PM_{2.5}$ in Beijing, China. *Atmos. Environ.* **2001**, *35*, 4959–4970.

29. Hughes, L.S.; Cass, G.R.; Gone, J.; Ames, M.; Olmez, I. Physical and chemical characterization of atmospheric ultrafine particles in the Los Angeles area. *Environ. Sci. Technol.* **1998**, *32*, 1153–1161.

30. Katrinak, K.A.; Anderson, J.R.; Buseck, P.R. Individual particles types in the aerosol of Phoenix Arizona. *Environ. Sci. Technol.* **1995**, *29*, 321–329.

31. Chow, J.C.; Watson, J.G.; Lowenthal, D.H.; Solomon, P.A.; Magliano, K.L.; Ziman, S.D.; Richards, L.W. PM_{10} and $PM_{2.5}$ composition in California's San Joaquin Valley. *Aerosol Sci. Technol.* **1993**, *18*, 105–128.

32. Tazaki, K.; Wakimoto, R.; Minami, Y.; Yamamoto, M.; Miyata, K.; Sato, K.; Saji, I.; Chaerun, S.K.; Zhou, G.; Morishita, T.; *et al.* Transport of carbon-bearing dusts from Iraq to Japan during Iraq's war. *Atmos. Environ.* **2004**, *38*, 2091–2109.

33. Posfai, M.; Anderson, J.R.; Buseck, P.R.; Shattuck, T.W.; Tindale, N.W. Constituents of a remote Pacific marine aerosol: A TEM study. *Atmos. Environ.* **1994**, *28*, 1747–1756.

34. Murr, L.E.; Bang, J.J. Electron microscope comparisons of fine and ultra-fine carbonaceous and non-carbonaceous, airborne particulates. *Atmos. Environ.* **2003**, *37*, 4795–4806.

35. Linak, W.P.; Wendt, J.O.L. Trace metal transformation mechanisms during coal combustion. *Fuel Process. Technol.* **1994**, *39*, 173–198.

36. Pitchford, M.; Flocchini, R.G.; Draftz, R.G.; Cahill, T.A.; Ashbaugh, L.L.; Eldred, R.A. Silicon in submicron particles in the Southwest. *Atmos. Environ.* **1981**, *15*, 321–333.

37. Gomes, L.; Bergametti, G.; Coude-Gaussen, G.; Rognon, P. Submicron desert dusts: A sandblasting process. *J. Geophys. Res.* **1990**, *95*, 13927–13935.

38. Hindy, K.T. Silicon, aluminum, iron, copper, and zinc levels in desert soil-related dust in Cairo. *Atmos. Environ.* **1991**, *25*, 213–217.

39. Paoletti, L.; de Berardis, B.; Diociaiuti, M. Physico-chemical characterization of the inhalable particulate matter (PM_{10}) in an urban area: An analysis of the seasonal trend. *Sci. Total Environ.* **2002**, *292*, 265–275.

40. Taylor, S.R. Abundance of chemical elements in the continental crust: A new Table. *Geochim. Cosmochim. Acta* **1964**, *28*, 1273–1285.

41. Tang, X. *Atmospheric Chemistry*; Higher Education Publisher: Beijing, China, 1990. (In Chinese)

42. Kulmala, M.; Vehkamaki, H.; Petaya, T.; dal Maso, M.; Lauri, A.; Kerminem, V.M.; Birmili, W.; McMurry, P.H. Formation growth rates of ultrafine atmospheric particles: A review of observations. *Aerosol Sci.* **2004**, *35*, 143–176.

43. Seinfeld, J.H.; Pandis, S.N. *Atmospheric Chemistry and Physics*; Wiley: New York, NY, USA, 1998.

44. Buseck, P.R.; Posfai, M. Airborne minerals and related aerosol particles: Effects on climate and the environment. *Proc. Natl. Acad. Sci. USA* **1999**, *96*, 3372–3379.

45. Zhang, D.; Shi, G.-Y.; Iwasaka, Y.; Hu, M. Mixture of sulfate and nitrate in coastal atmospheric aerosols: Individual particle studies in Qingdao. *Atmos. Environ.* **2000**, *34*, 2669–2679.

46. Posfai, M.; Simonics, R.; Li, J.; Hobbs, P.V.; Buseck, P.R. Individual aerosol particles from biomass burning in southern Africa: 1. Compositions and size distributions of carbonaceous particles. *J. Geophys. Res. Atmos.* **2003**, *104*, 15941–15954.

47. Zhang, D.; Tang, X.; Qin, Y.; Iwasaka, Y.; Gai, X. Tests for individual sulfate containing particles in urban atmosphere in Beijing. *Adv. Atmos. Sci.* **1995**, *12*, 343–350.

48. Tsitouridou, R.; Voutsa, D.; Kouimtzis, T. Ionic composition of PM_{10} in the area of Thessaloniki, Greece. *Chemosphere* **2003**, *52*, 883–891.

49. Horng, C.L.; Cheng, M.T. Distribution of $PM_{2.5}$, acidic and basic gases near highway in Central Taiwan. *Atmos. Res.* **2008**, *88*, 1–12.

50. Bassett, M.E.; Seinfeld, J.H. Atmospheric equilibrium model of sulfate and nitrate aerosols—II: Particle size analysis. *Atmos. Environ.* **1984**, *18*, 1163–1170.

51. Ganor, E.; Levin, Z.; Grieken, R.V. Composition of individual aerosol particles above the Israelian Mediterranean coast during the summer time. *Atmos. Environ.* **1998**, *32*, 1631–1642.

52. Gilmour, P.S.; Brown, D.M.; Lindsay, T.G.; Beswick, P.H.; MacNee, W.; Donaldson, K. Adverse health effects of PM_{10} particles: Involvement of iron in generation of hydroxyl radical. *Occup. Environ. Med.* **1996**, *53*, 817–822.

53. Draxler, R.R.; Rolph, G.D. *HYSPLIT (HYbrid Single-Particle Lagrangian Integrated Trajectory)*; NOAA Air Resources Laboratory: College Park, MD, USA, 2013.

Characteristics of PM_{10} and $PM_{2.5}$ at Mount Wutai Buddhism Scenic Spot, Shanxi, China

Zhihui Wu, Fenwu Liu and Wenhua Fan

Abstract: A survey was conducted to effectively investigate the characteristics of airborne particulate pollutants PM_{10} and $PM_{2.5}$ during the peak tourist season at Mount Wutai Buddhism scenic spot, Shanxi, China. Characteristics of the PM_{10} and $PM_{2.5}$ in Wu Ye Temple (core incense burners), Manjusri Temple (a traffic hub), Yang Bai Lin Village (a residential district), and Nan Shan Temple (located in a primitive forest district), were determined. The results showed that the PM_{10} concentration was more than 1.01–1.14 times higher than the threshold (50 $\mu g/m^3$) of World Health Organization Air Quality Guidelines (2005), and the $PM_{2.5}$ concentration was 1.75–2.70 times higher than the above standard (25 $\mu g/m^3$). Particle size analysis indicated that the distribution of fine particulate matter in Wu Ye Temple ranged from 0 to 3.30 μm. In other sampling points, the fine particulate was mainly distributed in the range of 0–5.90 μm. The particulates in Wu Ye Temple were mainly characterized by spherical, rod-like, and irregular soot aggregates (PM_{10}) and spherical particles of dust ($PM_{2.5}$). Manjusri Temple and Yang Bai Lin Village predominantly exhibited irregular soil mineral particulate matter (PM_{10}), and amorphous ultrafine soot particulate matter ($PM_{2.5}$).

Reprinted from *Atmosphere*. Cite as: Wu, Z.; Liu, F.; Fan, W. Characteristics of PM_{10} and $PM_{2.5}$ at Mount Wutai Buddhism Scenic Spot, Shanxi, China. *Atmosphere* **2015**, *6*, 1195–1210.

1. Introduction

Particulate matter impacts the environment, climate, and human health and has become one of the most important pollutants affecting air quality. However, particulate matter, including PM_{10} with an aerodynamic equivalent diameter less than 10 μm and $PM_{2.5}$ with an aerodynamic equivalent diameter less than 2.5 μm, has significant effect on human health in the last decades [1]. PM_{10} can be absorbed by the human body, thereby causing diseases following deposition in the respiratory tract and alveoli, and a number of studies have been conducted on this issue. For example, Wu *et al.* and Amador-Muñoz *et al.* studied the concentration and potential threats of PM_{10} in coal-fired industrial cities in China [2] and in Southwestern Mexico [3]. Makra *et al.* and Malandrino *et al.* studied the distribution of PM_{10} under different traffic patterns in Szeged and Bucharest [4], and the yearly and quarterly changes of the trace elements in PM_{10} in Turin, Italy [5], respectively. However, few studies have been conducted on PM_{10} distribution in Buddhist scenic spots, where a large number of incenses and candles are burned during the peak tourist season.

Studies on human health, particularly respiratory diseases, have increased in recent years. $PM_{2.5}$ is likely to adsorb toxic substances and to be inhaled into the alveoli and blood [6], which has increased people's awareness regarding this matter. Compared with PM_{10}, $PM_{2.5}$ is characterized by small particle size, large area, and high activity. It is also likely to absorb toxic and harmful substances, have a long residence time in the air, and travel long distances. The majority of recent reports on $PM_{2.5}$ have focused on the analysis of chemical components in organic pollutants, inorganic chemical components, and microbial components in particulate matter. For instance, Mikuška et al. analyzed all organic pollutants contained in the $PM_{2.5}$ of Ostrava, an industrial city in the Czech Republic during winter [7]. Salameh et al. studied the seasonal and spatial distribution characteristics of the chemical components of $PM_{2.5}$ in five cities around the Mediterranean Sea in Europe [8]. Cao et al. extracted DNA from some microorganisms in $PM_{2.5}$ in Beijing for metagenomic analysis [9].

Work-related stress, fast-paced lifestyles, and severe atmospheric and environmental pollution have caused the degradation of living standards in urban areas. This has led to the popularity of tourism as it allows urban dwellers to get "close to nature". However, increasing vehicle flow and emissions from fossil fuel combustion have resulted in anthropogenic pollution in scenic spots and these effects cannot be underestimated any longer. In scenic spots, some studies on PM_{10} and $PM_{2.5}$ focused on particulate matter concentration and analysis of chemical components. For instance, Zhang, et al. [10] studied the samples of aerosol from $PM_{2.5}$ in the Yulong Scenic Spot southeast of the Qinghai Tibet Plateau, China, during the winter to determine the chemical composition and source. Lee, et al. [11] analyzed the impact of soot caused by biomass burning, the dispersion of $PM_{2.5}$ concentrations in Southeast Asia, and the presence of water-soluble ions in Lulin Mountain Region, Taiwan.

Few studies have been carried out on single-particle morphology of $PM_{2.5}$ in a natural scenic spot. Single-particulate matter research techniques can provide single-particle morphology, chemical composition, particle size distribution, and other pertinent information. Analytical data related to single-particulate matter can be used as "fingerprint information", to some extent, to show the corresponding natural or anthropogenic source. Unlike urban areas and other scenic spots, Mount Wutai Scenic Spot, located in Shanxi, China, a sacred place for Buddhists, is visited by many pilgrims who burn candles while praying. During tourists praying at Buddhist tourist spots at Mount Wutai, they burning a lot of incense and candle, which consisted with some combustible biomass materials, such as natural sticky powder, elm wood powder, and fragrant material. Strong pungent odor can be produced when incense, candles, and its plastic package materials burning accompanied with intense black smoke. Sometimes, the pungent odor can be easy-nosed by tourists in 30–50 m away. In addition, transportation and catering business are also expansion with the increasing of tourists. Furthermore, some tourists carry out their tourism activities in primitive forest district. During these activities, large amounts of particulate matter may be produced from incense and candles burning in praying, automobile exhaust, coal burning in catering business, and fugitive dust come from bared land caused by the tourists trample in primitive forest district. Unfortunately, to the best of our knowledge, none studies have investigated the PM_{10} or $PM_{2.5}$ pollution at Buddhist scenic areas, such as Mount Wutai Scenic Spot.

The initial objective of the present work was to investigate the concentrations of PM_{10} and $PM_{2.5}$ in the four different areas, including Wu Ye Temple (core incense burners), Manjusri Temple (a traffic hub), Yang Bai Lin Village (a residential district), and Nan Shan Temple (located in a primitive forest district), at Mount Wutai scenic spot during the peak season. In order to explore: (1) the typical pattern of particulate matter produced in Mount Wutai Scenic Spot and (2) the main elements, which may enter the human body by breathing these particulate matter. The morphology and elemental composition of PM_{10} and $PM_{2.5}$ in the different areas were also investigated. The outcomes will provide scientific data for atmospheric pollution control at Mount Wutai scenic spot.

2. Results and Discussion

2.1. Sampling Points

The locations of the different sampling areas are shown in Table 1. Three sampling points were included in each area using a triangle model. The arrangement of sampling points in each area is shown in Figure 1.

Table 1. Information on the sampling areas in Mount Wutai, Shanxi, China.

Sampling Point	Area	Geographical Position	Altitude (m)
Wu Ye Temple	a candle burning gathering site in the scenic spot	113°35′22.9″E 39°03′23.3″N	1 672
Manjusri Temple	a traffic hub in the scenic spot	113°35′42.5″E 39°02′31.2″N	1 664
Yang Bai Lin Village	a residential district in the scenic spot	113°35′21.5″E 39°00′9.4″N	1 680
Nan Shan Temple	a primitive forest district	113°34′18.4″E 38°58′54.8″N	1 678

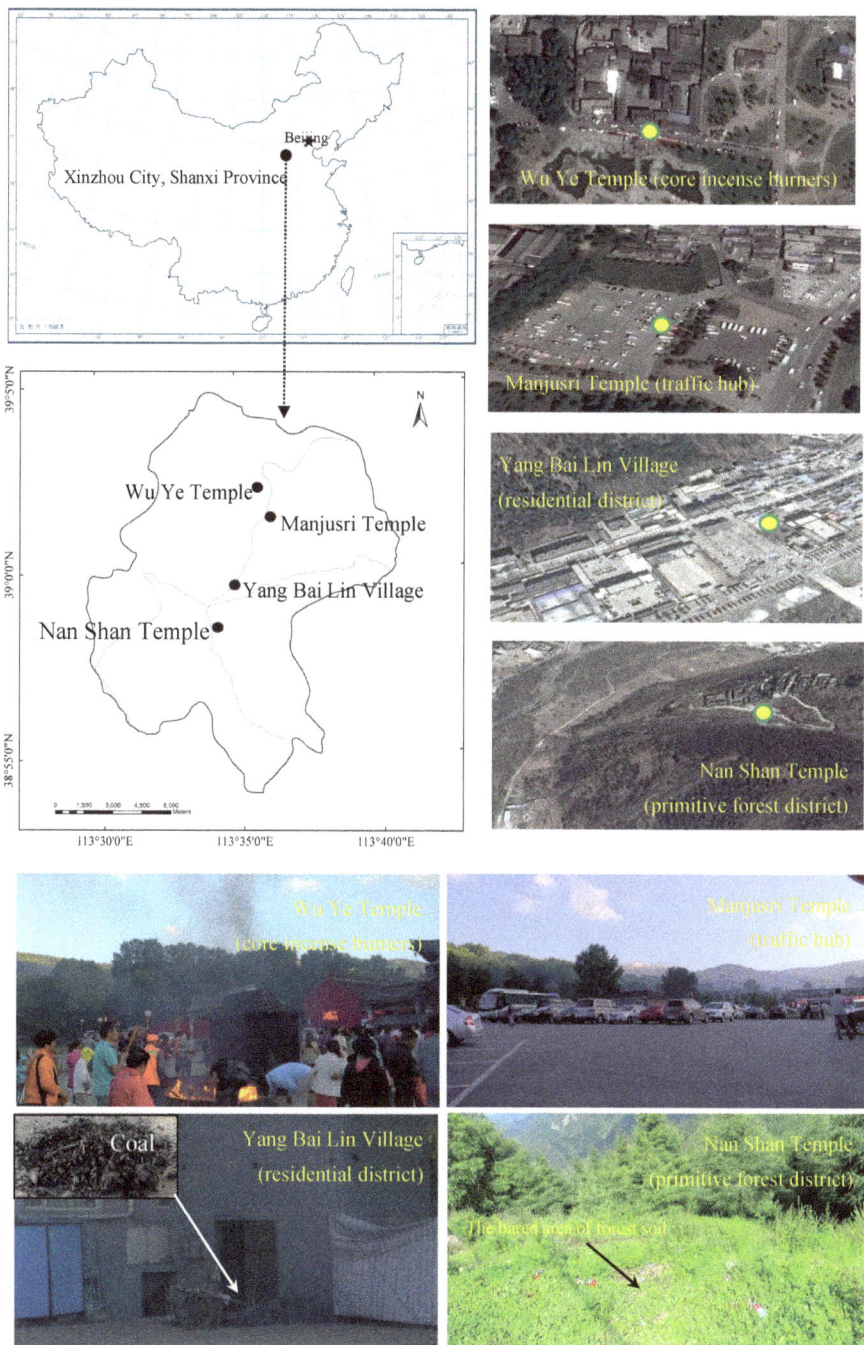

Figure 1. Distribution of sampling sites and its real scene pictures in the four different areas at Mount Wutai, Shanxi, China.

129

2.2. Sampling Methods

August is the peak tourist time at Mount Wutai. Atmospheric pollution caused by tourism activities during this period is more than that observed in the other seasons. Samples of particulate matter were collected from the four different areas at Mount Wutai scenic spot (Manjusri Temple, Wu Ye Temple, Yang Bai Lin Village, and Nan Shan Temple Forest District) from 20–27 August. These four different areas represent the main area of Mount Wutai. A KB-6120 four-channel moderate-flow integrated atmospheric sampler, provided by Qingdao Xuyu Environmental Co., Ltd. (Qingdao, China), was used with PM_{10} and $PM_{2.5}$ cutters to sample the airborne particulates, with a sampling flow of 100 L/min. A filter membrane of glass fiber (diameter: 47 mm; bore diameter: 0.2 μm) was also used. During sampling, the daily mean temperature was 13.31°C and the daily mean wind speed was 2.6 m/s. Blank samples of collection and transportation were included during sampling. Samples were collected at each sampling point from 3:00 to 23:00 for seven consecutive days with measurements taken every 2 h. To eliminate heterogeneity in the particulate matter settling rate at different altitudes, four different areas were selected at a similar altitude. In this study, eighty particulate matter samples for PM_{10} monitoring and eighty particulate matter samples for $PM_{2.5}$ monitoring have been collected at each sampling point.

2.3. Sample Preparation and Analytical Methods

The weight of PM_{10} or $PM_{2.5}$ was measured using a precision electronic balance. The concentration of PM_{10} or $PM_{2.5}$ was calculated by the weight of PM_{10} or $PM_{2.5}$ and the sampling volume. The micrograph or morphology of PM_{10} or $PM_{2.5}$ were analyzed at the Shanxi Institute of Coal Chemistry of the Chinese Academy of Sciences. The detailed steps are as follows: (1) choose and cut a 5 mm × 5 mm filter membrane for random sampling; (2) paste the membrane to a circular sample stage of aluminum with appropriate amounts of conductive adhesive; (3) under certain high-vacuum conditions, coat the filter membrane surface with a very thin Platinum (Pt) membrane by ion sputtering; (4) place the membrane in a JEOL JSM-7001F field-emission scanning electron microscope (SEM) for morphology analysis and to identify the pattern of particulate matter. A Bruker QX200 energy dispersive X-ray spectroscope (EDS, Bruker, Germany) was used for element analysis of the particulate matter. Each sample had five randomly selected micrographs that were obtained using the same optical zoom distance. Image-Pro Plus 6.0, micrograph analysis software, was used to process the particulate matter micrographs, obtain the number and equivalent spherical diameter of the particulate matter of different particle sizes, and calculate the average frequency of occurrence of each type of particulate matter of different particle sizes in every micrograph.

2.4. Statistical Methods

Statistical software (SAS 9.2) was used to analyze the experimental data. All results shown in figures are mean values with their standard deviations to show reproducibility and reliability. In other words, the data are represented by means and error bars where the error bars are the standard deviation in all figures. All figures in this paper were drawn using Origin 7.5 software. Two numbers were connected by "–" in this article means the data ranges from minimum value (before "–") to maximum value (after "–"). Two numbers were connected by "±" means the mean value (before "±") plus or minus standard deviation (after "±"). Significant differences among treatment means were determined using the Student-Newman-Keuls test ($p < 0.05$).

3. Results and Discussions

3.1. Concentration Distribution of PM$_{10}$ and PM$_{2.5}$ at Mount Wutai Scenic Spot

The mean daily concentrations of PM$_{10}$ and PM$_{2.5}$ are shown in Figure 2. The results showed that the PM$_{10}$ concentration at the four different areas were within the range of 17.40 µg/m^3 to 161.45 µg/m^3, whereas the mean daily value at Wu Ye Temple was 1.01–1.14 times higher than the threshold (50 µg/m^3) of the World Health Organization (WHO) Air Quality Guidelines (2005). The daily value for Nan Shan Temple Forest District was 0.99–1.46 times higher than the threshold of the above standard, respectively. However, at Manjusri Temple and Yang Bai Lin Village sampling points, the PM$_{10}$ concentrations did not exceed the threshold of the WHO standard. The PM$_{2.5}$ concentration at the four different areas was 1.43 µg/m^3 to 59.20 µg/m^3; the mean daily values of PM$_{2.5}$ at sampling points Wu Ye Temple and Nan Shan Temple Forest District were 1.75–2.70 times and 0.99–1.87 times higher than the WHO standard (25 µg/m^3), respectively. However, the PM$_{10}$ and PM$_{2.5}$ concentrations at the Manjusri Temple and Yang Bai Lin Village sampling points were lower than the WHO standard.

In addition, the results of this study show that the mean concentration ratios of PM$_{2.5}$ to PM$_{10}$ at Wu Ye Temple, Manjusri Temple, Yang Bai Lin Village, and Nan Shan Temple Forest District were 0.367, 0.435, 0.083 and 0.760, respectively. Thus, the mean concentration ratio of PM$_{2.5}$ to PM$_{10}$ in the four areas was 0.387. It is concluded that the particulate matter around Manjusri Temple was mostly sub-fine particles with an equivalent spherical diameter range of 2.5–10 µm. The sub-fine particle concentrations at Wu Ye Temple and Yang Bai Lin Village were lower that the level at Manjusri Temple.

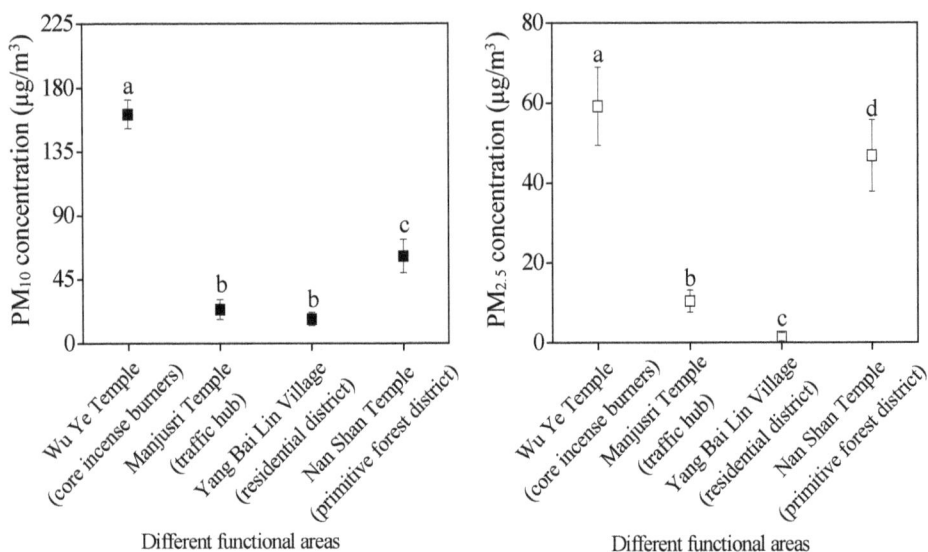

Figure 2. PM_{10} and $PM_{2.5}$ concentration distribution in the four different areas at Mount Wutai, Shanxi, China. Data labeled in the point with the same letter are not significantly different at $p < 0.05$. The vertical T-bars indicated standard deviation about the means.

3.2. Particle Size Distribution of Airborne Particulate Matter at Mount Wutai Scenic Spot

It is now generally accepted that fine particles are more harmful to health effect than larger particles because fine particles offer a larger surface area and hence potentially larger concentrations of adsorbed or condensed toxic air pollutants per unit mass [12] or fine particles are more efficiently retained in the peripheral lung [13]. Based on equivalent spherical diameter, the particle size distributions of particulate matter in the four different areas are shown in Figure 3.

The main particulate matter at the Wu Ye Temple sampling point had an equivalent spherical diameter of 0–3.30 μm. The particulate numbers in the range of 0–3.30 μm accounted for 80.0% of the particulate numbers of PM_{10} at this point, with the peak occurring at 1.10–2.20 μm (Figure 3a). However, no particulate matter in the range between 7.70 and 9.90 μm was observed. Furthermore, the particulate numbers of $PM_{2.5}$ accounted for 65.0% of the PM_{10} particulate numbers at this point. The particle size distribution of particulate matter in Manjusri Temple assumed an unimodal pattern (Figure 3b), with particulate matter mainly within the 0–4.60 μm range and accounted for 85.6% of the total particulate numbers in PM_{10} at this point. The peak interval was within 1.08–2.16 μm. Unlike Wu Ye Temple, the number ratio of $PM_{2.5}$ to PM_{10} was 37.1%. Therefore, the proportion of $PM_{2.5}$ that can penetrate into the lungs was smaller than the particulate matter at Wu Ye Temple (65.0%).

The particle size distribution of particulate matter at Yang Bai Lin Village assumed a bimodal pattern (Figure 3c), which was different to the other sampling points, and the individual peak-to-peak interval at this point was within 1.7–2.4 µm and 4.5–5.2 µm Furthermore, the particulate matter was mainly distributed within the range of 0–5.90 µm. The particulate number in the range of 0–5.90 µm accounted for 86.9% of the total particulate numbers of PM_{10} at this point. In addition, the number ratio of $PM_{2.5}$ to PM_{10} was 32.6%. The particle size distribution of particulate matter in Nan Shan Temple Forest District assumed a unimodal pattern, with the particulate matter mainly distributed within the range of 0–4.32 µm, the particulate numbers in this scale accounted for 77.9% of the total PM_{10} particulate numbers. The peak occurred within the range of 1.08–2.16 µm (Figure 3d). The number ratio of $PM_{2.5}$ to PM_{10} was 42.5% at this sampling point. Nan Shan Temple Forest District is located in a primitive forest district, and previous studies have shown that the plant leaf surfaces have a relatively strong absorptive capacity for fine particles, thus effectively reducing the airborne particulate matter concentration [14]. However, compared with Manjusri Temple and Yang Bai Lin Village sampling points, the particulate matter concentrations of PM_{10} and $PM_{2.5}$ in this area were relatively high (Figure 2). The forest coverage is relatively small or a portion of the primitive forest has been damaged by human activities and may be the main reason for these results.

In conclusion, the number ratios of $PM_{2.5}$ to PM_{10} were 65.0%, 37.1%, 32.6% and 42.5% at Wu Ye Temple, Manjusri Temple, Yang Bai Lin Village, and Nan Shan Temple Forest District sampling points. $PM_{2.5}$ was the main air pollutant in Wu Ye Temple (core incense and candle burners). The size of particulate matter in Yang Bai Lin Village (a residential district), Manjusri Temple (a traffic hub), and Nan Shan Temple Forest District (located in a primitive forest district) were mainly distributed between 2.5 and 10 µm.

3.3. Single-Particle Morphology and Elemental Analysis of PM_{10} in the Four Different Areas at Mount Wutai, Shanxi, China

On the basis of field-emission scanning electron microscopy and energy dispersive X-ray spectroscopy (EDS) results of the collected samples (Figures 4 and 5), the single-particle morphology of PM_{10} was divided into two categories: soot particles and soil dust particles.

On the basis of the distribution of the major elements in particulate matter, the atomic percentages of O, Si, C, Na, Al, Ca, and K were 54.78%, 21.40%, 9.81%, 4.63%, 3.54%, 1.96%, and 1.19%, respectively; among these elements, O and Si were the major components (Figure 5d). It is concluded that soot aggregates are found in Wu Ye Temple and Yang Bai Lin Village, and soil particles are found in Manjusri Temple, Yang Bai Lin Village, and Nan Shan Temple Forest District.

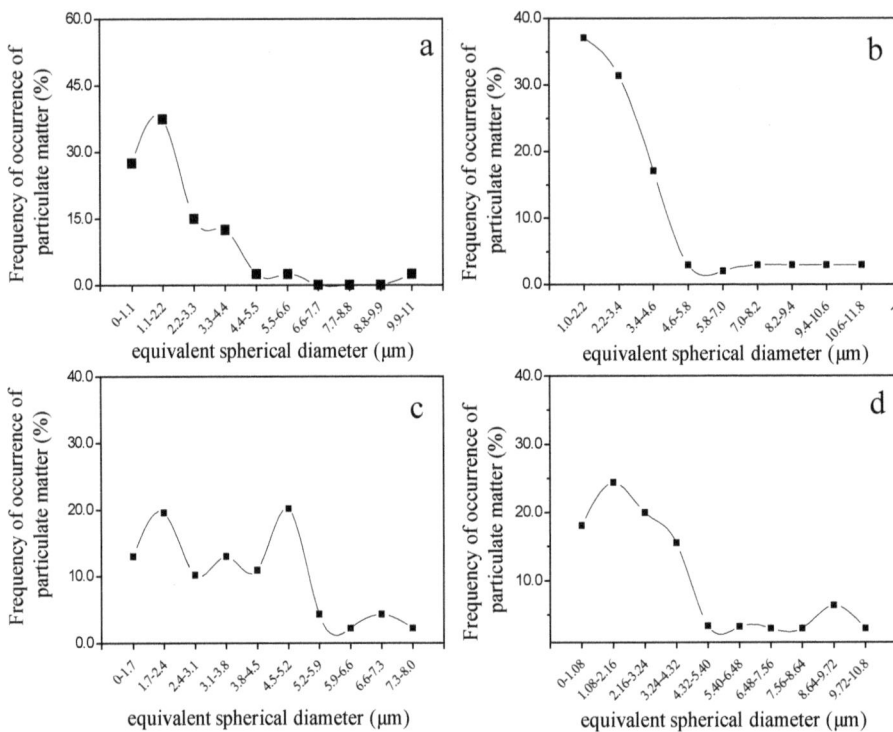

Figure 3. Equivalent spherical diameter distribution of atmospheric fine particles in the four different areas at Mount Wutai, Shanxi, China. ((**a**) Sampling points in the Wu Ye Temple; (**b**) sampling points in the Manjusri Temple; (**c**) sampling points in the Yang Bai Lin Village; (**d**) sampling points in the Nan Shan Temple Forest District).

Soot particles are formed by scaly fine particle aggregates, with a bitty appearance, loose structure, and particle size of 5.35–8.36 μm Most of the soot particles appeared near the primary tourist areas of Wu Ye Temple and Yang Bai Lin Village. Soot particles can be roughly divided into three categories on the basis of morphology: near-spherical (Figure 4a), rod-like (Figure 4b), and flocculent amorphous bodies (Figure 4c). The SEM-EDS results of particulate matter collected in the area of incense burners in Wu Ye Temple (Figure 5a) indicated that the distribution of major elements in such aggregates included C, O, N, Cl, Na, and K of which the atomic percentages were 60.98%, 17.67%, 8.93%, 6.46%, 4.88%, and 1.08%, respectively, with high contents of C, O, and N. The main sources of particulate matter were anthropogenic combustion. This phenomenon was described by Carrico *et al.* [15], who found that an increase in the particle size of soot aerosol particulate matter from biomass burning was mostly dependent upon its hygroscopic growth. The large number of soot particle aggregates distributed at the Mount Wutai scenic spot was closely associated with high rainfall and humidity during the summer. The surfaces of most soil dust particles have impurities. Thus, dust particles are not single-mineral particulate matter and had a relatively high frequency of occurrence in the samples collected in Manjusri Temple and from Yang Bai Lin Village. The SEM-EDS results in Figure 5b,c show that the distribution of the major elements in ellipsoidal particles collected from Manjusri Temple (Figure 4d) and Yang Bai Lin Village (Figure 4e) sampling points, which included O, Si, Ca, Al, and Na with atomic percentages of 57.87%, 16.60%, 16.32%, 4.71%, and 4.49% (Figure 5b), and of 50.43%, 14.58%, 12.10%, 22.42%, and 0.47% (Figure 5c), respectively. Particulate matter containing aluminosilicate minerals was likely formed during secondary atmospheric reactions. This result was verified previously by Dada *et al.* [16]. Another type of ellipsoidal particle (Figure 4f) was found in the Nan Shan Temple Forest District sampling point, and its surface had a slight depression and greater clastic particle adhesion.

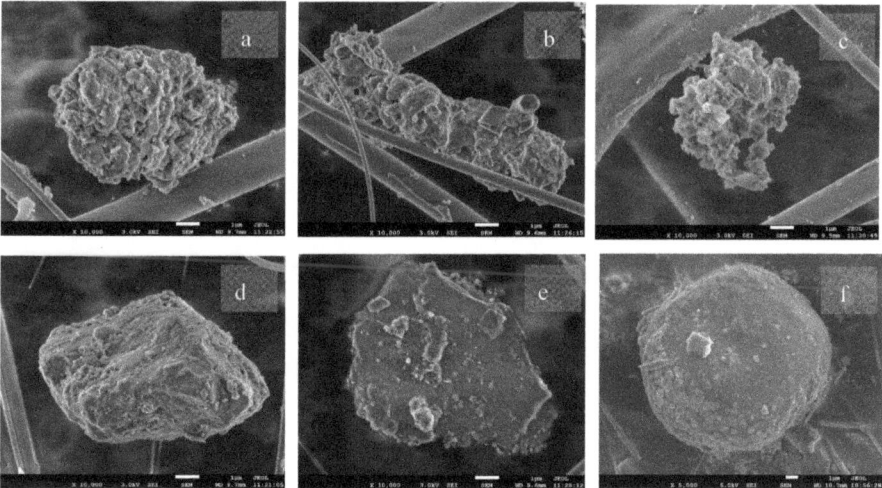

Figure 4. Scanning electron microscopy (SEM) images of atmospheric particles in PM_{10} at Wutai Mount scenic spot ((**a**) in the Wu Ye Temple sampling point; (**b**) in the Wu Ye Temple sampling point; (**c**) in the Yang Bai Lin Village sampling point; (**d**) in the Manjusri Temple sampling point; (**e**) in the Yang Bai Lin Village sampling point; (**f**) in the Nan Shan Temple Forest District sampling point).

a

Element	C	O	Cl	N	Na	K
Atomic percentage	60.98	17.67	6.46	8.93	4.88	1.08

b

Element	O	Si	C	Al	Na
Atomic percentage	57.87	16.6	16.32	4.71	4.49

c

Element	O	Si	Al	C	Na
Atomic percentage	50.43	14.58	12.1	22.42	0.47

d

Element	O	Si	C	Na	Al	Ca	K
Atomic precentage	57.48	21.40	9.81	4.63	3.54	1.96	1.19

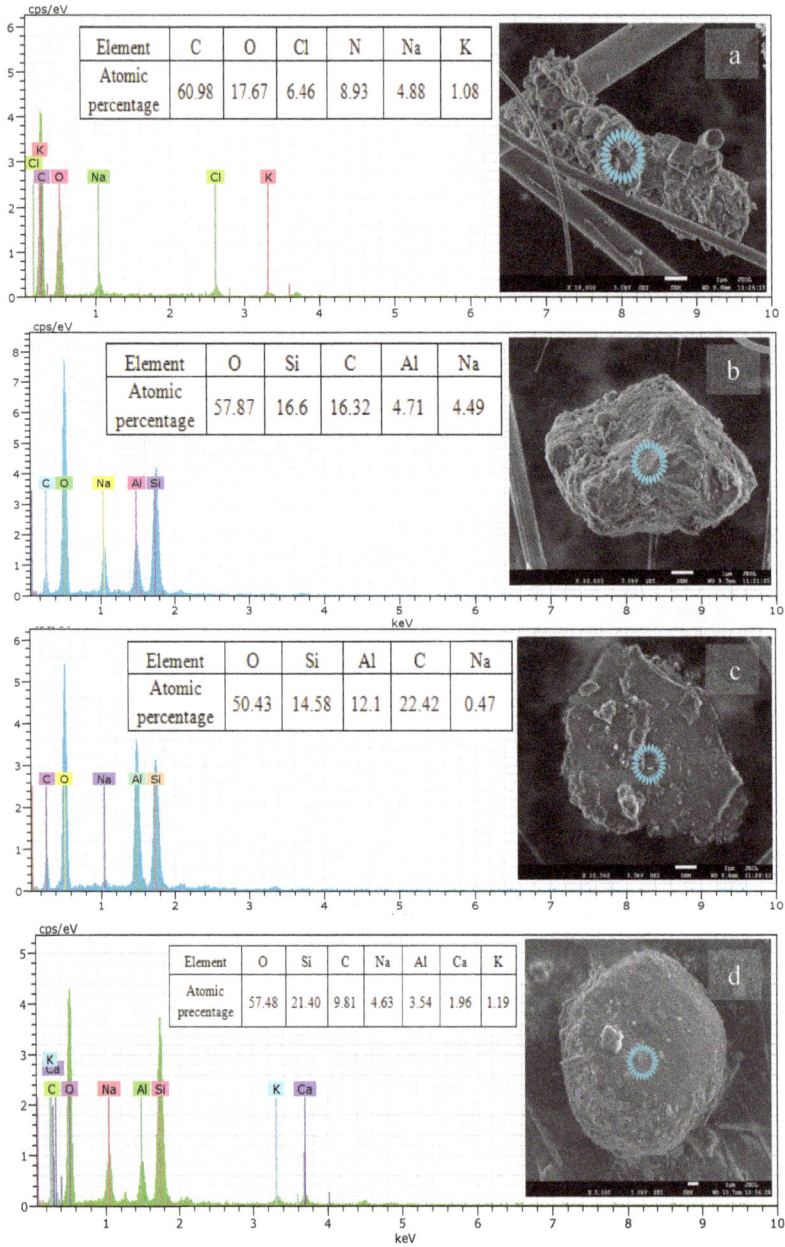

Figure 5. Element analysis of atmospheric particles in PM_{10} at Wutai Mount scenic spot ((**a**) in the Wu Ye Temple sampling point; (**b**) in the Manjusri Temple sampling point; (**c**) in the Yang Bai Lin Village sampling point; (**d**) in the Nan Shan Temple Forest District sampling point).

136

3.4. Single-Particle Morphology and Elemental Analysis of PM$_{2.5}$ in the Four Different Areas at Mount Wutai, Shanxi, China

Similar to PM$_{10}$, the single-particle morphologies of PM$_{2.5}$ were also divided into two categories: soot particles and soil dust particles. Figures 6 and 7 show the single-particle morphology and elemental analysis of PM$_{2.5}$ in the four different areas at Mount Wutai. The shape of PM$_{2.5}$ was roughly divided into the following three types: ultrafine soot particles, soil dust particles, and spherical soot particles. Soot particles of PM$_{2.5}$ collected at Wu Ye Temple and Yang Bai Lin Village were characterized by scaly aggregates with a particle size distribution of 0.06–2.37 µm (Figure 6a,b). The elemental composition of these particles included C, O, N, Cl, Na, and K with atomic percentages of 60.98%, 17.67%, 8.93%, 6.46%, 4.88%, and 1.08%, respectively (Figure 7a). The elemental composition of PM$_{2.5}$ was approximately consistent with that of PM$_{10}$ (Figure 5a–c); therefore, PM$_{2.5}$ and PM$_{10}$ had similar emission sources.

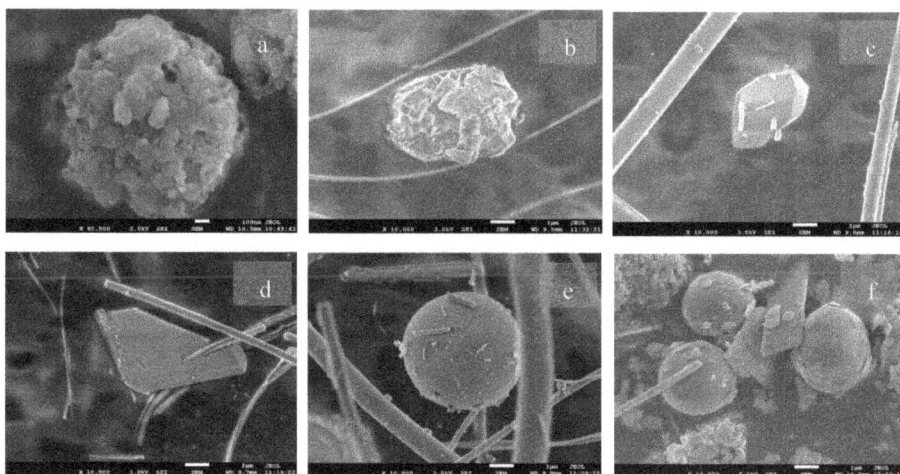

Figure 6. Scanning electron microscopy (SEM) images of atmospheric particles in PM$_{2.5}$ at Wutai Mount scenic spot ((**a**) in the Wu Ye Temple sampling point; (**b**) in the Yang Bai Lin Village sampling point; (**c**) in the Manjusri Temple sampling point; (**d**) in the Nan Shan Temple Forest District sampling point; (**e**) in the Wu Ye Temple sampling point; (**f**) in the Yang Bai Lin Village sampling point).

Consistent with the findings of Baumgartner *et al.* [17], black carbon emissions from the area with the incense burners at the Mount Wutai scenic spot, as well as the emissions from six villages in Yunnan and Tibet, are primarily produced by biomass burning.

Soil dust particles of PM$_{2.5}$ collected at Manjusri Temple and Nan Shan Temple Forest District were mostly characterized by an irregular morphology with distinct edges and corners and a particle size distribution of 0.03–1.49 μm (Figure 6c,d). Most fine dust debris were from crust-derived minerals and had elemental compositions of O, Ca, Mg, and C with atomic percentages of 58.47%, 17.71%, 12.33%, and 11.49%, respectively (Figure 7b). The debris is presumed to be the mineral particulate matter of dolomite ($CaMg(CO_3)_2$) [18] or magnesite ($MgCO_3$). [19] reporting on in-site observations in Hengshan, China, confirmed that gritty dust with a particle size of 15–50 μm easily becomes the core of dust under light conditions, thus allowing photochemical isomerization with moisture and dust in the atmosphere, promoting the formation of new composite particles, and providing the carrier for the long-distance transport of pollutants. Reducing soil dust particles is important for the reduction of atmospheric particulate matter.

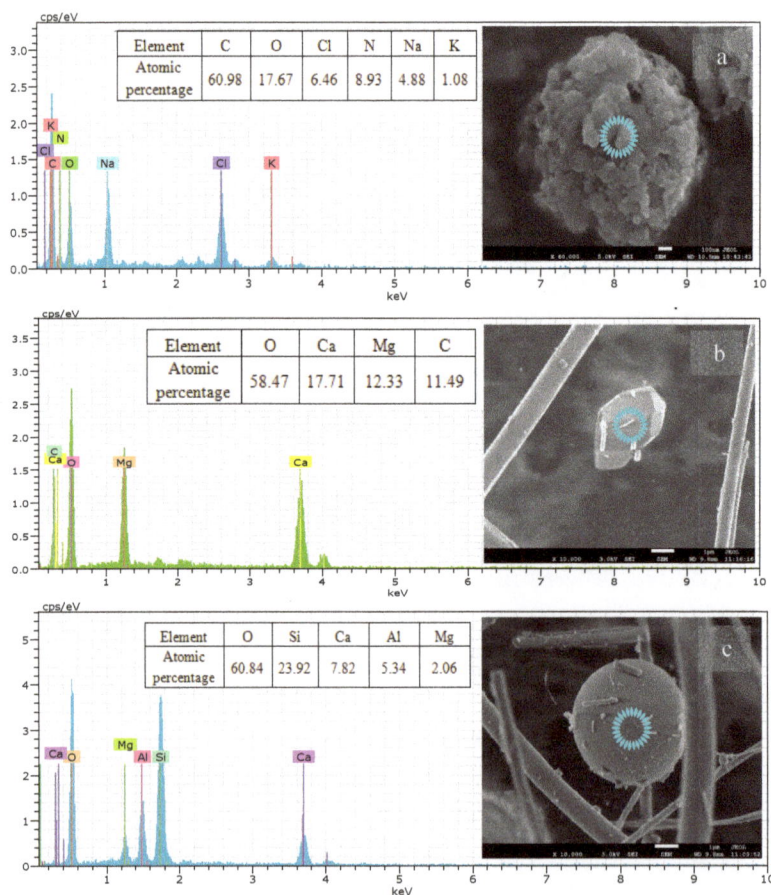

Figure 7. Element analysis of atmospheric particles in PM$_{2.5}$ at Wutai Mount scenic spot ((**a**) in the Wu Ye Temple sampling point; (**b**) in the Manjusri Temple sampling point; (**c**) in the Wu Ye Temple sampling point).

However, with the exception of scaly aggregates mentioned above, spherical particles with a smooth surface collected at Wu Ye Temple and Yang Bai Lin Village were another type of soot particle at these sampling points. SEM images combined with micrograph analysis method (Image-Pro Plus 6.0) showed that the average roundness of such particulate matter was 0.37 ± 0.30 µm, and that the particle size distribution was 2.99–5.04 µm. EDS revealed that the distribution of major elements in the particulate matter included O, Si, Ca, Al, and Mg, with atomic percentages of 60.84%, 23.92%, 7.82%, 5.34%, and 2.06%, respectively (Figure 7c). Shi et al. [20] found that the particulate matter in the emissions from coal and other high-temperature combustion sources included spherical particulate matter with a smooth surface and a spherical particulate matter with the surface covered with particles. The particulate matter had the same morphology as that of fly ash particles determined by Geng et al. [21] in Tokchok Island, South Korea, during the summer; an 80% similarity in the distribution of the major elements was found in this study and the current research. Considering the actual situation at the scenic spot, it is concluded that the particulate matter in the scenic spot was mainly sourced from the secondary atmospheric reactions of the emission from biomass burning, such as the fire coal or candles used in catering the primary tourist area.

In conclusion, the main single-particulate morphology of airborne particulate matter at Mount Wutai scenic spot included soot particles and soil dust mineral particles. Huang et al. [22] found that the main emission sources of soot particles are fossil fuel and biomass burning. Reducing the emission of such particulate matter can effectively reduce the concentration of secondary organic aerosol precursors and help avoid the occurrence of haze.

This research, which is consistent with other published literature, suggest implementing pollution control measures in the different functional areas of the Mount Wutai Buddhism Scenic Spot. For Wu Ye temple area, the incense and candles burning furnace should be covered by a smoke collector, in which the strong black smoke can be treated efficiency through absorption and precipitation devices. For Yang Bai Lin Village residential area, the natural gas, electromagnetic or solar energy types should be advocate use to replace coal as the main energy resource for heating and catering. For Manjusri Temple transport hub area, the green belt should be set at the both sides of the road. In addition, the water sprinkler working times and road cleaning should be increased. The vehicles numbers should be restricted in this area. The electric energy or other clean fuel should be employed for vehicles running. For Nanshan Temple next to the primeval forest area, it is should facilitate afforestation as more as possible. Some broad-leaved plants should be cultivated because the broad-leaved plants can intercept the dust produced from bare land and in further reduce the concentration of particulate matter in the atmosphere in this area.

It is noted that in the process of this research, we focused on the elements kinds and contents in the typical particulate matters, but not the total particulate matters, in the different sampling points. In order to identify potential local origin or long-range transport of the elements, the enrichment factor (EF) of the elements in total particulate matter in different survey areas will be the focus of our further research.

4. Conclusions

Characteristics of PM_{10} with an aerodynamic equivalent diameter less than 10 μm and $PM_{2.5}$ with an aerodynamic equivalent diameter less than 2.5 μm in four different areas, including Wu Ye Temple (core incense burners), Manjusri Temple (a traffic hub), Yang Bai Lin Village (a residential district), and Nan Shan Temple (located in a primitive forest district), were determined at Mount Wutai, Shanxi, China, a Buddhist scenic spot. To the best of our knowledge, this study is the first to present the characteristics of PM_{10} and $PM_{2.5}$ in a Buddhist scenic spot. It was found that the daily average concentrations of PM_{10} and $PM_{2.5}$ were 17.40–161.45 μg/m^3 and 1.43–59.20 μg/m^3, respectively. The number ratio of $PM_{2.5}$ to PM_{10} was 65.0% at the Wu Ye Temple sampling point, and was 37.1%–42.5% at the other three sampling points. The particulates in Wu Ye Temple were mainly characterized by spherical, rod-like, and irregular soot aggregates (PM_{10}) and spherical particles of dust ($PM_{2.5}$). Manjusri Temple and Yang Bai Lin Village predominantly exhibited irregular soil mineral particulate matter (PM_{10}), and amorphous ultrafine soot particulate matter ($PM_{2.5}$). It is concluded that soot particles with high content of C, O, and N in PM_{10} or with high content of O, Si, and Ca in $PM_{2.5}$, and soil dust particles with high content of O, Si, Ca, and Mg elements in PM_{10} and $PM_{2.5}$ are present at Mount Wutai. The findings in this study have great significance in developing measures to control atmospheric pollution at Mount Wutai scenic spot.

Acknowledgments: This work was supported by the Science and Technology Key Project of Shanxi Province (funded social development) (No.20120313011-1) and Shanxi Agricultural University Talent recruitment project (No. 2011017).

Author Contributions: Wenhua Fan conceived and designed the experiments. Zhihui Wu and Fenwu Liu collected and analyzed the data. Zhihui Wu and Fenwu Liu wrote the paper.

Conflicts of Interest: The authors declare no conflict of interest.

References

1. Pui, D.Y.H.; Chen, S.C.; Zuo, Z. $PM_{2.5}$ in China: Measurements, sources, visibility and health effects, and mitigation. *Particuology* **2014**, *13*, 1–26.

2.	Wu, D.; Wang, Z.S.; Chen, J.H.; Kong, S.F.; Fu, X.; Deng, H.B.; Shao, G.F.; Wu, G. Polycyclic aromatic hydrocarbons (PAHs) in atmospheric $PM_{2.5}$ and PM_{10} at a coal-based industrial city: Implication for PAH control at industrial agglomeration regions, China. *Atmos. Res.* **2014**, *149*, 217–229.

3.	Amador-Muñoz, O.; Bazán-Torija, S.; Villa-Ferreira, S.A.; Villalobos-Pietrini, R.V.; Bravo-Cabrera, J.L.; Munive-Colín, Z.; Hernández-Mena, L.; Saldarriaga-Noreña, H.; Murillo-Tovar, M.A. Opposing seasonal trends for polycyclic aromatic hydrocarbons and PM_{10}: Health risk and sources in southwest Mexico City. *Atmos Res.* **2013**, *122*, 199–212.

4.	Makra, L.; Ionel, I.; Csépe, Z.; Matyasovszky, I.; Lontis, N.; Popescu, F.; Sümeghy, Z. The effect of different transport modes on urban PM_{10} levels in two European cities. *Sci. Total Environ.* **2013**, *458*, 36–46.

5.	Malandrino, M.; Martino, M.D.; Ghiotti, G.; Geobaldo, F.; Grosa, M.M.; Giacomino, A.; Abollino, O. Inter-annual and seasonal variability in PM_{10} samples monitored in the city of Turin (Italy) from 2002 to 2005. *Microchem. J.* **2013**, *107*, 76–85.

6.	Maté, T.; Guaita, R.; Pichiule, M.; Linares, C.; Díaz, J. Shortterm effect of fine particulate matter $PM_{2.5}$ on daily mortality due to diseases of the circulatory system in Madrid (Spain). *Sci. Total Environ.* **2010**, *408*, 570–575.

7.	Mikuška, P.; Křůmal, K.; Večeřa, Z. Characterization of organic compounds in the $PM_{2.5}$ aerosols in winter in an industrial urban area. *Atmos. Environ.* **2015**, *105*, 97–108.

8.	Salameh, D.; Detournay, A.; Pey, J.; Pérez, N.; Liguori, F.; Saraga, D.; Bove, M.C.; Brotto, P.; Cassola, F.; Massabò, D.; *et al.* $PM_{2.5}$ chemical composition in five European Mediterranean cities: A 1-year study. *Atmos. Res.* **2015**, *155*, 102–117.

9.	Cao, C.; Jiang, W.; Wang, B.; Fang, J.; Lang, J.; Tian, G.; Jiang, J.; Zhu, T.F. Inhalable microorganisms in Beijing's $PM_{2.5}$ and PM_{10} pollutants during a severe smog event. *Environ. Sci. Technol.* **2014**, *48*, 1499–1507.

10.	Zhang, N.N.; Cao, J.J.; Liu, S.X.; Zhao, Z.Z.; Xu, H.M.; Xiao, S. Chemical composition and sources of $PM_{2.5}$ and TSP collected at Qinghai Lake during summertime. *Atmos. Res.* **2014**, *138*, 213–222.

11.	Lee, C.T.; Chuang, M.T.; Lin, N.H.; Wang, J.L.; Sheu, G.R.; Chang, S.C.; Wang, S.H.; Huang, H.; Chen, H.W.; Liu, Y.L.; *et al.* The enhancement of $PM_{2.5}$ mass and water-soluble ions of biosmoke transported from Southeast Asia over the Mountain Lulin site in Taiwan. *Atmos. Environ.* **2011**, *45*, 5784–5794.

12.	Chiu, H.F.; Peng, C.Y.; Wu, T.N.; Yang, C.Y. Short-term effects of fine particulate air pollution on ischemic heart disease hospitalizations in Taipei: A case-crossover study. *Aerosol Air Qual. Res.* **2013**, *13*, 1563–1569.

13.	Calcabrini, A.; Meschini, S.; Marra, M.; Falzano, L.; Colone, M.; De, B.B.; Paoletti, L.; Arancia, G.; Fiorentini, C. Fine environmental particulate engenders alterations in human lung epithelial A549 cells. *Environ. Res.* **2004**, *95*, 82–91.

14.	Song, Y.S.; Maher, B.A.; Li, F.; Wang, X.K.; Sun, X.; Zhang, H.X. Particulate matter deposited on leaf of five evergreen species in Beijing, China: Source identification and size distribution. *Atmos. Environ.* **2015**, *105*, 53–60.

15. Carrico, C.M.; Kreidenweis, S.M.; Malm, W.C.; Day, D.E.; Lee, T.; Carrillo, J.; McMeeking, G.R.; Collett, J.L., Jr. Hygroscopic growth behavior of a carbon-dominated aerosol in Yosemite National Park. *Atmos. Environ.* **2005**, *39*, 1393–1404.

16. Dada, L.; Mrad, R.; Siffert, S.; Salib, N.A. Atmospheric markers of African and Arabian dust in an urban eastern Mediterranean environment, Beirut, Lebanon. *J. Aerosol Sci.* **2013**, *66*, 187–192.

17. Baumgartner, J.; Zhang, Y.; Schauer, J.J.; Huang, W.; Wang, Y.; Ezzati, M. Highway proximity and black carbon from cookstoves as a risk factor for higher blood pressure in rural China. *P. Natl. Acad. Sci. USA.* **2014**, *111*, 13229–13234.

18. Krueger, B.J.; Grassian, V.H.; Cowin, J.P.; Laskin, A. Heterogeneous chemistry of individual mineral dust particles from different dust source regions: The importance of particle mineralogy. *Atmos. Environ.* **2004**, *38*, 6253–6261.

19. Nie, W.; Ding, A.J.; Wang, T.; Kerminen, V.M.; George, C.; Xue, L.K.; Wang, W.X.; Zhang, Q.Z.; Petaja, T.; Qi, X.M.; *et al.* Polluted dust promotes new particle formation and growth. *Sci. Rep.* 2014.

20. Shi, Z.B.; Shao, L.Y.; Jones, T.P.; Whittaker, A.G.; Richards, R.J.; Zhang, P.F. Oxidative stress on plasmid DNA induced by inhalable particles in the urban atmosphere. *Sci. Bull.* **2004**, *49*, 692–697.

21. Geng, H.; Jung, H.J.; Park, Y.M.; Hwang, H.J.; Kim, H.K.; Kim, Y.J.; Sunwoo, Y.; Ro, C. Morphological and chemical composition characteristics of summertime atmospheric particles collected at Tokchok Island, Korea. *Atmos. Environ.* **2009**, *43*, 3364–3373.

22. Huang, R.J.; Zhang, Y.; Bozzrtti, C.; Ho, K.F.; Cao, J.J.; Han, Y.; Daellenbach, K.R.; Slowik, J.G.; Platt, S.M.; Canonaco, F.; *et al.* High secondary aerosol contribution to particulate pollution during haze events in China. *Nature.* **2014**, *514*, 218–222.

Analysis of PAHs Associated with Particulate Matter PM$_{2.5}$ in Two Places at the City of Cuernavaca, Morelos, México

Hugo Saldarriaga-Noreña, Rebecca López-Márquez, Mario Murillo-Tovar,
Leonel Hernández-Mena, Efrén Ospina-Noreña, Enrique Sánchez-Salinas,
Stefan Waliszewski and Silvia Montiel-Palma

Abstract: This study was carried out between January and February 2013, at two sites in the city of Cuernavaca, México, using low-volume equipment. Fifteen Polycyclic aromatic hydrocarbons (PAHs), were identified by gas chromatography coupled with mass spectrometry. The total average concentration observed for PAHs was 24.0 ng·m^{-3}, with the high molecular weight compounds being the most abundant. The estimated equivalent concentration for Benzo (a) P (BaPE) was 4.05 ng·m^{-3}. Diagnostic ratios together with the principal components analysis (PCA) allowed for establishing coal burning and vehicle emissions as being the main sources of these compounds in the area. The PAHs used to calculate this index account for 51% of the 15 PAHs identified, which probably involves a risk to the exposed population.

Reprinted from *Atmosphere*. Cite as: Saldarriaga-Noreña, H.; López-Márquez, R.; Murillo-Tovar, M.; Hernández-Mena, L.; Ospina-Noreña, E.; Sánchez-Salinas, E.; Waliszewski, S.; Montiel-Palma, S. Analysis of PAHs Associated with Particulate Matter PM$_{2.5}$ in Two Places at the City of Cuernavaca, Morelos, México. *Atmosphere* **2015**, *6*, 1259–1270.

1. Introduction

Polycyclic aromatic hydrocarbons (PAHs) are one of the most studied families of organic compounds present in the atmosphere, given their proven negative effects on health and persistence in the environment. Different PAHs, specifically those with high molecular weight, which predominate in the particulate phase of atmospheric aerosols, have shown to be carcinogenic [1].

It is also known that PAHs accumulated in particles are mainly associated with fine particulate matter (PM$_{2.5}$). These fine particles are better able to penetrate the respiratory system, which increases their potential health effects [2,3].

PAHs are produced by incomplete combustion and pyrolysis of fossil fuels such as oil and coal, and from other organic materials from natural and anthropogenic sources [4–6], including vehicle emissions, wood burning, waste incineration, coke production, and metal production [7–9].

Their concentrations in ambient air are lower in rural than in urban and suburban areas. Moreover, they can have different profiles depending on the geographical

143

location, the type of sources and the dominant atmospheric characteristics [10]. To identify possible sources of these compounds, diagnostic ratios were used. However, these should be used with caution, because sometimes the discrimination between these compounds is difficult, taking into account the reactivity of some of the PAHs with species such as NO_x, O_3, etc. [11].

For the city of Cuernavaca, there is little information on the chemical characterization of $PM_{2.5}$, specifically in determining PAHs. Therefore, the present study has as its main objective to determine ambient levels of these compounds and their possible sources. The results will serve as a basis for mitigation programs of air pollution and/or generation of regulatory standards for the environmental authorities responsible for monitoring of air quality.

2. Results and Discussion

2.1. Wind Trajectories

During the study period, the winds came mainly from the southeast of the state (66%), which probably suggests that part of air pollutants identified in the Autonomous University of State of Morelos (in the north) had its origin in the industrial zone of the city (located to the south) and the downtown area of the city. However, other possible sources could be the emissions from the Popocatepetl volcano, located to the south east of the State (Figure 1).

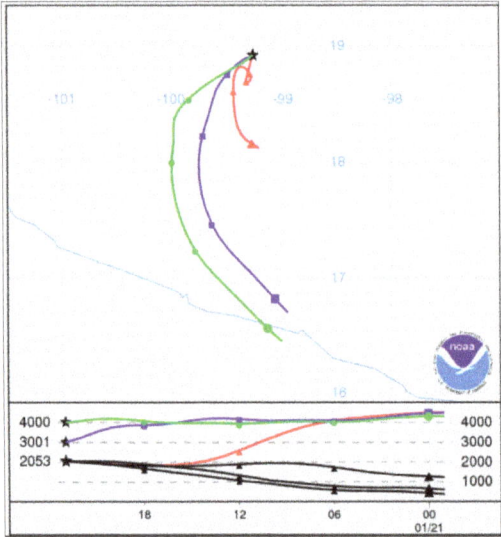

Figure 1. Behavior of wind trajectories in Cuernavaca during January and February 2013. Source: 18°55′N, 99°13′W, heights: 1000, 2000 and 3000 m above mean sea level (a.m.s.l.).

144

2.2. Concentration of PAHs

PAHs with 2 and 3 aromatic rings (Acen, Ace, Flu, Phe and Ant) showed the lowest levels of concentration at both sites, whereas those with 4–6 rings (Flt, Pyr, BaA, Chr, BbF, BkF, BaP, IP, DBahA, BghiP) were the most abundant. This suggests that the combustion of gasoline constitutes one of the main sources of these compounds in the study area, which is consistent with that observed by Zielinska *et al.* 2004 [12].

At the downtown site the average total PAHs concentration was 22.57 $ng \cdot m^{-3}$ ($\Sigma15$ PAHs), while that at the CIQ site was 25.43 $ng \cdot m^{-3}$ ($\Sigma15$ PAHs), indicating that there were no statistically significant differences in the concentration of this family of compounds between sites ($p < 0.05$). Such behavior suggests relatively stable atmospheric conditions or similar emission sources in the study area (Table 1). Also the proximity of the University Campus to the México-Cuernavaca highway, probably is affecting the concentration of PAHs in this area, as well as transportation from the southern part of the city, such as was seen in the wind field.

2.3. Comparison with Other Studies

Comparisons with other studies, were carried out using the average between the two study sites. The average total concentration in Cuernavaca (24.0 $ng \cdot m^{-3}$, $\Sigma15$ PAHs) was similar to that reported in Taichung, China (22.29 $ng \cdot m^{-3}$, $\Sigma15$ PAHs) [13] and in Seoul, Korea (26.14 $ng \cdot m^{-3}$, $\Sigma14$ PAHs) [14]. It was lower than that reported for Nanjing (62.58 $ng \cdot m^{-3}$, $\Sigma15$ PAHs) [6] and Jinzhou, China (190.86 $ng \cdot m^{-3}$, $\Sigma13$ PAHs) [15]. All these places have the characteristic of being large urban areas with high vehicular activity. However, the average total concentration was higher than that reported in Atlanta, USA (1.52 $ng \cdot m^{-3}$, $\Sigma15$ PAHs) [16], Wanqingsha, Hong Kong (19.3 $ng \cdot m^{-3}$, $\Sigma17$ PAHs) [17] Porto, Portugal (13.3 $ng \cdot m^{-3}$, $\Sigma15$ PAHs) [18] and Mount Taishan, China (7.05 $ng \cdot m^{-3}$ $\Sigma15$ PAHs) [19]. These results are consistent with the type of industrial and/or vehicular activities, which occur in this urban area.

2.4. Possible Sources of PAHs in Cuernavaca

2.4.1. Diagnostic Ratios

Given the variability in the concentration of PAHs emitted into the atmosphere, often diagnostic ratios between them are used in order to estimate the possible sources [20]. However, they should be used with caution, if one considers, for example, reactivity with some of the PAH species such as NO_x, O_3, *etc.* [9].

Diagnostic ratios used in this study, were calculated and reported by various authors, which established ranges or values indicating possible sources of origin of polycyclic aromatic hydrocarbons (Table 2).

Table 1. Average concentration of PAHs associated with particulate matter $PM_{2.5}$ in two places of the city of Cuernavaca, Morelos during the winter season of 2013 $(ng \cdot m^{-3})$.

–	CIQ					GB (Downtown)				
	Media	Min	Max	SD	Media	Media	Min	Max	SD	Median
Acen	0.81	0.72	0.90	0.06	0.80	1.29	0.67	6.17	1.62	0.82
Ace	0.96	0.79	1.25	0.16	0.94	0.93	0.63	1.46	0.23	0.89
Flu	0.95	0.77	1.09	0.12	0.98	0.94	0.72	1.15	0.14	0.99
Phe	1.27	1.00	1.45	0.14	1.34	1.18	0.79	1.46	0.25	1.23
Ant	1.07	0.87	1.24	0.10	1.10	1.14	0.77	1.64	0.24	1.15
Flt	1.73	1.07	2.76	0.51	1.61	1.24	0.63	1.69	0.34	1.32
Pyr	1.39	0.99	1.82	0.24	1.45	1.22	0.63	1.56	0.32	1.29
BaA	1.57	1.17	2.11	0.26	1.57	1.33	0.77	1.79	0.30	1.32
Chr	1.54	0.99	2.04	0.27	1.55	1.39	0.69	1.78	0.36	1.44
BbF	2.68	1.83	3.49	0.49	2.73	2.36	1.04	3.34	0.77	2.53
BkF	2.75	1.82	3.58	0.46	2.78	2.45	1.17	3.32	0.74	2.52
BaP	2.63	1.53	3.54	0.50	2.55	2.27	0.93	3.13	0.73	2.47
IP	2.14	1.56	2.81	0.35	2.14	1.70	0.89	2.56	0.56	1.64
DBahA	1.68	1.11	2.16	0.26	1.72	1.46	0.92	2.00	0.32	1.56
BghiP	2.25	1.57	3.36	0.50	2.16	1.68	0.53	3.13	0.83	1.72
Total (ΣHAP)	25.43	–	–	–	–	22.57	–	–	–	–

n = 11; Min: minimum value; Max: maximal value; SD: standard deviation.

Table 2. Diagnostic ratios.

Diagnostic Ratios	Gasoline Engines	Diesel Engines	Coal	Wood Combustion	This Study
Flt/ (Flt + Pyr) [21,22]	0.40–0.50	0.40–0.50	>0.50	>0.50	0.42
IP/(IP + BghiP) [21,23–25]	0.18	0.37–0.70	0.56	0.62	0.50
Phe/ (Phe + Ant) [23,26]	0.50	0.65	0.76	–	0.52
IP/ BghiP [27,28]	0.22	0.50	1.30	–	1.05

The ratio Flt/(Flt + Pyr) was between 0.40 and 0.50, suggesting that the study area is influenced by emissions from gasoline and diesel [21,22]. Meanwhile the IP/(IP + BghiP) ratio obtained in this study was 0.5. Some previous studies that were done in other places have reported values between 0.18 and 0.40 corresponding to vehicle emissions, 0.56 corresponding to coal burning and 0.62 to wood combustion [21–25], indicating that in the study area coal burning is an important source. The Phe/(Ant + Phe) ratio was also estimated, giving a value of 0.52. This value is associated with vehicle emissions, specifically gasoline [23,26]. Meanwhile the IP/BghiP (1.05) ratio indicated coal burning as a major source in this zone [27,28]. These results are congruent with the dimensions, anthropogenic activities and population density of the Cuernavaca city. Recently, Ortíz-Hernández et al. [29] performed an emissions inventory, this study suggests vehicle emissions, coal burning and forest activities as the principal sources of greenhouse gases in this region.

2.5. Principal Component Analysis

A principal component analysis (PCA) with varimax rotation was performed to determine the possible sources of these compounds in the city of Cuernavaca. The principal components (PC) were selected with eigenvalues greater than 1.0. In total 3 PC were extracted. The PC1 explained 75.6% of the total variance. In this component the compounds loads were similar. Although the BaA, Chr and BaP stand out slightly, these compounds are considered as markers of pyrogenic sources (combustion of oil and coal) [30]. While the PC2 represented 8.5% of the remaining variance, the compounds with the greatest burden in this component were BkF, IP and BghiP, which are associated with vehicle emissions [30,31]. Meanwhile, the PC3 explained 7.0%, and the greatest burden was provided by Ace and Flu, considered as indicators of coke production (Table 3).

The results show that the study area is being specifically affected by emissions from gasoline and diesel.

The combined results of the PCA and diagnostic ratios suggests that vehicle emissions and coal combustion are probably the main sources of PAHs associated with the $PM_{2.5}$ in the study area.

Table 3. Principal Component Analysis for Cuernavaca.

–	PC1	PC2	PC3
Eigenvalues	11.30	1.30	1.10
% Variance accumulated	75.60	84.10	91.10
Acen	0.23	0.15	−0.05
Ace	0.17	−0.24	**0.66**
Flu	0.25	0.17	−0.14
Phe	0.26	0.10	−0.31
Ant	0.26	−0.26	−0.15
Flt	0.24	0.04	**0.41**
Pyr	0.27	0.14	0.27
BaA	**0.28**	0.06	0.20
Chr	**0.29**	0.12	0.02
BbF	0.27	0.16	−0.02
BkF	0.26	**0.37**	−0.19
BaP	**0.29**	0.12	−0.13
IP	0.27	**0.29**	0.05
DBahA	0.27	−0.06	−0.27
BghiP	0.26	**0.28**	−0.11

2.6. Evaluation of Health Risk

The value obtained for the BaPE in this study was 4.05 ng·m^{-3}. This value is lower than that reported for Nanging, China, (7.1 ng·m^{-3}) [6] and Zonguldak, Turkey (14.1 ng·m^{-3}) [32]. However, it was higher than that found in Hamilton, Canada (0.84 ng·m^{-3}) [33] and in Florence, Italy (0.92 ng·m^{-3}) [34]. The compounds used to calculate BaPE (BaA, BaP, BbF, BkF, IP, DBah) represent 51% of the PAHs identified in Cuernavaca, which constitutes a risk factor for the exposed population, taking into account the high degree of penetration in the respiratory system that have the PM$_{2.5}$ particles.

3. Experimental Section

3.1. Sampling Sites

The study was conducted at two sites in the city of Cuernavaca. The first one is located in the downtown of the city (Government Building, GB), which is characterized by high commercial and vehicular activity. The second is located north of the city, at the Autonomous University of State of Morelos Campus ((Center of Chemical Research, CIQ), for its acronym in Spanish). This location is characterized as being surrounded by mountains, having a great variety of vegetation, and being influenced by the Mexico-Cuernavaca highway. The city has a total area of 207.5 km^2 and a population of 338,620 inhabitants. During the study period, the average temperature was 23.1 °C (minimum 9.6 °C, maximum 26.7 °C), the relative humidity was 36.2% and the average wind speed was 5.3 m·s^{-1} (Figure 2).

The particles were collected on quartz filters of 47 mm diameter and in a pore size of 2.0 μm, baked to 180 °C for at least 24 h to remove adsorbed organics, after which they were transferred to a chamber with relative humidity of <40% at 20–23 °C for another 24 h for conditioning. Afterward, the filters with particles were equilibrated in the chamber for an additional 24 h. Low volume equipment (Mini-Vol) were used at a constant flow of 5.0 L·min^{-1}, equipped with impactors for PM$_{10}$ and PM$_{2.5}$. Sampling was carried out in periods of 24 h (12:00–12:00), from January 25 to 28 February 2013, every two days. In total, 11 samples were collected for each site. After sampling, the filters were refrigerated (4 °C) until extraction.

3.2. Wind Trajectories

To determine the behavior of winds in the study area, HYSPLIT4 model trajectories obtained from the NOAA (National Oceanic and Atmospheric Administration) were used. The trajectories were made for the entire study period [35].

148

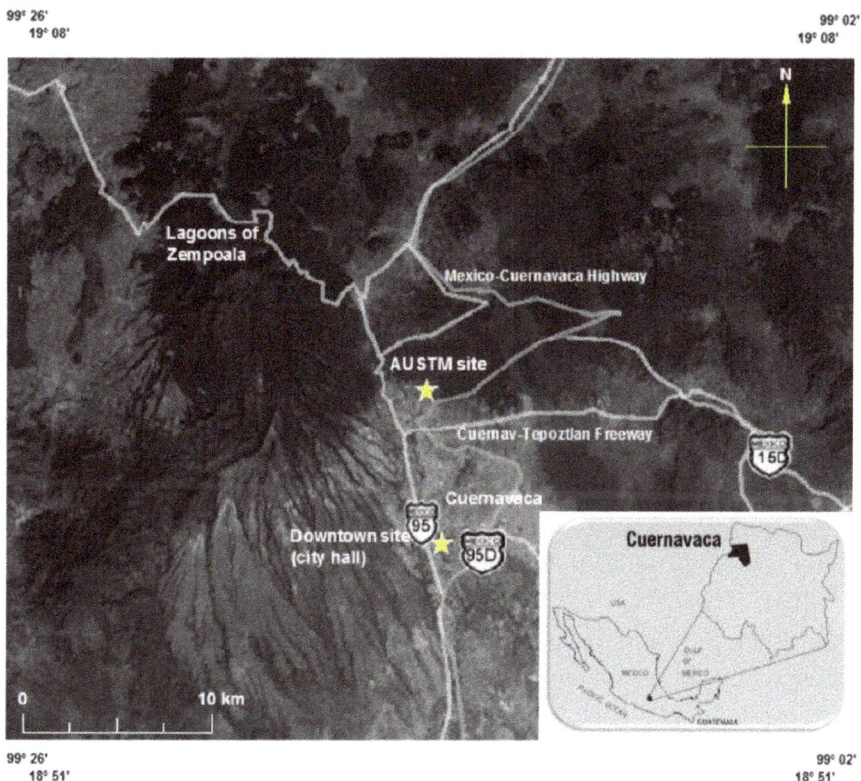

Figure 2. Sampling sites at the city of Cuernavaca, Morelos.

3.3. Extraction of Organic Matter

Sampled filters were extracted using 10 mL of dichloromethane (Burdick & Jackson, HPLC grade) in an ultrasound bath (Branson 3210) for 30 min. This procedure was performed twice. The Erlenmeyer flasks in which the filters were introduced, were fitted with cooled condensers with water at 10 °C. The extracts were filtered through a polytetrafluoroethylene (PTFE) membrane (0.22 μm, Millipore), and then concentrated on a rotary evaporator (Buchi R-3) to about 1.0 mL and stored under refrigeration until analysis.

3.4. Quality Control of the Analytical Method

Field blanks (glass fiber filters) were performed once a week. Field blanks were stored in Petri dishes and returned to the laboratory until analysis. Further, blanks of laboratory material were used to discard the presence of analytes of interest during each one of the stages. The efficiency of the extraction methodology was evaluated by the enrichment of sampled filters with airborne particles with a solution containing 16 PAHs:

Acenaphthylene (Acen), Acenaphthene (Ace), Fluorene (Flu), Phenanthrene (Phe), Anthracene (Ant), Fluoranthene (Flt), Pyrene (Pyr), Benzo[a]Anthracene (BaA), Chrysene (Chr), Benzo[b]Fluoranthene (BbF), Benzo[k]Fluoranthene (BkF), Benzo[a]Pyrene (BaP), Indeno[1,2,3-cd]Pyrene (IP), Dibenzo[a,h]Anthracene (DBahA) and Benzo[g,h,i]Perylene (BghiP), at a concentration of 250 parts per billion (ppb). The amount recovered for each compound was compared to the amount added to calculate the recoveries. Once the percent recovery was determined on the enriched filters, the repeatability was calculated by estimating the coefficient of variation (% CV). The recoveries ranged from 60% to 90%, for Fluorene and IP, respectively. The repeatability in terms of % CV was less than 10%, which is consistent with the variation (30%) set by the United States Environmental Protection Agency [36], for these samples.

Instrumental detection limits were performed based on a weight regression, proposed by Miller and Miller 2002 [37]; these ranged between 16.71 and 61.70 ng·mL^{-1} for BghiP and BaP, respectively (Table 4). The concentrations of the compounds in ambient air were corrected using the recovery percentages.

Table 4. Detection limits ng·mL^{-1}.

Compound	LOD
Acen	33.62
Ace	32.20
Flu	31.19
Phe	35.17
Ant	37.30
Flt	46.85
Pyr	50.00
BaA	26.00
Chr	44.00
BbF	26.04
BkF	22.50
BaP	61.70
IP	59.22
DahA	44.30
BghiP	16.71

LOD: limit of detection.

3.5. Instrumental Analysis

The extracts obtained were concentrated under a gentle stream of nitrogen to near dryness and subsequently resuspended with 150 μL of dichloromethane in an insert injection. To this volume, 50 µL of a solution containing 6 PAH-d (internal standard) were added to a final volume of 200 µL (naphthalene-d8, acenaphthene-d10, phenanthrene-d10, pyrene-d10, chrysene-d12 and perylene-d12) (Chemservice West Chester, PA), at a final concentration of 2500 ppb of internal standard.

The chromatographic analysis was performed on an Agilent Technologies, Model 6890 gas chromatograph (GC) coupled to a mass spectrometer (MS) 5973N with quadrupole mass filter. The separation of the compounds was carried out on a capillary column (J & W Scientific, USA) with an internal diameter of 0.25 mm with 5% phenyl stationary phase and 95% dimethyl polysiloxane and a film thickness of 0.25 µm. The oven temperature program was as follows: 60 °C for 10 min at a gradient of 5 °C per minute to 300 °C for 10 min. The injector temperature was 300 °C in the splitless mode (purge time 30 sec) and the injection volume of 2.0 µL. Helium was used as carrier gas at a flow of 1.0 mL·min^{-1}. The MS was operated in the electron impact (70 eV) mode and the temperature of the ion source and the quadrupole filter were 230 °C and 150 °C, respectively.

The mass-charge ratios (m/z) monitored were: Acenaphthylene (Acen, 152), Acenaphthene (Ace, 153), Fluoranthene (Flt, 202), Pyrene (Pyr, 202), Benzo[a]Anthracene (BaA, 228), Chrysene (Chr, 228), Benzo[a]Pyrene (BaP, 252), Benzo[b]Fluoranthene (BbF, 252), Benzo[k]Fluoranthene (BkF, 252), Indeno[1,2,3-cd]Pyrene (IP, 276), Dibenzo[a,h]Anthracene (DahA, 278), Benzo[g,h,i]Perylene (BghiP, 276). Multicomponent calibration curves were made for all PAHs in a concentration range of 12.5 to 1600 ng·mL^{-1} (r > 0.99, p < 0.001).

3.6. Evaluation of Health Risk

BaP is considered one of the most powerful mutagens, in many cases is used as a general indicator of PAHs and regarded by the World Health Organization (WHO) as a good index for whole PAH carcinogenicity. However, BaP it degrades easily in the presence of sunlight and some oxidants [38,39]. For this reason, BaP concentration alone does not give a good indication of the hazard presented by all the PAHs, and the PAHs' carcinogenic character, could be underestimated under determined conditions if only this compound is taken as the representative of carcinogenicity.

Therefore, an equivalent index for BaP (BaPE), was created with the objective of allowing a better estimation of the carcinogenic potential of PAHs associated with atmospheric particles. The concentration of BaPE for each PAH, was calculated by multiplying their concentration by its corresponding toxic equivalent factor (TEF), which represents the relative carcinogenic potency of the corresponding PAH [40]:

$$BaPE = BaA \times 0.06 + BF \times 0.07 + BaP + DBahA \times 0.6 + IP \times 0.08 \qquad (1)$$

where BF includes all the isomers of benzofluoranthene. The BaPE index tries to parameterize the health risk for human health related to ambient PAH exposure, and was calculated by multiplying the concentrations of each carcinogenic congener with its carcinogenic factor obtained by laboratory studies.

3.7. Statistical Analysis

The statistical analysis, including Kruskal-Wallis test, correlation analysis and principal component analysis (PCA), were realized using STATGRAPHICS program (Statistical Graphics Corp.).

4. Conclusions

The most abundant PAHs were those of high molecular weight, suggesting that in this area the combustion processes significantly contribute to atmospheric emissions. The diagnostic ratios and PCA revealed that coal burning and vehicle emissions are the main source of PAHs in the studied sites. The value obtained for the BaPE suggests a risk to the population, taking into account the degree of penetration of the $PM_{2.5}$. It is important to mention that this result must be taken with caution, because the study was performed just for one month in the winter season, what suggests longer sampling periods in this urban zone for a better estimation of the risk.

The results obtained reflect the deterioration in air quality that is happening in the urban area of Cuernavaca as a result of high population growth, vehicle fleet growth and development in the industrial park that this region has experienced in recent years. This suggests that establishing strategies oriented to the reduction of atmospheric emissions is required, including the modernization of road transport and implementation of mass transportation systems, among other strategies.

Acknowledgments: The authors would like to express their appreciation to Winston Smith of the Peace Corps for the revision of this paper, to María Gregoria Medina Pintor for her support in the chromatographic analysis, to the Air Quality Monitoring Network of the State of Morelos (RAMAMOR, for its acronym in Spanish) for allowing the installment of the equipment in their locations. Special thanks also Program for the Professional Development of Teachers (PRODEP, for its acronym in English), for the financial support in this project. The authors gratefully acknowledge the NOAA Air Resources Laboratory (ARL) for the provision of the HYSPLIT transport and dispersion model and/or READY website (http://www.ready.noaa.gov) used in this publication. Also to National Laboratory of Macromolecular Structures (LANEM, for its acronym in Spanish, project 251613).

Author Contributions: Saldarriaga-Noreña and López-Márquez wrote the article, Murillo-Tovar, Montiel-Palma and Hernández-Mena designed the study, Waliszewski realized the PAHs analysis, Ospina-Noreña and Sánchez-Salinas performed the statistical analysis.

Conflicts of Interest: The authors declare no conflict of interest.

References

1. Monographs on the Evaluation of Carcinogenic Risks to Humans. Available online: http://monographs.iarc.fr/ENG/Monographs/vol32/volume32.pdf (accessed on 11 July 2015).

2. Aceves, M.; Grimalt, J.O. Seasonally dependent size distributions of aliphatic and polycyclic aromatic hydrocarbons in urban aerosols from densely populated areas. *Environ. Sci. Technol.* **1993**, *27*, 2896–2908.

3. Kiss, G.; Varga-Puchony, Z.; Rohrbacher, G.; Hlavay, J. Distribution of polycyclic aromatic hydrocarbons on atmospheric aerosol particles of different sizes. *Atmos. Res.* **1998**, *46*, 253–261.

4. Mazquiaran, M.A.B.; de Pinedo, L.C.O. Organic composition of atmospheric urban aerosol: Variations and sources of aliphatic and polycyclic aromatic hydrocarbons. *Atmos. Res.* **2007**, *85*, 288–299.

5. Ravindra, K.; Bencs, L.; Wauters, E.; de Hoog, J.; Deutsch, F.; Roekens, E.; Bleux, N.; Berghmans, P.; Grieken, R. Seasonal and site-specific variation in vapour and aerosol phase PAHs over Flanders Belgium and their relation with anthropogenic activities. *Atmos. Environ.* **2006**, *40*, 771–785.

6. Wang, G.; Huang, L.; Zhao, X.; Niu, H.; Dai, Z. Aliphatic and polycyclic aromatic hydrocarbons of atmospheric aerosols in five locations of Nanjing urban area, China. *Atmos. Res.* **2006**, *81*, 54–66.

7. Galarneau, E. Source specificity and atmospheric processing of airborne PAHs: Implications for source apportionment. *Atmos. Environ.* **2008**, *42*, 8139–8149.

8. Sklorz, M.; Kreis, J.S.; Liu, Y.B.; Orasche, J.; Zimmermann, R. Daytime resolved analysis of polycyclic aromatic hydrocarbons in urban aerosol samples—Impact of sources and meteorological conditions. *Chemosphere* **2007**, *67*, 934–943.

9. Liu, Y.; Liu, L.B.; Lin, J.M.; Tang, N.; Hayakawa, K. Distribution and characterization of polycyclic aromatic hydrocarbon compounds in airborne particulates of East Asia. *China Partic.* **2006**, *6*, 283–292.

10. Cotham, W.E.; Bidleman, T.F. Polycyclic aromatic hydrocarbons and polychlorinated biphenyls in air at an urban and a rural site near Lake Michigan. *Environ. Sci. Technol.* **1995**, *29*, 2782–2789.

11. Mantis, J.; Chaloulakou, A.; Samara, C. PM_{10}-bound polycyclic aromatic hydrocarbons (PAHs) in the Greater Area of Athens, Greece. *Chemosphere* **2005**, *59*, 593–604.

12. Zielinska, B.; Sagebiel, J.; McDonald, J.; Whitney, K.; Lawson, D. Emission rates and comparative chemical composition from selected in-use diesel and gasoline-fueled vehicles. *J. Air Waste Manag. Assoc.* **2004**, *54*, 1138–1150.

13. Fang, G.C.; Wu, Y.S.; Chen, J.C.; Fu, P.P.C.; Chang, C.N.; Ho, T.T.; Chen, M.H. Characteristic study of polycyclic aromatic hydrocarbons for fine and coarse particulates at Pastureland near Industrial Park sampling site of central Taiwan. *Chemosphere* **2005**, *3*, 427–433.

14. Park, S.S.; Kim, Y.J.; Kang, C.H. Atmospheric polycyclic aromatic hydrocarbons in Seoul, Korea. *Atmos. Environ.* **2002**, *36*, 2917–2924.

15. Kong, S.; Ding, X.; Bai, Z.; Han, B.; Chen, L.; Shi, J.; Li, Z.Y. A seasonal study of polycyclic aromatic hydrocarbons in $PM_{2.5}$ and $PM_{2.5-10}$ in five typical cities of Liaoning Province, China. *J. Hazard Mater.* **2010**, *183*, 70–80.

16. Li, Z.; Porter, E.N.; Sjödin, A.; Needham, L.L.; Lee, S.; Russell, A.G. Characterization of $PM_{2.5}$-bound polycyclic aromatic hydrocarbons in Atlanta Seasonal variations at urban, suburban, and rural ambient air monitoring sites. *Atmos. Environ.* **2009**, *43*, 4187–4193.

17. Huang, B.; Liu, M.; Bi, X.; Chaemfa, C.; Ren, Z.; Wang, X.; Sheng, G.; Fu, J. Phase distribution, sources and risk assessment of PAHs, NPAHs and OPAHs in a rural site of Pearl River Delta region, China. *Atmos. Pollut. Res.* **2014**, *5*, 210–218.

18. Slezakova, K.; Castro, D.; Pereira, M.C.; Morais, S.; Delerue-Matos, C.; Alvim-Ferraz, M.C. Influence of Traffic Emissions on the Carcinogenic Polycyclic Aromatic Hydrocarbons in Outdoor Breathable Particles. *J. Air Waste Manage. Assoc.* **2010**, *60*, 393–401.

19. Li, P.H.; Wang, Y.; Li, Y.H.; Wang, Z.F.; Zhang, H.Y.; Xu, P.J. Characterization of polycyclic aromatic hydrocarbons deposition in $PM_{2.5}$ and cloud/fog water at Mount Taishan (China). *Atmos. Environ.* **2010**, *44*, 1996–2003.

20. Kendall, M.; Hamilton, R.; Watt, J.; William, I.D. Characterization of selected speciated organic compounds associated with particulate matter in London. *Atmos. Environ.* **2001**, *35*, 2483–2495.

21. Kavouras, I.; Koutrakis, P.; Tsapakis, M.; Lagoudaki, E.; Stephanou, E.; Stephanou, E.G.; von Baer, D.; Oyola, P. Source apportionment of urban particulate aliphatic and polynuclear aromatic hydrocarbons (PAHs) using multivariate methods. *Environ. Sci. Technol.* **2001**, *35*, 2288–2294.

22. Yunker, M.; Macdonald, R.; Vingarzan, R.; Mitchell, R.; Goyette, D.; Sylvestre, S. PAHs in the Fraser River basin: A critical appraisal of PAH ratios as indicators of PAH source and composition. *Org. Geochem.* **2002**, *33*, 489–515.

23. Khalili, N.; Scheff, P.; Holsen, T. PAH source fingerprints for coke ovens, diesel and gasoline engines, highway tunnels, and wood combustion emissions. *Atmos. Environ.* **1995**, *29*, 533–542.

24. Sicre, M.A.; Saliot, A.; Aparicio, X.; Grimalt, J.; Albaiges, J. Aliphatic and aromatic hydrocarbons in different sized aerosols over the Mediterranean Sea: occurrence and origin. *Atmos. Environ.* **1987**, *21*, 2247–2259.

25. Rogge, W.F.; Hildemann, L.; Mazurek, M.A.; Cass, G.R.; Simoneit, B.R.T. Sources of fine organic aerosol: Non-catalyst and catalyst equipped automobiles and heavy duty diesel trucks. *Environ. Sci. Technol.* **1993**, *27*, 636–651.

26. Alves, C.; Pio, C.; Durate, A. Composition of extractable organic matter of air particles from rural and urban Portugese areas. *Atmos. Environ.* **2001**, *35*, 5485–5496.

27. Grimmer, G.; Jacob, J.; Naujack, K. Profile of the polycyclic aromatic compounds from crude oils-inventory by GC GC-MS. PAH in environmental materials: Part 3. *Fresen. J. Anal. Chem.* **1983**, *316*, 29–36.

28. Caricchia, A.M.; Chiavarini, S.; Pezza, M. Polycyclic Aromatic Hydrocarbons in the urban atmospheric particulate matter in the city of Naples (Italy). *Atmos. Environ.* **1999**, *33*, 3731–3738.

29. Ortíz-Hernández, M.L.; Quiroz-Castañeda, R.E.; Sánchez-Salinas, E.; Castrejón-Godínez, M.L.; Macedo-Abarca, B. *Emissions of Greenhouse Gases in the State of Morelos*, 1st. ed.; Environmental Management Degree Program: Morelos, Mexico, 2013. (In Spanish)

30. Simcik, M.F.; Eisenreich, S.J.; Lioy, P.J. Source apportionment and source/sink relationships of PAHs in the coastal atmosphere of Chicago and Lake Michigan. *Atmos. Environ.* **1999**, *33*, 5071–5079.

31. Kulkarni, P.; Venkataraman, C. Atmospheric polycyclic aromatic hydrocarbons in Mumbai, India. *Atmos. Environ.* **2000**, *34*, 2785–2790.

32. Akyüz, M.; Cabuk, H. Meteorological variations of $PM_{2.5}/PM_{10}$ concentrations and particle-associated polycyclic aromatic hydrocarbons in the atmospheric environment of Zonguldak, Turkey. *J. Hazard Mater.* **2009**, *170*, 13–21.

33. Anastasopoulos, A.T.; Wheeler, A.J.; Karman, D.; Kulka, R.H. Intraurban concentrations, spatial variability and correlation of ambient polycyclic aromatic hydrocarbons (PAH) and $PM_{2.5}$. *Atmos. Environ.* **2012**, *59*, 272–283.

34. Lodovici, M.; Venturini, M.; Marini, E.; Grechib, D.; Dolara, P. Polycyclic aromatic hydrocarbons air levels in Florence, Italy, and their correlation with other air pollutants. *Chemosphere* **2003**, *50*, 377–382.

35. Draxler, R.R.; Rolph, G.D. HYSPLIT—Hybrid Single-Particle Lagrangian Integrated Trajectory Model. Available online: http://www.arl.noaa.gov/HYSPLIT.php (accessed on 11 July 2015).

36. Environmental Protection Agency (EPA) of the United States. *Compendium Method TO-13A, Determination of Polycyclic Aromatic Hydrocarbons (PAHs) in Ambient Air Using Gas Chromatography/Mass Spectrometry (GC/MS)*; EPA: Cincinnati, OH, USA, 1999.

37. Miller, J.N.; Miller, J.C. *Estadística y Quimiometría para Química Analítica*, 6th ed.; Prentice Hall: Harlow, UK, 2002.

38. Daisey, J.M.; Cheney, J.L.; Lioy, P.J. Profiles of organic particulate emissions from air pollution sources: status and needs for receptor source apportionment modelling. *J. Air. Pollut. Control. Assoc.* **1986**, *36*, 17–33.

39. Muller, J.F.; Hawker, D.F.; Connel, D.F. Polycyclic aromatic hydrocarbons in the atmospheric environment of Brisbane, Australia. *Chemosphere* **1998**, *37*, 1369–1383.

40. Nisbet, C.; LaGoy, P. Toxic equivalency factors (TEFs) for polycyclic aromatic hydrocarbons (PAHs). *Regul. Toxicol. Pharmacol.* **1992**, *16*, 290–300.

Inland Concentrations of Cl_2 and $ClNO_2$ in Southeast Texas Suggest Chlorine Chemistry Significantly Contributes to Atmospheric Reactivity

Cameron B. Faxon, Jeffrey K. Bean and Lea Hildebrandt Ruiz

Abstract: Measurements of molecular chlorine (Cl_2), nitryl chloride ($ClNO_2$), and dinitrogen pentoxide (N_2O_5) were taken as part of the DISCOVER-AQ Texas 2013 campaign with a High Resolution Time-of-Flight Chemical Ionization Mass Spectrometer (HR-ToF-CIMS) using iodide (I-) as a reagent ion. $ClNO_2$ concentrations exceeding 50 ppt were regularly detected with peak concentrations typically occurring between 7:00 a.m. and 10:00 am. Hourly averaged Cl_2 concentrations peaked daily between 3:00 p.m. and 4:00 p.m., with a 29-day average of 0.9 ± 0.3 (1σ) ppt. A day-time Cl_2 source of up to 35 ppt·h^{-1} is required to explain these observations, corresponding to a maximum chlorine radical ($Cl^\bullet$) production rate of 70 ppt·h^{-1}. Modeling of the Cl_2 source suggests that it can enhance daily maximum O_3 and $RO_2^\bullet$ concentrations by 8%–10% and 28%–50%, respectively. Modeling of observed $ClNO_2$ assuming a well-mixed nocturnal boundary layer indicates O_3 and $RO_2^\bullet$ enhancements of up to 2.1% and 38%, respectively, with a maximum impact in the early morning. These enhancements affect the formation of secondary organic aerosol and compliance with air quality standards for ozone and particulate matter.

Reprinted from *Atmosphere*. Cite as: Faxon, C.B.; Bean, J.K.; Ruiz, L.H. Inland Concentrations of Cl_2 and $ClNO_2$ in Southeast Texas Suggest Chlorine Chemistry Significantly Contributes to Atmospheric Reactivity. *Atmosphere* **2015**, *6*, 1487–1506.

1. Introduction

Research over the past several decades has shown that the presence of reactive gas phase chlorine species, particularly chlorine radicals ($Cl^\bullet$), can contribute significantly to atmospheric reactivity [1–7]. High concentrations of $Cl^\bullet$ can lead to increased rates of Volatile Organic Compound (VOC) oxidation [8–13] and enhanced rates of O_3 production in the troposphere [1,3,7,14–16]. Photochemical sources of $Cl^\bullet$ include Cl_2 and HOCl [17], which are the primary forms of anthropogenic chlorine emissions [7,18,19]. Chlorine radicals, once produced, can participate in reactions with VOCs to produce alkyl radicals. These reactions typically proceed via hydrogen abstraction as shown in Reaction (R1), where R represents a generic VOC, and $R^\bullet$ represents the resulting alkyl radical [17,20].

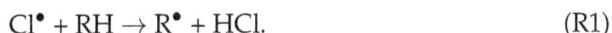

$$Cl^\bullet + RH \rightarrow R^\bullet + HCl. \qquad (R1)$$

Previous studies have reported field observations of elevated tropospheric Cl_2 at various locations across North America. Many of these observations were made near coastal areas, and maximum observed concentrations range from 20 ppt to 250 ppt [21–24]. Additionally, concentrations of up to 400 ppt Cl_2 were observed in the Arctic marine boundary layer in Barrow, Alaska [25]. Cl_2 source strengths on the order of 10–100 ppt $Cl_2 \cdot h^{-1}$ are required to explain such concentrations due to the rapid photolysis of Cl_2 [22,23,25].

Anthropogenic sources [1,26], aqueous reaction pathways [27,28] and naturally occurring heterogeneous or surface reaction routes of Cl_2 production [25,29,30] have been proposed to explain these observations. Previous experiments have observed that Cl_2 production from irradiated mixtures of O_3 and particulate chloride proceeds at a rate that is too rapid to be explained unless heterogeneous chemistry is active [30]. Numerous mechanisms have been proposed to explain the exact pathway for the heterogeneous formation of Cl_2. For example, the reaction of gas phase O_3 at the particle surface has been suggested [31,32].

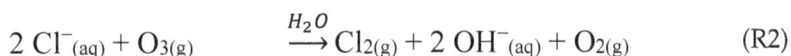

$$2\ Cl^-_{(aq)} + O_{3(g)} \xrightarrow{H_2O} Cl_{2(g)} + 2\ OH^-_{(aq)} + O_{2(g)} \qquad (R2)$$

Other work suggests that the reaction of hydroxyl radicals ($OH^\bullet$) at the particle surface more accurately depicts the chemistry active during heterogeneous Cl_2 production [30,33,34], where the formation of a surface complex between particulate Cl^- and gas phase $OH^\bullet$ was suggested as the rate limiting step, as is shown in Reactions (R3) and (R4).

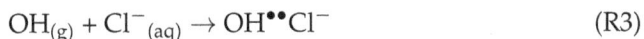

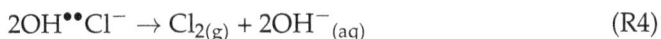

$$OH_{(g)} + Cl^-_{(aq)} \rightarrow OH^{\bullet\bullet}Cl^- \qquad (R3)$$

$$2OH^{\bullet\bullet}Cl^- \rightarrow Cl_{2(g)} + 2OH^-_{(aq)} \qquad (R4)$$

Regardless of the exact mechanism, the presence of a heterogeneous route for significant Cl_2 production from particulate chloride has implications for air quality through the production of O_3 [16] and particulate matter.

Heterogeneous production of gas phase $ClNO_2$ from particulate chloride has also been shown to occur and can lead to the generation of $Cl^\bullet$ in the presence of sunlight [35–38]. $ClNO_2$ chemistry also decreases the loss of reactive nitrogen via N_2O_5 deposition by producing a photolytic form of reactive nitrogen that is reintroduced into the gas phase [39–42]. Heterogeneous routes for the production of Cl_2 from particulate chloride have also been shown to exist [30,42–44].

Observations [24,37,45–47], modeling work [39–42] and laboratory studies [36–38,48–52] in the recent past have identified heterogeneous $ClNO_2$ formation as a major route for the production of reactive gas phase chlorine. The mechanism of $ClNO_2$ production is initiated by the reactive uptake of N_2O_5 on chloride-containing aerosol particles, as seen in Reaction (R5). This reaction competes with the heterogeneous hydrolysis of N_2O_5 (Reaction (R6)), which produces HNO_3.

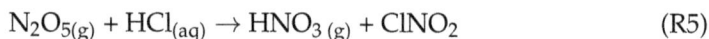

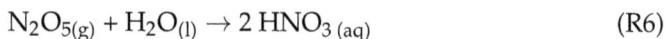

$$N_2O_{5(g)} + HCl_{(aq)} \rightarrow HNO_{3\,(g)} + ClNO_2 \qquad (R5)$$

$$N_2O_{5(g)} + H_2O_{(l)} \rightarrow 2\,HNO_{3\,(aq)} \qquad (R6)$$

This results in relatively unreactive particulate chloride being transformed into reactive gas phase chlorine. The $ClNO_2$ produced by the mechanism in Reaction (5) is photolytic and will decompose to produce gas phase NO_2 and $Cl^{\bullet}$ [17].

Aside from coastal [24,37,53,54] and oceanic [35] observations, several studies have recently reported significant $ClNO_2$ concentration in inland and mid-continental regions [45–47,55]. Previous measurements and modeling work have indicated that ppb-level concentrations of $ClNO_2$ are present in Houston, TX, USA and along the coast near the Houston Ship Channel [37,40,41]. The measurements reported here provide further insight into the formation and concentrations of Cl_2 and $ClNO_2$ at an inland location in southeast Texas and their implications to atmospheric reactivity.

2. Results and Discussion

The data presented in this work were collected during the DISCOVER-AQ 2013 campaign [56] in Houston, TX, USA for the period of 1 September 2013–1 October 2013. The data were obtained at an air quality monitoring ground site in Conroe, TX, USA (30.350278°N, 95.425000°W) situated next to the Lone Star Executive Airport in Montgomery county. The site is located approximately 60 km north northwest from the Houston urban center and approximately 125 km northwest of the nearest coastline. The gas phase species Cl_2, N_2O_5 and $ClNO_2$ were measured using an iodide High Resolution Time-of-Flight Chemical Ionization mass Spectrometer (HR-ToF-CIMS) and were identified by adduct ions Cl_2I^-, $N_2O_5I^-$, and $ClNO_2I^-$, respectively. The concentration and bulk composition of particulate matter smaller than 1 µm in diameter (PM_1) was measured using an Aerosol Chemical Speciation Monitor (ACSM, Aerodyne Research) [57]. Particle size distributions were measured using a Scanning Electrical Mobility System (SEMS, Brechtel Manufacturing).

Observations over the entire month of September 2013 revealed average peak concentrations of Cl_2, $ClNO_2$, and N_2O_5 of 1.6, 25, and 1.5 ppt, respectively. The highest concentrations of chlorine species were detected during the period of 11–20 September, and this time period is the focus of the following analysis. During this period, daily peak concentrations of Cl_2, $ClNO_2$, and N_2O_5 frequently exceeded 2 ppt, 60 ppt, and 20 ppt, respectively. Average nocturnal (8 p.m.–2 a.m.) concentrations of N_2O_5 were approximately 5 ppt. $ClNO_2$ morning (5 a.m.–11 a.m.) and nocturnal concentrations averaged 24 and 21 ppt, respectively. Cl_2 typically peaked in the afternoon, and Figure 1 shows the time series for 11–20 September.

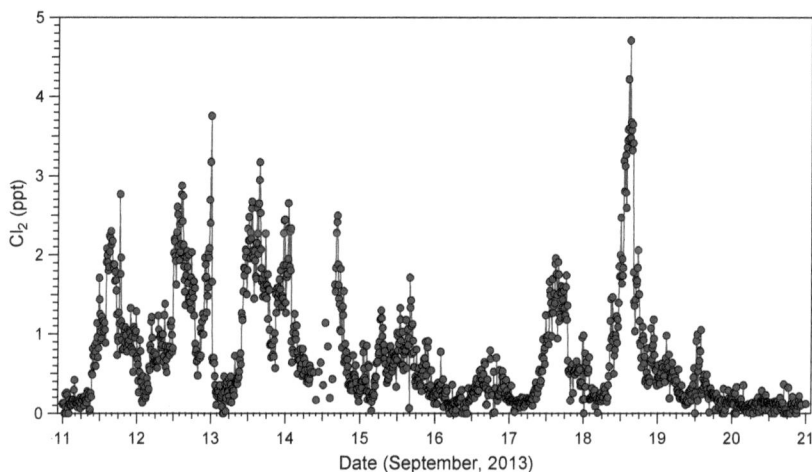

Figure 1. Time series of Cl_2 (10-min averages) at the Conroe, Texas measurement site from 11–21 September.

N_2O_5 exhibited peak nocturnal concentrations exceeding 20 ppt on the nights between 11–13 September. However, concentrations on most other nights rarely exceeded 2 ppt, even when increases in $ClNO_2$ concentrations were observed during the same time periods. Concentration time series of N_2O_5 and $ClNO_2$ from 11–21 September are shown in Figure 2. The implications of these observations are discussed further in the following sections.

2.1. Cl_2 Measurements

Concentrations of Cl_2 typically peaked in the afternoon at concentrations ranging 1–10 ppt. During the afternoon hours (12:00–6:00 p.m.) on many days, Cl_2 concentrations remained above 1–2 ppt. On most days, Cl_2 peaked between 3:00 p.m. and 4:00 p.m. (Figure 1). Figure 3 shows the diurnal pattern of Cl_2 concentrations between 11–21 September.

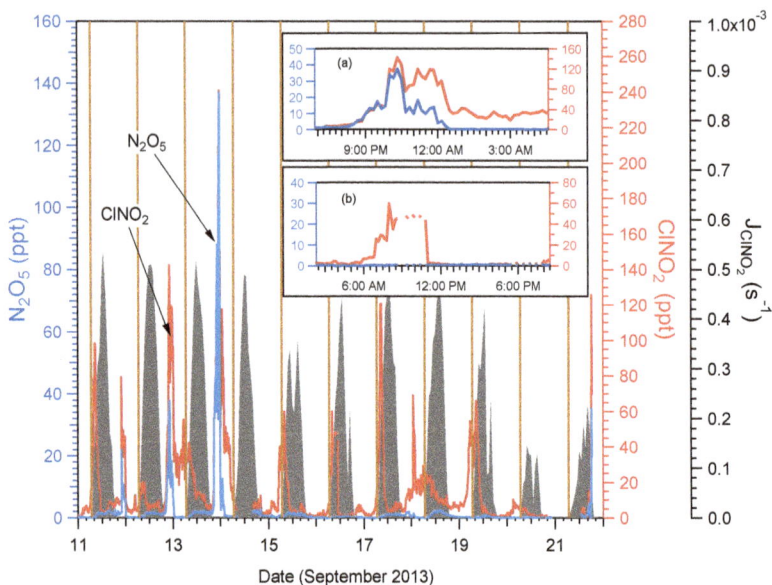

Figure 2. Time series of 10-min average $ClNO_2$ and N_2O_5 concentrations at the Conroe, Texas site from 11–21 September. Two distinct patterns were observed with respect to N_2O_5 and $ClNO_2$. Inset (**a**) shows a time period (9/13) when $ClNO_2$ and N_2O_5 concentrations correlated. Inset (**b**) shows a time period (9/16) when little N_2O_5 concentrations remained below the detection limit despite elevated $ClNO_2$ concentrations. A time series of the $ClNO_2$ photolysis rate is shown in grey, and vertical lines represent sunrise.

These observed concentrations are slightly lower than concentrations observed in previous studies [22–24,58], and significantly lower than concentrations that have been detected in the arctic marine boundary layer [24,25]. However, the higher concentrations of Cl_2 in these previous studies would be expected since they were taken in coastal areas, in close proximity to a large sea salt source. Although the Cl_2 concentrations reported here are relatively low compared to other studies, the observation of a sustained concentration in spite of the rapid day-time photolysis of Cl_2 suggests a significant source of Cl_2 is present. For example, the average Cl_2 concentration between 2:00 p.m. and 2:30 p.m. on 18 September was 3.8 ppt. The calculated photolysis rate constant for this same period, using the coordinates of the measurement site, was 2.5×10^{-3} s^{-1}. This implies an approximate Cl_2 photolysis rate of 35 ppt $Cl_2 \cdot$h^{-1} for a mid-day Cl_2 concentration of 4 ppt Cl_2; calculated 30-min average photolysis rates between 10:00 a.m. and 4:00 p.m. ranged from 7–35 ppt $Cl_2 \cdot$h^{-1}. Considering that concentrations on 18 September steadily increased from 10:00 a.m. and peaked at 2:00 p.m., this suggests that a minimum source strength of approximately 10–30 ppt·h^{-1} during the mid-day is necessary to explain the observations. This is similar in strength to sources suggested by previous studies to explain Cl_2 observations elsewhere [22,23].

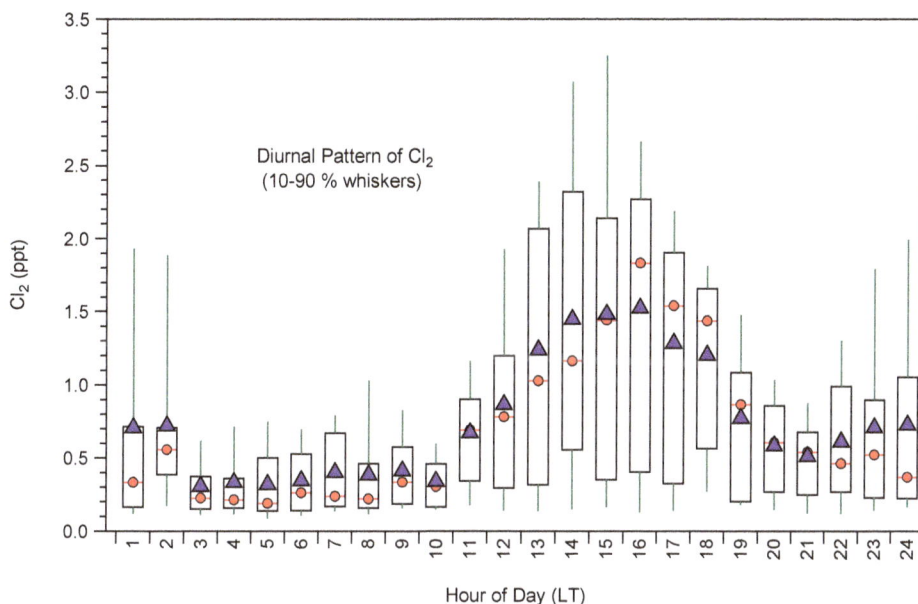

Figure 3. Diurnal pattern of Cl_2 concentrations observed from 11–21 September. Concentrations were typically elevated in the mid to late afternoon, with a lower enhancement late at night due to late night peaks such as the night of 12 and 13 September. Boxes indicate 25th and 75th percentiles, and whiskers indicate 10th and 90th percentiles.

Similar daytime Cl_2 concentrations were observed for several other days during the campaign (Figure 1), suggesting that a similar source was present during those times as well. Such a Cl_2 source could potentially lead to the enhanced oxidation of CH_4 and other VOCs [23,25,58], and therefore enhanced formation of secondary organic aerosol. Additionally, enhanced O_3 production can result from the presence of a $Cl^•$ source [1,33]. The implications of this are discussed further in Section 2.4.

On several occasions (13–15 September, for example), increases in Cl_2 were observed late at night or in the early morning, coinciding with increases in $ClNO_2$. No significant anthropogenic emissions of gas phase Cl_2 were known to be present in the area at the time. It is possible that the source of Cl_2 in these situations was the same as $ClNO_2$ since N_2O_5 has been shown to oxidize directly to Cl_2 when particles are acidic [43]. Recent measurements in inland regions suggest that soil deflation could possibly play a larger role as a particulate chloride source than previously

161

assumed [59–61]. Considering the diurnal pattern of Cl_2, with peak concentrations occurring in the mid-afternoon, it is more likely that Cl_2 is produced inland. Gas phase Cl_2 transported from the coast would be degraded by photolysis by the time it reaches the measurement site, which could take several hours. Additionally, the heterogeneous production of reactive chlorine is thought to be driven by O_3 or $OH^\bullet$, species which peak during the mid-day. The timing of peak Cl_2 concentrations thus suggest a heterogeneous mechanism as a possible source of observed Cl_2.

2.2. $ClNO_2$ and N_2O_5 Measurements

For the month of September, 2013, $ClNO_2$ was frequently detected at concentrations exceeding 50 ppt (approximately 40% of the days in the month). Observations for the period spanning 11–21 September are shown in Figure 2. Although the timing of peak $ClNO_2$ concentration varied between days (e.g., Figure 2a compared to Figure 2b), the typical pattern was a peak $ClNO_2$ concentration between 25–60 ppt occurring between 7:00–10:00 a.m. Daily 1-hour maximum concentrations averaged 24 ppt over the course of the month. On several days, however, $ClNO_2$ concentrations peaked at night between 23:00–24:00, reaching concentrations over 100 ppt by midnight. Examples of such episodes can be seen in Figure 2 on the nights of 13 and 14 September. Observations of N_2O_5 were also made, and during periods when the concentration was high (11–13 September), N_2O_5 correlated well with $ClNO_2$. However, there were many days (e.g., 14–21 September) during which elevated concentrations of $ClNO_2$ were observed while N_2O_5 remained low or was below the detection limit, perhaps suggesting that $ClNO_2$ production was limited by N_2O_5 on these days.

Observed concentrations of $ClNO_2$ were lower than previous measurements made in coastal areas. During the Texas Air Quality Study II (TexAQS II), $ClNO_2$ concentrations reaching 1200 ppt were detected in the Houston Ship Channel [37]. At another coastal location, in Los Angeles, CA, concentrations up to 1500 ppt were detected off the California coast in the Los Angeles region [24]. The lower magnitude of $ClNO_2$ concentrations reported here is likely due to the fact that measurements were taken further inland (~125 km from the coast) and therefore further from a sea salt chloride source. However, the $ClNO_2$ concentrations observed in Conroe, TX, USA, are also lower than recent observations much further from the coast. For example, $ClNO_2$ concentrations of up to 800 ppt and 450 ppt were detected at Kohler Mesa, CO (~1400 km from coast) [46] and Hessen, Germany (~380 km from coast) [45], respectively. The most likely explanation for these observations is that an inland source of chloride was present at these locations, and more recent work [59] suggest the presence of a significant soil source for chloride in Colorado. In this study, the measurements are consistent with the influence of a moderate particulate chloride source, and the observation that N_2O_5 was never present in the absence of $ClNO_2$ suggests that heterogeneous production was limited by N_2O_5 availability. The concentrations reported here are within the range predicted by previous modeling studies of $ClNO_2$ production in the region [39,40,42]. Concentrations of this magnitude were predicted to result in an enhancement of O_3 production up to several ppb [42].

On nights when $ClNO_2$ and N_2O_5 correlated well, the concentrations typically peaked around midnight (Figure 2a). However, when elevated $ClNO_2$ was present in the absence of N_2O_5, the concentrations typically peaked in the early to mid-morning (Figure 2b). Figure 4 shows a diurnal pattern of $ClNO_2$ and $PM_{2.5}$ surface area (as measured by the SEMS). Within the diurnal cycle of $ClNO_2$ at the measurement site, two distinct patterns of $ClNO_2$ concentrations are clearly visible. One pattern peaks early in the morning between 8 and 9 a.m., shortly after sunrise. In the other, late night $ClNO_2$ peaks are observed around midnight, coinciding with peak N_2O_5 concentrations (11–13 September). The peak 1-h average particulate surface area also occurs around the time of peak $ClNO_2$ concentrations in the early mornings, suggesting that conditions are favorable for the heterogeneous production of $ClNO_2$ from N_2O_5 in air masses that are advected to the site in the early mornings.

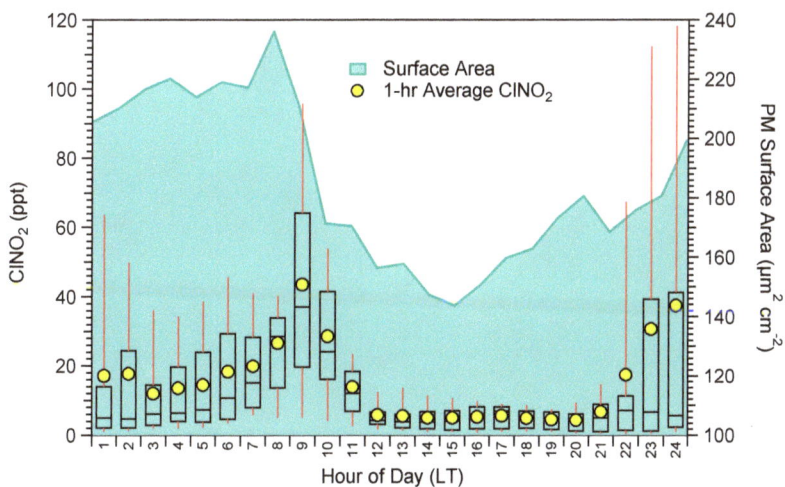

Figure 4. Diurnal cycle of $ClNO_2$ concentrations, showing two distinct patterns that are present during the campaign: (1) an early morning peak concentrations where $ClNO_2$ is advected to the site, and (2) late night peak concentrations, when $ClNO_2$ and N_2O_5 concentrations correlate significantly, suggesting more local inland production of $ClNO_2$.

Analysis of ACSM data collected during DISCOVER-AQ suggests high concentrations of particulate organic nitrates [62], which are formed either from photo-oxidation of hydrocarbons in the presence of NO_x or from the reactions of hydrocarbons with NO_3 radicals. Nitrate radical chemistry acts as a NO_y sink, reducing the formation of N_2O_5 and therefore $ClNO_2$ and could be responsible for relatively low N_2O_5 concentrations observed throughout most of the campaign.

2.3. Air Source Regions and Back Trajectories

On several days during the campaign, differences in the correlation of $ClNO_2$ and N_2O_5 concentrations and the timing of $ClNO_2$ peak concentrations suggest possible differences in the production and transport of $ClNO_2$. To investigate this difference further, HYSPLIT back trajectories were calculated for each case [63]. Starting height was set to 70 m, the end point of the back trajectories was set to the coordinates of the Conroe field site, and the end time was set to the time of maximum observed $ClNO_2$ concentration on each night. The model was used to generate a new back trajectory every two hours from the start time for a total of six 24-h back trajectories for each time period. Figure 5 shows a comparison of calculated back trajectories for the 13th and 16th of September.

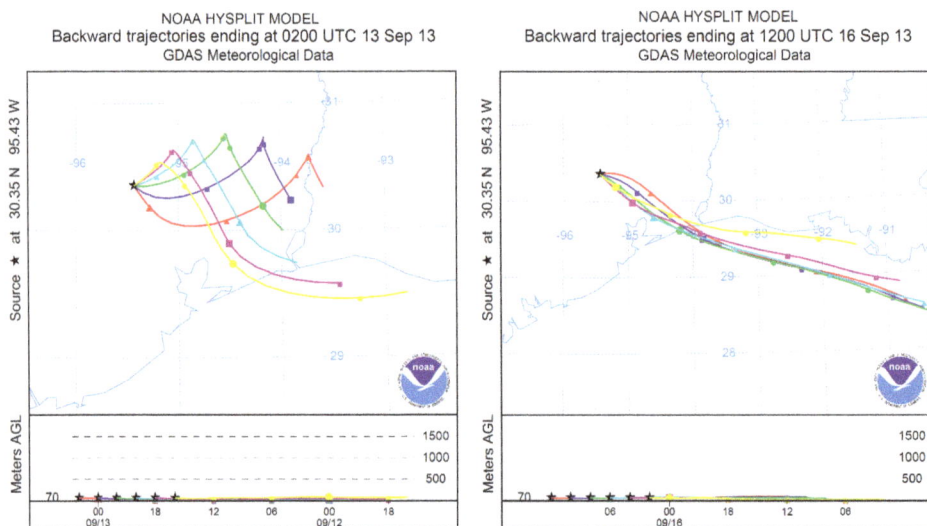

Figure 5. A comparison of HYSPLIT back trajectories for the days shown in the insets of Figure 2. Left: Back trajectories for the morning of 13 September 2013. The back trajectories ending at the time of peak $ClNO_2$ concentrations (10–11 p.m.) show air that originated inland near the TX-LA border 24 hours prior. Right: Back trajectories for the morning of 16 September 2013. The back trajectory at the time of peak $ClNO_2$ concentration (7–8 a.m.) indicates that the air mass originated in the Gulf of Mexico 24 hours prior, passing between the Houston, TX region and the LA-TX border.

One major difference between the two back trajectories is the amount of time that the air mass spent over land or over sea prior to arriving at the site. For example, air that was sampled at the site on 15–16 September originated in the Gulf of Mexico, resulting in a lower residence time over industrialized regions relative to the air sampled on the night of 12–13 September. Observations on the night of the 15th of September indicated very low N_2O_5. Elevated morning-time concentrations of $ClNO_2$ in these scenarios (e.g., 15–17 September) most likely resulted from transport of $ClNO_2$ to the measurement site from the Gulf Coast. $ClNO_2$ concentrations of up to 1.2 ppb were previously detected along the coast [37] where HYSPLIT back trajectories indicate the air originated on 16 September (Figure 5). The lack of a corresponding increase in N_2O_5 concentrations during these early morning episodes also suggests that the $ClNO_2$ observed at these times is produced elsewhere. The total amount of $ClNO_2$ transported to the site on these mornings was probably limited by N_2O_5 availability since SEMS data indicate that the highest measured particulate surface area concentrations were observed during these times (Figure 4).

In contrast to the early morning peak concentrations, the air that was sampled on the night of 12 September originated inland, and the majority of its path was over land. On the night of 12 September, it might be expected that the observed air mass had been exposed to additional inland NO_x sources that would explain the higher N_2O_5 concentration relative to the night of 15–16 September.

On several occasions, $ClNO_2$ was found to be elevated in the afternoon during daylight hours. Examples of this include the afternoons of 12, 13 and 17 September. Back trajectories for the afternoon hours on these days indicate that air transported to the site originated along the Gulf coast. On the afternoon of 12 September between the hours of 1 and 5 p.m., incoming air originated along the gulf coast 18 hours prior to arriving at the site. The back trajectories for 13 and 17 September indicate that the air masses originated from the same region, frequently crossing the region between Lake Charles, LA and Houston, TX. The amount of $ClNO_2$ that would need to be produced in the transported air can be estimated. For example, $ClNO_2$ concentrations on the afternoons of 12 and 13 September were approximately 10–15 ppt. An air mass arriving at 2 p.m. would have been exposed to daylight for approximately 7 hours. Taking the average observed photolysis rate of $ClNO_2$ between 7 a.m. and 2 p.m. for these dates, this corresponds to an average $ClNO_2$ photolysis rate of $2.8 \times 10^{-4} \text{ s}^{-1}$. This suggests that approximately 1–2 ppb of $ClNO_2$ would need to be produced in the incoming air masses during transport to the site in order for concentrations on the order of 10 ppt to be observed. Previous measurements and modeling predictions in this region indicate that elevated concentrations of $ClNO_2$ can reach concentrations of this magnitude [37,39,40,42].

Back trajectories in the afternoons when Cl_2 concentrations were typically elevated indicate that air is transported directly from the Gulf of Mexico, often passing over the metropolitan area of Houston, TX, USA. For example, air sampled at the site between 12–5 p.m. on the afternoon of 12, 13, 16 and 18 September consistently originated from the Gulf of Mexico 12 to 18 hours before being sampled. The transport of these air masses from the coast could suggest that long range sea salt transport contributed as a source of heterogeneously produced Cl_2 at the site.

2.4. Box Modeling Results

The contribution of observed Cl2 and ClNO2 to $Cl^\bullet$ production and resulting atmospheric reactivity were analyzed using ambient box modeling. The Statewide Air Pollution Research Center (SAPRC) software was used in combination with an updated version of the Carbon Bond 6 (CB6r2) chemical kinetics mechanism [64,65]. The CB6r2 mechanism was modified to include basic gas phase chlorine chemistry [65,66] in addition to Cl2 and ClNO2 photolysis in a manner similar to Sarwar et al. [39]. A list of reactions that were added is provided in the supplementary material. The photolysis rate of Cl2 was calculated internally to the mechanism for a latitude and longitude matching the location of the measurement site. For the dates used in the simulations, this resulted in a peak j_{Cl_2} of 2.8×10^{-2} s^{-1}, occurring at 12:30 p.m. Deposition was not included and boundary layer height was fixed for the duration of the model run.

Conditions on 18 September were modeled in order to assess the effects of Cl_2 on atmospheric reactivity. Although this day did not have the highest concentration of Cl_2 observed throughout the campaign (Figure 1), sustained concentrations over 2 ppt were observed for the entire period between 12:00–6 p.m., suggesting a continuous source of Cl_2 was present. Emissions of Cl_2 were added to the model so that the observed Cl_2 concentration profile was replicated in the simulations. On this day, relative humidity was 98% at 6:00 a.m., dropping to a minimum of 43% by 2:00 p.m., and temperature rose from 294 K at 6:00 am to 308 K at 2:00 p.m. Initial concentrations included 0.4 ppt Cl_2, 8 ppb NO_2, and 10 ppb O_3, consistent with early morning observations at the site on 18 September. Using these initial conditions and meteorological inputs, the production of OH radicals was calculated Equation (1), and day-time OH production ranged 10^6–10^7 molecules·cm^{-3} s^{-1}. A typical urban VOC mixture [67] was included in simulations at a concentration of 20 ppbC. Four scenarios were simulated: (1) no Cl_2 and no VOCs, (2) including Cl_2 emissions, no VOCs, (3) no Cl_2 emissions with 20 ppbC VOCs and (4) including Cl_2 emissions and 20ppbC VOCs. $Cl^\bullet$ production was calculated from the photolysis of measured Cl_2 concentrations Equation (2); an average emissions rate of 15 ppt·h^{-1} between 8:00 a.m.–6:00 p.m. was found to be necessary to replicate the observed concentrations. A comparison of modeled $OH^\bullet$ and $Cl^\bullet$ production rates and concentrations are shown in Figure 6.

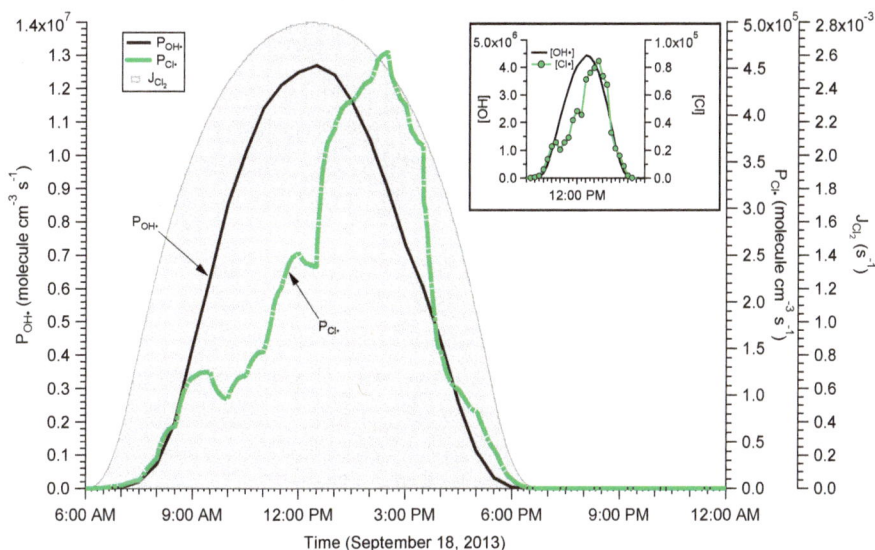

Figure 6. Production of OH and Cl radicals in a box modeling simulations for the conditions on 18 September 2013. The inset shows calculated concentrations of OH and Cl radicals in molecules·cm^{-3}·s^{-1}.

Although Cl$^\bullet$ production rates and concentrations are approximately 1–2 orders of magnitude lower than those of OH$^\bullet$, the calculated concentrations are significant when the relative reactivity of the two radicals is taken into account. The rate constants for the reactions of Cl$^\bullet$ with many VOCs is significantly (up to 100 times) faster than the corresponding reactions of OH$^\bullet$ [9–11,68,69]. Thus, Cl$^\bullet$ concentrations on the order of 10^5 molecules·cm^{-3} s^{-1} represent a significant radical source with respect to the oxidation of VOCs. For example, taking the maximum Cl$^\bullet$ concentration (8.5×10^4 molecules·cm^{-3}) and the corresponding OH$^\bullet$ concentration at the same time (3.6×10^6 molecules·cm^{-3}), the rate constant for the reaction of Cl$^\bullet$ and OH$^\bullet$ with CH$_4$ at 298 K are approximately 1.0×10^{-13} and 6.4×10^{-15} cm^3 molecules^{-1} s^{-1}, respectively. Assuming a CH$_4$ concentration of 1700 ppb, this corresponds to rates of reaction with CH$_4$ for Cl$^\bullet$ and OH$^\bullet$ of 3×10^5 and 8×10^5 molecules·cm^{-3} s^{-1}, respectively. These rates of reaction indicate that the Cl$^\bullet$ from observed concentrations of Cl$_2$ could significantly contribute to atmospheric reactivity with respect to VOC oxidation. It is also important that Cl$_2$ and Cl$^\bullet$ peak in the mid to late afternoon when OH$^\bullet$ production begins to wane.

Measured O$_3$ concentrations on the morning of 18 September were low with concentrations reaching around 5 ppb at 6:00 a.m. Simulations under-predicted day-time O$_3$ concentrations by over 25% compared to measurements. Transport of O$_3$ to the site in the afternoon (which is not included in this box modeling simulation) likely leads to higher observed O_3 concentrations. O_3 production was found to be enhanced by Cl_2. The addition of Cl_2 without VOC (model scenario 1 *vs.* 2) increased maximum modeled O_3 concentration between 8:00 a.m.–6:00 p.m. by 10% (1 ppb) and $RO_2^\bullet$ concentration by up to 50% (1.1 ppt). The addition of Cl_2 in the presence of VOC (model scenario 3 *vs.* 4) increased the maximum day-time O_3 concentration by 8% (2 ppb) and $RO_2^\bullet$ concentration by 28% (10 ppt). These results are summarized in Table 1.

Conditions on the morning of 13 September (scenario 5 and 6) and 16 September (scenarios 7 and 8) were modeled in order to assess the effects of $ClNO_2$ on atmospheric reactivity. Relative humidity and temperature in the model were set to approximate conditions present at the Conroe site. The simulation start time was set to midnight in order to capture the changes in radical production between nocturnal and early morning conditions. Model inputs for 13 September also included initial concentrations of 35 ppb O_3 and 20 ppb NO_x (consistent with measurements), as well as 20 ppbC VOC. A base case model run was performed without $ClNO_2$ (scenario 5) and compared to the model run with 40 ppt initial $ClNO_2$ (scenario 6). Model inputs for 16 September included initial concentrations of 15 ppb O_3 and 10 ppb NO_x (consistent with measurements), as well as 20 ppbC VOC. A base case with no $ClNO_2$ (scenario 7) was compared to a separate model run (scenario 8), where initial $ClNO_2$ concentrations were set to 0, and $ClNO_2$ emissions were added to gradually increase the total amount of $ClNO_2$ from 0 to 50 ppt between 6–8 a.m. (Figure 2b).

A comparison of the radical production rates of $OH^\bullet$ and $Cl^\bullet$ resulting from measured $ClNO_2$, O_3 and H_2O concentrations is shown in Figure 7. Although $Cl^\bullet$ production reaches a peak rate that is approximately 2 orders of magnitude lower than peak $OH^\bullet$ production, its production in the early morning increases atmospheric reactivity during a time when $OH^\bullet$ concentrations are low. Concentrations of $Cl^\bullet$ peak in the very early morning (Figure 7 inset), approximately two hours before $OH^\bullet$ concentrations begin to increase. Using the peak modeled $Cl^\bullet$ concentration (2.5×10^4 molecules·cm^{-3} and an approximate $OH^\bullet$ concentration from the same time period (1.5×10^6 molecules·cm^{-3}), the reaction rates with 1700 ppb CH_4 are similar at 1.1×10^5 and 4.1×10^5 molecules·cm^{-3} s^{-1} for $Cl^\bullet$ and $OH^\bullet$, respectively.

The inclusion of 40 ppt $ClNO_2$ enhanced O_3 and $RO_2^\bullet$ concentrations between 7:00 a.m.–3:00 p.m. by a maximum of 6.5% and 0.4%, respectively, when compared to a simulation with no $ClNO_2$ (scenarios 5 and 6 in Table 1). For simulations on the morning of 16 September, which included emissions of 50 ppt $ClNO_2$ in the early morning, similar enhancements in O_3 and $RO_2^\bullet$ concentrations were observed: in the presence of 20 ppbC VOC, O_3 and RO_2 concentrations were enhanced by 11.5% and 1.0%, respectively, when compared to a base case simulation without $ClNO_2$ (scenarios 7 and 8 in Table 1). In the absence of VOCs the addition of $ClNO_2$ did not significantly increase O_3 or $RO_2^\bullet$ concentrations (not listed in

Table 1). The increases in $RO_2^\bullet$ exemplify how $ClNO_2$ at the observed concentrations can contribute significantly to atmospheric reactivity, particularly in the morning. A summary of these modeling results and the results of Cl_2 simulations are listed in Table 1.

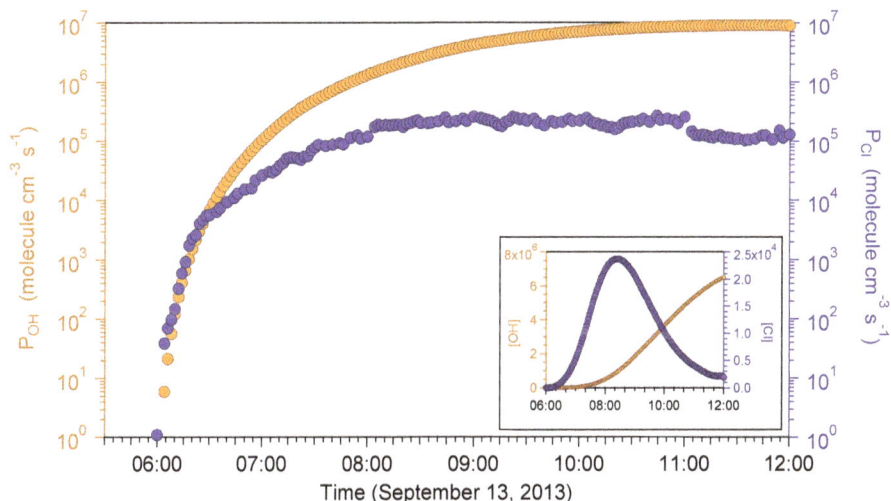

Figure 7. $OH^\bullet$ production resulting from the photolysis of O_3 compared to $Cl^\bullet$ production from observed concentrations of $ClNO_2$ on the morning of 13 September 2013. The inset shows a comparison of modeled $Cl^\bullet$ and $OH^\bullet$ radical concentrations (molecules·cm^{-3}) during the same time period, assuming an initial $ClNO_2$ concentration of 40 ppt (model scenario 6).

Table 1. Summary of Modeling Results.

Scenario	Chlorine Species	Comparison	[VOC] (ppbC)	$\Delta[O_3]_{max}$ (%)	$\Delta[RO_2^\bullet]_{max}$ (%)
1	–	1	0	10	50
2	Cl_2		0		
3	–	2	20	8	28
4	Cl_2		20		
5	–	3	20	6	0.4
6	$ClNO_2$		20		
7	–	4	20	12	1
8	$ClNO_2$		20		

3. Experimental Section

The area surrounding the air-quality monitoring station at Conroe, TX is primarily affected by pollution in the outflow of air from Houston, which hosts significant energy and petrochemical industries in addition to a large urban population. The regional atmospheric chemistry is also influenced by marine air from the Gulf of Mexico. The site itself is located in the middle of a field adjacent to the airport, with a gravel parking lot nearby and bordered by trees approximately 200 m to the North.

A permanent Texas Commission on Environmental Quality (TCEQ) ambient measurement station exists at this site and provided continuous meteorological data (wind speed, wind direction, temperature and relative humidity) for the duration of the campaign. NO_x and O_3 monitors were also present at the TCEQ site. During DISCOVER-AQ a temporary ground site was set up adjacent to the permanent station. Chlorine species were detected using a High Resolution Time-of-Flight Chemical Ionization Mass Spectrometer (HR-ToF-CIMS, henceforth referred to as CIMS) [70–73], which was operated in negative ionization mode utilizing the iodide reagent ion. Similar CIMS techniques have been used in previous studies for the detection of $ClNO_2$, N_2O_5 and Cl_2 [24,35,37,45–47,74,75], and other studies have described in detail the operation and application of the Aerodyne CIMS. Species reported here that were detected with the iodide CIMS included $ClNO_2$, Cl_2 and N_2O_5; the ions monitored for these species were $IClNO_2{}^-$, $ICl_2{}^-$, and $IN_2O_5{}^-$.

A sample flow of 2 LPM was introduced into the instrument through a $\frac{1}{4}$" outer diameter perfluoroalkoxy (PFA) Teflon®sample line connected to the stainless steel inlet by a $\frac{1}{4}$" to 10 mm stainless steel Swagelok union. The sample flow then passed into the ion-molecule reaction (IMR) chamber of the instrument, where it was mixed with a reagent ion flow of 2 LPM. Ionization in the IMR took place at a pressure of 200 mbar and temperature of 34°C (controlled by an IMR heater). $I(H_2O)_n{}^-$ reagent ions were generated by passing a flow of ultra-high purity (UHP) N_2 over a methyl iodide (CH_3I, Sigma Aldrich, 99%) permeation tube. The N_2 flow first passed through purified water (Milli-Q, model Advantage-A10) upstream of the CH_3I permeation tube in order to provide the humidity necessary for sufficient production of H_2OI^- ions. This also helped to stabilize the relative humidity within the IMR and resulting variations in sensitivity that have been noted for iodide chemical ionization [35,72]. Inside of the inlet, the sample flow then passed through 44.5 mm of the 10 mm outer diameter stainless steel tubing before passing through a critical orifice into the IMR. After mixing in the IMR the combined reagent and sample flow passed through two quadrupoles before entering the Time-of-Flight (ToF) mass analyzer. During the entire measurement period, the ToF was operated in V mode, providing higher sensitivity but lower resolution than W mode.

In order to test for measurement artifacts a flow of ultra-high purity (UHP) N_2 was periodically introduced into the sampling inlet. This flushing of the sampling line resulted in a rapid drop to background levels in the signals for Cl_2, N_2O_5 and $ClNO_2$, suggesting that adsorption and subsequent desorption of these species from the inlet line walls was not a significant source of measurement artifacts. When the sampling commenced afterwards ion signals returned rapidly to ambient levels, suggestive of a high transmission efficiency. No explicit tests were performed to assess the N_2O_5 transmission efficiency in the field introducing uncertainty in the quantification of N_2O_5 which could be lost irreversibly to the inlet walls. The N_2O_5 concentrations in this study are reported assuming that transmission efficiency was equivalent to that observed during calibrations. We estimate that this introduces a measurement uncertainty of 20%, a conservative estimate compared to previous work [76] which cited a $\pm$ 7% accuracy for N_2O_5 measured using a different instrument.

Equations (1) and (2) were used to calculate the production of $Cl^{\bullet}$ and $OH^{\bullet}$ consistent with box modeling simulation results. The production of $OH^{\bullet}$ was calculated from the reactions of $O(^1D)$ with H_2O, N_2, and O_2 (Equation (1)), in a manner similar to a previous observational study [45]. The rate of O_3 photolysis was used to calculate the rate of $O(^1D)$ generation, and additional details on the derivation of Equation (1) are described in the supplementary material. The production of $Cl^{\bullet}$ proceeded directly from the photolysis of Cl_2 (Equation (2)).

$$P_{OH} = 2J_{(O_3)} [O_3] k_{H_2O} [H_2O] / (k_{H2O} [H_2O] + k_{N2} [N_2] + k_{O2} [O_2]) \quad (1)$$

$$P_{Cl} = 2 \times j_{Cl_2} [Cl_2] \quad (2)$$

Additional simulations were performed to assess the impact of observed $ClNO_2$ concentrations on $Cl^{\bullet}$ production and atmospheric reactivity, and Equation (3) was used to quantify the production of $Cl^{\bullet}$ from $ClNO_2$.

$$P_{Cl} = j_{ClNO2} [ClNO_2] \quad (3)$$

4. Conclusions

During the month of September 2013, concentrations of Cl_2 and $ClNO_2$ regularly reached or exceeded 2 ppt and 60 ppt, respectively, at an inland monitoring site in Southeast Texas. The concurrent presence or absence of N_2O_5 at the site depended on the source of the sampled air, and longer paths over land corresponded to higher concentrations of N_2O_5. Peak $ClNO_2$ concentrations in the early mornings when little to no N_2O_5 was present on some days suggest that the production and transport of $ClNO_2$ from a non-local source is occurring. Contemporaneous measurements of PM surface area suggest conditions favorable to heterogeneous conversion of N_2O_5 in the advected air masses on these mornings. The presence of Cl_2 and $ClNO_2$ at the site are significant due to their role as Cl• sources. Although Cl_2 concentrations are lower in magnitude, the fact that Cl_2 concentrations consistently peak in the afternoon when photolysis rates are highest suggests a Cl_2 source of approximately 30 $ppt·h^{-1}$ Cl_2. Box modeling simulations suggest that such concentrations can contribute to enhanced O_3 and RO_2• production during the day. The source of $ClNO_2$ is likely the result of reactions between anthropogenic NO_x sources and particulate chloride originating along the Gulf Coast, and the reported values are within the range of previous model predictions in the region. Assuming that the surface measurements made in the early morning are representative of the entire nocturnal boundary layer, $ClNO_2$ was also found to contribute to Cl• production and atmospheric reactivity, particularly in the early morning before significant OH• production begins. Overall, the results suggest that Cl_2 and $ClNO_2$ affect atmospheric reactivity and can impact the formation of ozone and secondary organic aerosol. In a region where attainment of the National Ambient Air Quality Standard for ozone has been an issue in the past decades, quantifying the effects of reactive chlorine chemistry is important for assuring future attainment.

Acknowledgments: This work was funded in part through a grant from the Texas Commission on Environmental Quality (TCEQ), administered by The University of Texas through the Air Quality Research Program (Project 12-012). The contents, findings opinions and conclusions are the work of the authors and do not necessarily represent findings, opinions or conclusions of the TCEQ. The work was also funded in part with funds from the State of Texas as part of the program of the Texas Air Research Center. The contents do not necessarily reflect the views and policies of the sponsor nor does the mention of trade names or commercial products constitute endorsement or recommendation for use. The authors would like to acknowledge the support of the entire DISCOVER-AQ Texas 2013 team. Code for the SAPRC box modeling software can be found at http://www.engr.ucr.edu/~carter/SAPRC. Meteorological data used for model input can be obtained via the TCEQ website (http://www.tceq.state.tx.us).

Author Contributions: Cameron Faxon, Jeffrey Bean, and Lea Hildebrandt Ruiz planned, prepared and conducted the ambient measurements. Cameron Faxon conducted the box model simulations and analyzed ambient data and box model results presented in this work. The manuscript was written by Cameron Faxon with the guidance of Lea Hildebrandt Ruiz.

Conflicts of Interest: The authors declare no conflict of interest.

References

1. Tanaka, P.L.; Riemer, D.D.; Chang, S.; Yarwood, G.; McDonald-Buller, E.C.; Apel, E.C.; Orlando, J.J.; Silva, P.J.; Jimenez, J.L.; Canagaratna, M.R.; *et al.* Direct evidence for chlorine-enhanced urban ozone formation in Houston, Texas. *Atmos. Environ.* **2003**, *37*, 1393–1400.

2. Wang, L.; Thompson, T.; McDonald-Buller, E.C.; Allen, D.T. Photochemical modeling of emissions trading of highly reactive volatile organic compounds in Houston, Texas. 2. Incorporation of chlorine emissions. *Environ. Sci. Technol.* **2007**, *41*, 2103–2107.

3. Faxon, C.B.; Allen, D.T. Chlorine chemistry in urban atmospheres: A review. *Environ. Chem.* **2013**, *10*, 221–233.

4. Wingenter, O.W.; Sive, B.C.; Blake, N.J.; Blake, D.R.; Rowland, F.S. Atomic chlorine concentrations derived from ethane and hydroxyl measurements over the equatorial Pacific Ocean: Implication for dimethyl sulfide and bromine monoxide. *J. Geophys. Res.* **2005**, *110*.

5. Graedel, T.E.; Keene, W.C. Tropospheric budget of reactive chlorine. *Glob. Biogeochem. Cycles* **1995**, *9*, 47–77.

6. Keene, C.; Aslam, M.; Khalil, K.; Erickson, J.; Archie, I.I.I.; Graedel, E.; Loberr, M.; Aucott, M.L.; Ling, S.; Harper, D.B.; *et al.* Composite global emissions of reactive chlorine from anthropogenic and natural sources: Reactive chlorine emissions inventory. *J. Geophys. Res.* **1999**, *104*, 8429–8440.

7. Sarwar, G.; Bhave, P.V. Modeling the effect of chlorine emissions on ozone levels over the eastern United States. *J. Appl. Meteorol. Climatol.* **2007**, *46*, 1009–1019.

8. Liu, C.L.; Smith, J.D.; Che, D.L.; Ahmed, M.; Leone, S.R.; Wilson, K.R. The direct observation of secondary radical chain chemistry in the heterogeneous reaction of chlorine atoms with submicron squalane droplets. *Phys. Chem. Chem. Phys.* **2011**, *13*, 8993–9007.

9. Wang, L.; Arey, J.; Atkinson, R. Reactions of chlorine atoms with a series of aromatic hydrocarbons. *Environ. Sci. Technol.* **2005**, *39*, 5302–10.

10. Nelson, L.; Rattigan, O.; Neavyn, R.; Sidebottom, H. Absolute and relative rate constants for reactions of hydroxyl radicals and chlorine atoms with series of aliphatic alcohols and ethers at 298K. *Int. J. Chem. Kinet.* **1990**, *22*, 1111–1126.

11. Aschmann, S.M.; Atkinson, R. Rate Constants for the Gas-Phase Reactions of alkanes with Cl atoms at 296. *Int. J. Chem. Kinet.* **1995**, *27*, 613–622.

12. Canosa-Mas, C.E.; Hutton-Squire, H.R.; King, M.D.; Stewart, D.J.; Thompson, K.C.; Wayne, R.P. Laboratory Kinetic Studies of the Reactions of Cl Atoms with Species of Biogenic Origin: Δ3-Carene, Methacrolein and Methyl Vinyl Ketone. *J. Atmos. Chem.* **1999**, *34*, 163–170.

13. Ragains, M.L.; Finlayson-Pitts, B.J. Kinetics and mechanism of the reaction of Cl atoms with 2-Methyl-1,3-butadiene (Isoprene) at 298 K. *J. Phys. Chem. A* **1997**, *5639*, 1509–1517.

14. Chang, S.; McDonald-Buller, E.; Kimura, Y.; Yarwood, G.; Neece, J.; Russell, M.; Tanaka, P.; Allen, D. Sensitivity of urban ozone formation to chlorine emission estimates. *Atmos. Environ.* **2002**, *36*, 4991–5003.

15. Tanaka, P.L.; Allen, D.T.; Mullins, C.B. An environmental chamber investigation of chlorine-enhanced ozone formation in Houston, Texas. *J. Geophys. Res.* **2003**, *108*.

16. Knipping, E.M.; Dabdub, D. Impact of chlorine emissions from sea-salt aerosol on coastal urban ozone. *Environ. Sci. Technol.* **2003**, *37*, 275–284.

17. Sander, S.P.; Friedl, R.R.; Barker, J.R.; Golden, D.M.; Kurylo, M.J.; Sciences, G.E.; Wine, P.H.; Abbatt, J.P.D.; Burkholder, J.B.; Kolb, C.E.; *et al.* Chemical Kinetics and Photochemical Data for Use in Atmospheric Studies: Evaluation Number 17. Available online: https://jpldataeval.jpl.nasa.gov/pdf/JPL%2010-6%20Final%2015Jun8011.pdf (accessed on 31 August 2015).

18. Chang, S.; Allen, D.T. Chlorine chemistry in urban atmospheres: Aerosol formation associated with anthropogenic chlorine emissions in southeast Texas. *Atmos. Environ.* **2006**, *40*, 512–523.

19. Chang, S.; Tanaka, P.; Mcdonald-buller, E.; Allen, D.T. Emission inventory for atomic chlorine precursors in Southeast Texas. Available online: http://www.tceq.state.tx.us/assets/public/implementation/air/am/contracts/reports/oth/EmissioninventoryForAtomicChlorinePrecursors.pdf (accessed on 31 August 2015).

20. Atkinson, R.; Baulch, D.L.; Cox, R.A.; Crowley, J.N.; Hampson, R.F.; Hynes, R.G.; Jenkin, M.E.; Rossi, M.J.; Troe, J. Evaluated kinetic and photochemical data for atmospheric chemistry: Volume III—Gas phase reactions of inorganic halogens. *Atmos. Chem. Phys.* **2007**, *7*, 981–1191.

21. Pszenny, A.A.P.; Keene, W.C.; Jacob, D.J.; Fan, S.; Maben, J.R.; Zetwo, M.P. Evidence of inorganic chlorine gases other than hydrogen chloride in marine surface air. *Geopys. Res. Lett.* **1993**, *20*, 699–702.

22. Spicer, C.; Chapman, E.; Finlayson-Pitts, B.; Plastrige, R.; Hubbe, J.; Fast, J.; Berkowitz, C. Unexpectedly high concentrations of molecular chlorine in coastal air. *Nature* **1998**, *394*, 353–356.

23. Finley, B.D.; Saltzman, E.S. Measurement of Cl_2 in coastal urban air. *Geophys. Res. Lett.* **2006**, *33*.

24. Riedel, T.P.; Bertram, T.H.; Crisp, T.A.; Williams, E.J.; Lerner, B.M.; Vlasenko, A.; Li, S.M.; Gilman, J.; de Gouw, J.; Bon, D.M.; *et al.* A Nitryl chloride and molecular chlorine in the coastal marine boundary layer. *Environ. Sci. Technol.* **2012**, *46*, 10463–10470.

25. Liao, J.; Huey, L.G.; Liu, Z.; Tanner, D.J.; Cantrell, C.A.; Orlando, J.J.; Flocke, F.M.; Shepson, P.B.; Weinheimer, A.J.; Hall, S.R.; *et al.* High levels of molecular chlorine in the Arctic atmosphere. *Nat. Geosci.* **2014**, *7*, 91–94.

26. Chang, S.; Allen, D.T. Atmospheric chlorine chemistry in southeast Texas: impacts on ozone formation and control. *Environ. Sci. Technol.* **2006**, *40*, 251–262.

27. Oum, K.W.; Lakin, M.J.; DeHaan, D.O.; Brauers, T.; Finlayson-Pitts, B.J. Formation of molecular chlorine from the photolysis of ozone and aqueous sea-salt particles. *Science* **1998**, *279*, 74–76.

28. Herrmann, H.; Majdik, Z.; Ervens, B.; Weise, D. Halogen production from aqueous tropospheric particles. *Chemosphere* **2003**, *52*, 485–502.

29. Sadanaga, Y.; Hirokawa, J.; Akimoto, H. Formation of molecular chlorine in dark condition: Heterogeneous reaction of ozone with sea salt in the presence of ferric ion. *Geophys. Res. Lett.* **2001**, *28*, 4433–4436.

30. Knipping, E.M. Experiments and simulations of ion-enhanced interfacial chemistry on aqueous NaCl aerosols. *Science* **2000**, *288*, 301–306.

31. Behnke, W.; Zetzsch, C. Heterogeneous formation of chlorine atoms from various aerosols in the presence of O_3 and HCl. *J. Aerosol Sci.* **1989**, *20*, 1167–1170.

32. Keene, W.C.; Pszenny, A.A.P.; Jacob, D.J.; Duce, R.A.; Galloway, J.N.; Schultz-Tokos, J.J.; Sievering, H.; Boatman, J.F. The geochemical cycling of reactive chlorine through the marine troposphere. *Glob. Biogeochem. Cycles* **1990**, *4*, 407–430.

33. Knipping, E.M. Modeling Cl_2 formation from aqueous NaCl particles: Evidence for interfacial reactions and importance of Cl_2 decomposition in alkaline solution. *J. Geophys. Res.* **2002**, *107*.

34. George, I.J.; Abbatt, J.P.D. Heterogeneous oxidation of atmospheric aerosol particles by gas-phase radicals. *Nat. Chem.* **2010**, *2*, 713–722.

35. Kercher, J.P.; Riedel, T.P.; Thornton, J.A. Chlorine activation by N_2O_5: Simultaneous, in situ detection of $ClNO_2$ and N_2O_5 by chemical ionization mass spectrometry. *Atmos. Measu. Tech.* **2009**, *2*, 193–204.

36. Thornton, J.A.; Abbatt, J.P.D. N_2O_5 reaction on submicron sea salt aerosol: Kinetics, products, and the effect of surface active organics. *J. Phys. Chem. A* **2005**, *109*, 10004–10012.

37. Osthoff, H.D.; Roberts, J.M.; Ravishankara, a. R.; Williams, E.J.; Lerner, B.M.; Sommariva, R.; Bates, T.S.; Coffman, D.; Quinn, P.K.; Dibb, J.E.; *et al.* High levels of nitryl chloride in the polluted subtropical marine boundary layer. *Nat. Geosci.* **2008**, *1*, 324–328.

38. Roberts, J.M.; Osthoff, H.D.; Brown, S.S.; Ravishankara, A.R.; Coffman, D.; Quinn, P.; Bates, T. Laboratory studies of products of N_2O_5 uptake on Cl-containing substrates. *Geophys. Res. Lett.* **2009**, *36*.

39. Sarwar, G.; Simon, H.; Bhave, P.; Yarwood, G. Examining the impact of heterogeneous nitryl chloride production on air quality across the United States. *Atmos. Chem. Phys.* **2012**, *12*, 6455–6473.

40. Simon, H.; Kimura, Y.; McGaughey, G.; Allen, D.T.; Brown, S.S.; Coffman, D.; Dibb, J.; Osthoff, H.D.; Quinn, P.; Roberts, J.M. Modeling heterogeneous $ClNO_2$ formation, chloride availability, and chlorine cycling in Southeast Texas. *Atmos. Environ.* **2010**, *44*, 5476–5488.

41. Simon, H.; Kimura, Y.; McGaughey, G.; Allen, D.T.; Brown, S.S.; Osthoff, H.D.; Roberts, J.M.; Byun, D.; Lee, D. Modeling the impact of $ClNO_2$ on ozone formation in the Houston area. *J. Geophys. Res.* **2009**, *114*.

42. Sarwar, G.; Simon, H.; Xing, J.; Mathur, R. Importance of tropospheric $ClNO_2$ chemistry across the Northern Hemisphere. *Geophys. Res. Lett.* **2014**, *41*, 4050–4058.

43. Roberts, J.M.; Osthoff, H.D.; Brown, S.S.; Ravishankara, A.R. N_2O_5 oxidizes chloride to Cl_2 in acidic atmospheric aerosol. *Science* **2008**, *321*, 1059–1059.

44. Thomas, J.L.; Jimenez-Aranda, A.; Finlayson-Pitts, B.J.; Dabdub, D. Gas-phase molecular halogen formation from NaCl and NaBr aerosols: when are interface reactions important? *J. Phys. Chem. A* **2006**, *110*, 1859–1867.

175

45. Phillips, G.J.; Tang, M.J.; Thieser, J.; Brickwedde, B.; Schuster, G.; Bohn, B.; Lelieveld, J.; Crowley, J.N. Significant concentrations of nitryl chloride observed in rural continental Europe associated with the influence of sea salt chloride and anthropogenic emissions. *Geophys. Res. Lett.* **2012**, *39*.

46. Thornton, J.A.; Kercher, J.P.; Riedel, T.P.; Wagner, N.L.; Cozic, J.; Holloway, J.S.; Dubé, W.P.; Wolfe, G.M.; Quinn, P.K.; Middlebrook, A.M.; *et al.* A large atomic chlorine source inferred from mid-continental reactive nitrogen chemistry. *Nature* **2010**, *464*, 271–274.

47. Mielke, L.H.; Furgeson, A.; Osthoff, H.D. Observation of $ClNO_2$ in a mid-continental urban environment. *Environ. Sci. Technol.* **2011**, *45*, 8889–8896.

48. Behnke, W.; George, C.; Scheer, V.; Zetzsch, C. Production and decay of $ClNO_2$ from the reaction of gaseous N_2O_5 with NaCl solution: Bulk and aerosol experiments. *J. Geophys. Res.* **1997**, *102*, 3795–3804.

49. Behnke, W.; Krüger, H.-U.; Scheer, V.; Zetzsch, C. Formation of atomic Cl from sea spray via photolysis of nitryl chloride: Determination of the sticking coefficient of N_2O_5 on NaCl aerosol. *J. Aerosol Sci.* **1991**, *22*, S609–S612.

50. Frenzel, A.; Scheer, V.; Sikorski, R.; George, C.; Behnke, W.; Zetzsch, C. Heterogeneous interconversion reactions of $BrNO_2$, $ClNO_2$, Br_2, and Cl_2. *J. Phys. Chem. A* **1998**, *102*, 1329–1337.

51. Schweitzer, F.; Mirabel, P.; George, C. Multiphase chemistry of N_2O_5, $ClNO_2$, and $BrNO_2$. *J. Phys. Chem. A* **1998**, *102*, 3942–3952.

52. Bertram, T.H.; Thornton, J.A. Toward a general parameterization of N_2O_5 reactivity on aqueous particles: The competing effects of particle liquid water, nitrate and chloride. *Atmos. Chem. Phys. Discuss.* **2009**, *9*, 15181–15214.

53. Mielke, L.H.; Stutz, J.; Tsai, C.; Hurlock, S.C.; Roberts, J.M.; Veres, P.R.; Froyd, K.D.; Hayes, P.L.; Cubison, M.J.; Jimenez, J.L.; *et al.* Heterogeneous formation of nitryl chloride and its role as a nocturnal NO_x reservoir species during CalNex-LA 2010. *J. Geophys. Res. Atmos.* **2013**, *118*, 10638–10652.

54. Tham, Y.J.; Yan, C.; Xue, L.; Zha, Q.; Wang, X.; Wang, T. Presence of high nitryl chloride in Asian coastal environment and its impact on atmospheric photochemistry. *Chin. Sci. Bull.* **2013**, *59*, 356–359.

55. Riedel, T.P.; Wagner, N.L.; Dubé, W.P.; Middlebrook, A.M.; Young, C.J.; Öztürk, F.; Bahreini, R.; VandenBoer, T.C.; Wolfe, D.E.; Williams, E.J.; *et al.* Chlorine activation within urban or power plant plumes: Vertically resolved $ClNO_2$ and Cl_2 measurements from a tall tower in a polluted continental setting. *J. Geophys. Res. Atmos.* **2013**, *118*, 8702–8715.

56. National Aeronautics and Space Administration. Deriving Information on Surface Conditions from Column and Vertically Resolved Observations Relevant to Air Quality (DISCOVER-AQ). Available online: http://discover-aq.larc.nasa.gov/ (accessed on 31 August 2015).

57. Ng, N.L.; Herndon, S.C.; Trimborn, a.; Canagaratna, M.R.; Croteau, P.L.; Onasch, T.B.; Sueper, D.; Worsnop, D.R.; Zhang, Q.; Sun, Y.L.; *et al.* An Aerosol Chemical Speciation Monitor (ACSM) for routine monitoring of the composition and mass concentrations of ambient aerosol. *Aerosol Sci. Technol.* **2011**, *45*, 780–794.

58. Finley, B.D.; Saltzman, E.S. Observations of Cl_2, Br_2, and I_2 in coastal marine air. *J. Geophys. Res.* **2008**, *113*, 1–14.

59. Jordan, C.E.; Pszenny, A.A.P.; Keene, W.C.; Cooper, O.R.; Deegan, B.; Maben, J.; Routhier, M.; Sander, R.; Young, A.H. Origins of aerosol chlorine during winter over north central Colorado, USA. *J. Geophys. Res. Atmos.* **2015**, *120*, 678–694.

60. Young, A.H.; Keene, W.C.; Pszenny, A.A.P.; Sander, R.; Thornton, J.A.; Riedel, T.P.; Maben, J.R. Phase partitioning of soluble trace gases with size-resolved aerosols in near-surface continental air over northern Colorado, USA, during winter. *J. Geophys. Res. Atmos.* **2013**, *118*, 9414–9427.

61. Pratt, K.A.; Prather, K.A. Aircraft measurements of vertical profiles of aerosol mixing states. *J. Geophys. Res. Atmos.* **2010**, *115*, 1–10.

62. Hildebrandt Ruiz, L.; Yarwood, G.; Koo, B.; Heo, G. Quality assurance project plan project 14-024—Sources of organic particulate matter in Houston: Evidence from DISCOVER-AQ data modeling and experiments. Available online: http://aqrp. ceer.utexas.edu/projectinfoFY14_15%5C14-024%5C14-024%20QAPP.pdf (accessed on 31 August 2015).

63. Draxler, R.R.; Hess, G. Description of the HYSPLIT_4 Modeling System. Available online: https://ready.arl.noaa.gov/HYSPLIT.php; 2010 (accessed on 31 August 2015).

64. Yarwood, G.; Jung, J.; Whitten, G.Z.; Heo, G.; Mellberg, J.; Estes, M. Updates to the Carbon Bond mechanism for version 6 (CB6). In Proceedings of the 9th Annual CMAS Conference, Chapel Hill, NC, USA, 11–13 October 2010; pp. 1–4.

65. Yarwood, G.; Rao, S. *Updates to the Carbon Bond Chemical Mechanism: CB05*; RT-0400675; Final report to the US EPA; US EPA: Washington, DC, USA, 2005.

66. Tanaka, P.L.; Allen, D.T.; McDonald-Buller, E.C.; Chang, S.; Kimura, Y.; Mullins, C.B.; Yarwood, G.; Neece, J.D. Development of a chlorine mechanism for use in the carbon bond IV chemistry model. *J. Geophys. Res.* **2003**, *108*.

67. Carter, W.P.L.; Luo, D.; Malkina, I.L.; Pierce, J.A. Environmental Chamber Studies of Atmospheric Reactivities of Volatile Organic Compounds: Effects of Varying Chamber and Light Source. 1995. Availavle online: https://www.cert.ucr.edu/~carter/pubs/explrept.pdf (accessed on 31 August 2015).

68. Sokolov, O.; Hurley, M.D.; Wallington, T.J.; Kaiser, E.W.; Platz, J.; Nielsen, O.J.; Berho, F.; Rayez, M.T.; Lesclaux, R. Kinetics and mechanism of the gas-phase reaction of Cl atoms with benzene. *J. Phys. Chem. A* **1998**, *102*, 10671–10681.

69. Wingenter, O.W.; Blake, D.R.; Blake, N.J.; Sive, B.C.; Rowland, F.S.; Atlas, E.; Flocke, F. Tropospheric hydroxyl and atomic chlorine concentrations, and mixing timescales determined from hydrocarbon and halocarbon measurements made over the Southern Ocean. *J. Geophys. Res.* **1999**, *104*, 21,819–21,828.

70. Bertram, T.H.; Kimmel, J.R.; Crisp, T.A.; Ryder, O.S.; Yatavelli, R.L.N.; Thornton, J.A.; Cubison, M.J.; Gonin, M.; Worsnop, D.R. A field-deployable, chemical ionization time-of-flight mass spectrometer. *Atmos. Meas. Tech.* **2011**, *4*, 1471–1479.

71. Yatavelli, R.L.N.; Lopez-Hilfiker, F.; Wargo, J.D.; Kimmel, J.R.; Cubison, M.J.; Bertram, T.H.; Jimenez, J.L.; Gonin, M.; Worsnop, D.R.; Thornton, J.A. A chemical ionization high-resolution time-of-flight mass spectrometer coupled to a Micro Orifice Volatilization Impactor (MOVI-HRToF-CIMS) for analysis of gas and particle-phase organic species. *Aerosol Sci. Technol.* **2012**, *46*, 1313–1327.

72. Lee, B.H.; Lopez-Hilfiker, F.D.; Mohr, C.; Kurtén, T.; Worsnop, D.R.; Thornton, J.A. An iodide-adduct high-resolution time-of-flight chemical-ionization mass spectrometer: Application to atmospheric inorganic and organic compounds. *Environ. Sci. Technol.* **2014**, *48*, 6309–6317.

73. Aljawhary, D.; Lee, A.K.Y.; Abbatt, J.P.D. High-resolution chemical ionization mass spectrometry (ToF-CIMS): Application to study SOA composition and processing. *Atmos. Meas. Tech.* **2013**, *6*, 3211–3224.

74. Slusher, D.L.; Huey, L.G.; Tanner, D.J.; Flocke, F.M.; Roberts, J.M. A Thermal Dissociation-Chemical Ionization Mass Spectrometry (TD-CIMS) technique for the simultaneous measurement of peroxyacyl nitrates and dinitrogen pentoxide. *J. Geophys. Res.* **2004**, *109*.

75. Zheng, W.; Flocke, F.M.; Tyndall, G.S.; Swanson, A.; Orlando, J.J.; Roberts, J.M.; Huey, L.G.; Tanner, D.J. Characterization of a thermal decomposition chemical ionization mass spectrometer for the measurement of Peroxy Acyl Nitrates (PANs) in the atmosphere. *Atmos. Chem. Phys.* **2011**, *11*, 6529–6547.

76. Fuchs, H.; Dube, W.P.; Ciciora, S.J.; Brown, S.S. Determination of inlet transmission and conversion efficiencies for in situ measurements of the nocturnal nitrogen oxides, NO_3, N_2O_5 and NO_2, via pulsed cavity ring-down spectroscopy. *Anal. Chem.* **2008**, *80*, 6010–6017.

CCN Properties of Organic Aerosol Collected Below and within Marine Stratocumulus Clouds near Monterey, California

Akua Asa-Awuku, Armin Sorooshian, Richard C. Flagan, John H. Seinfeld and Athanasios Nenes

Abstract: The composition of aerosol from cloud droplets differs from that below cloud. Its implications for the Cloud Condensation Nuclei (CCN) activity are the focus of this study. Water-soluble organic matter from below cloud, and cloud droplet residuals off the coast of Monterey, California were collected; offline chemical composition, CCN activity and surface tension measurements coupled with Köhler Theory Analysis are used to infer the molar volume and surfactant characteristics of organics in both samples. Based on the surface tension depression of the samples, it is unlikely that the aerosol contains strong surfactants. The activation kinetics for all samples examined are consistent with rapid $(NH_4)_2SO_4$ calibration aerosol. This is consistent with our current understanding of droplet kinetics for ambient CCN. However, the carbonaceous material in cloud drop residuals is far more hygroscopic than in sub-cloud aerosol, suggestive of the impact of cloud chemistry on the hygroscopic properties of organic matter.

Reprinted from *Atmosphere*. Cite as: Asa-Awuku, A.; Sorooshian, A.; Flagan, R.C.; Seinfeld, J.H.; Nenes, A. CCN Properties of Organic Aerosol Collected Below and within Marine Stratocumulus Clouds near Monterey, California. *Atmosphere* **2015**, *6*, 1590–1607.

1. Introduction

It is well established that organic compounds (especially water-soluble organic compounds, WSOC) are ubiquitous in marine aerosol; they can interact with water and affect aerosol hygroscopicity, droplet surface tension, and Cloud Condensation Nuclei (CCN) activity [1–6]. As a result, marine aerosol organic matter can affect cloud droplet number concentration as much as 15% [7–10] and may exert a climatically important impact on clouds.

Organic compounds, depending on their source, are classified as "primary" and "secondary". Primary organic marine aerosol (POMA) can include high molecular-weight compounds transferred onto sea-salt aerosol from the surfactant-rich surface ocean during the bubble-bursting process [9]. The presence of POMA is mostly attributed to biological activity, and its concentration varies with

179

season [9,11,12]. POMA can exhibit low hygroscopic growth factors but maintain high CCN activity [13] or *vice versa* [14]. Secondary organic marine aerosol (SOMA) can be produced during cloud processing [15–19]; perhaps the most studied chemical pathway for SOMA is glyoxylic acid oxidation [20] via several aqueous phase intermediates [21–23]. Continental biogenic emissions can also contribute to organic mass in marine clouds at high altitudes [24]. Anthropogenic emissions can substantially contribute to marine organics; for example, particulate emissions from ships are composed roughly of up to 10% carbon [25,26], in the form of sparingly soluble poly-aromatic hydrocarbons, ketones and quinones (PAHs, PAKs, and PAQs, respectively [13,27]). "Ship tracks" are a natural laboratory for studying aerosol indirect effects, as clouds with uniform dynamics, are exposed to a strong gradient in emissions concentrations and often exhibit droplet number, effective radius and drizzle rates responses that are consistent with local emission rates [28–32].

In this study, marine aerosol samples influenced by ship emissions are collected in-situ (in and out of cloudy regions) and studied for their cloud-droplet formation properties. Given the complexity of the water-soluble organic fraction, characterization is done by measuring the size-resolved CCN activity of the material, surface tension depression and using Köhler Theory Analysis (KTA) [33–35] to infer the thermodynamic properties (average molar volume, surfactant tension depression) of the organic fraction and its potential impact on droplet growth rate kinetics. These properties are then related to the influence of primary emissions and in-cloud oxidation processes on the CCN activity of the organic compounds.

2. Experimental Methods

2.1. Aerosol Sampling and Chemical Composition

The samples analyzed in this study were obtained during the Marine Stratus/Stratocumulus Experiments (MASE) that took place near the coast of Monterey, California, from July to August 2005. The airborne platform used the Center for Interdisciplinary Remotely-Piloted Aircraft Studies (CIRPAS) Twin Otter that sampled boundary layer air over 13 flights. A full description of the aerosol and cloud instrument payload aboard the plane during the MASE campaign can be found in Lu *et al.* [36]. Six of thirteen flights encountered strong, localized perturbations in aerosol concentration, size, and composition consistent with ship emissions. The data presented here were sampled on 13 July in the vicinity of ship tracks. An overview of airborne studies conducted in this region has been summarized by Coggon *et al.* [18].

Two sample types were collected aboard the aircraft in this study: cloud droplet residuals (CR) from evaporated cloud droplets, and sub-cloud (SC) aerosol sampled below cloud (and occasionally in clear-sky conditions). CR samples were collected with a counter-flow virtual impactor (CVI) [37,38], in which cloud droplets with diameter greater than 5 μm were inertially separated from interstitial (unactivated) aerosol and evaporated before collection. Analysis of cloud droplet residuals with this approach has been instrumental in understanding the origin of CCN in ambient clouds [18].

Table 1. Concentrations of Water Soluble Organic Carbon (WSOC) and ions (mg·L^{-1} sample), and, α (mN·m^{-1}·K^{-1}) and β (L·mg^{-1}) parameters of the Szyszkowski–Langmuir surface tension model * [45] fits for the aerosol samples analyzed.

Property	Cloud Residuals (CR)	Sub-Cloud (SC)
WSOC	220 ± 14	202 ± 7
Ca^{2+}	3.10	2.75
Mg^{2+}	0.07	0.58
Na^+	21.07	20.90
Cl^-	25.30	22.61
NH_4^+	21.00	36.38
NO_3^-	16.12	17.30
SO_4^{2-}	35.30	87.85
Oxalate	3.61	4.22
α	8.99×10^{-4}	2.91×10^{-3}
β	3.84×10^{-2}	1.63×10^{-2}

* Szyszkowski–Langmuir Equation (1): $\sigma = \sigma_w - \alpha T \ln(1 + \beta c)$.

A Brechtel Manufacturing Inc. Particle-Into-Liquid sampler (PILS; [20]) was used to collect the water-soluble fraction of SC and CR by exposing the aerosol to supersaturated steam and growing them into droplets that are subsequently collected by inertial impaction. The liquid stream was then collected in vials over 3.5 to 5 min each. The chemical composition of the ionic species in each sample was measured with a dual Ion Chromatography (IC) system (ICS-2000 with 25 μL sample loop, Dionex Inc.); the IC detection limit of aerosol species is less than 0.1 μg·m^{-3} for inorganic ions (Na^+, NH_4^+, K^+, Mg^{2+}, Ca^{2+}, Cl^-, NO_3^-, NO_2^-, and SO_4^{2-}) and less than 0.01 μg·m^{-3} air for the organic acid ions (dicarboxylic acids C_2–C_9, acetic, formic, pyruvic, glyoxylic, maleic, malic, methacrylic, benzoic, and methanesulfonic acids). The WSOC content was also measured off-line with a Total Organic Carbon (TOC) Analyzer (Sievers Model 800 Turbo, Boulder, CO) (Table 1). The contents of the vials were subsequently analyzed for their CCN activity and surfactant characteristics (Sections 2.2–2.4). The aerosol samples analyzed in the study were obtained on 13 July (when the highest organic acid concentrations were reported) from within and below cloud (cloud base and cloud top were measured at 101 and 450 m altitude, respectively) [36]. During this flight, the aircraft focused on sampling air in the vicinity of ship tracks. Backward trajectories calculated using the NOAA HYSPLIT (HYbrid Single-Particle Lagrangian Integrated Trajectory) model (http://www.arl.noaa.gov/ready/hysplit4.html) indicate that the air mass on the 13 July originated from the North Pacific; the vertical profile indicates that the aerosol masses sampled on 13 July originated from the free troposphere before descending into the marine boundary layer (Figure 1).

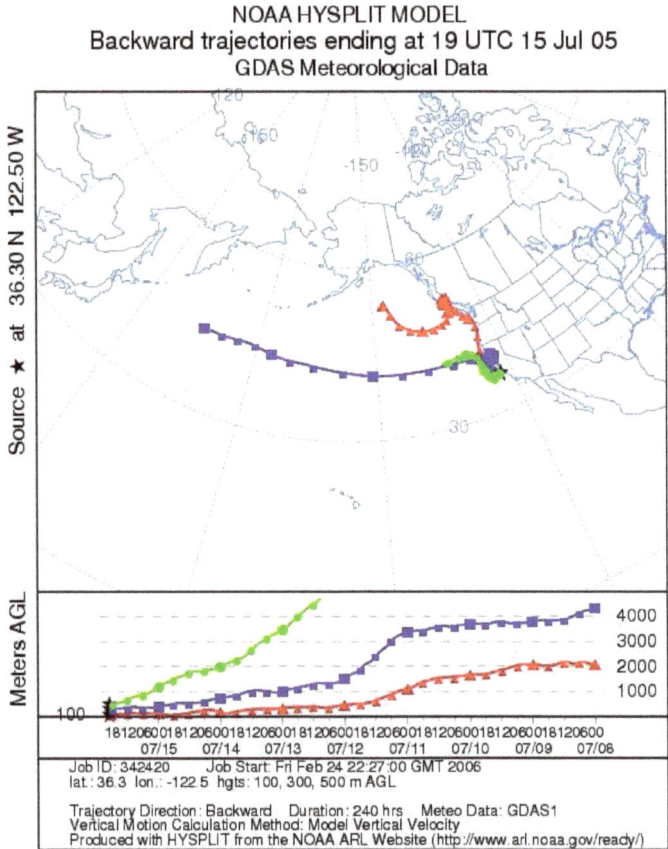

Figure 1. NOAA HYSPLIT (HYbrid Single-Particle Lagrangian Integrated Trajectory) Model Back Trajectories for air masses sampled aboard the Twin Otter on 13 July 2005.

2.2. CCN Activity of Soluble Material Collected

The aqueous contents of the PILS vials were atomized with a Collison-type atomizer (Figure 2) operated at 5 psig pressure. The atomized aerosol was then dried through two silica gel diffusional dryers, charged by a Kr-85 bipolar charger, and classified with a Differential Mobility Analyzer (DMA 3081) (Figure 2). The classified monodisperse aerosol was then split and passed through a TSI 3025A Condensation Particle Counter (CPC) to measure aerosol number concentration (CN); the other stream was sampled by a Droplet Measurement Technologies Continuous-Flow Streamwise Thermal Gradient CCN Counter (CFSTGC) [39–41]. Given the limited amount of sample, size-resolved CCN activity and growth kinetics measurements were obtained using scanning mobility CCN Analysis (SMCA) [42]. The SMCA process couples CFSTGC measurements with a scanning mobility particle sizer (SMPS). An inversion procedure was used to compute the ratio of CCN to CN as a function of aerosol size as the SMPS scans from 10 and 250 nm dry mobility diameter for a fixed supersaturation, s. The data were fit to a sigmoidal function, corrected for diffusion and multiple charges in the DMA [42]; the particle dry diameter size, d, for which 50% of the particles activated into droplets, represents the dry diameter of the particle with critical supersaturation, s_c, equal to the instrument supersaturation. The activation experiments were repeated (a minimum of four times) for each s level, which varied from 0.2% to 1.2%. The compositional data, aerosol surfactant behavior, and the dependence of d with respect to s_c, were used to infer the molar volume and surfactant characteristics of the organic fraction with Köhler Theory Analysis [33–35,43] (Section 3). The CFSTGC was calibrated using $(NH_4)_2SO_4$ (density = 1.77 g·cm^{-3}, and molar mass of 132 g·mol^{-1}) generated with the same experimental setup, and operating the DMA with a sheath:aerosol flow-rate ratio of 10:1; d for ammonium sulfate was then related to critical supersaturation by applying classical Köhler theory, using an effective van't Hoff factor of 2.5 [41,44]. The constant van't Hoff value is an approximation that can be improved with the use of the Pitzer Method. Multiple calibrations of the instrument were performed over the period of measurements, and each supersaturation was within 10% of the average.

2.3. Surface Tension Measurements

A CAM 200 pendant drop goniometer was used to measure bulk surface tension, σ, of the original samples and prescribed dilutions of them, following the approach of [33,35]. The instrument uses 5–6 mL of sample to form a drop at the end of a needle. The optical goniometer captures ~100 images of the droplet and computes droplet surface tension through application of the Young–Laplace equation; the standard deviations for $\sigma_{s/a}$ are <0.05 mN·m^{-1} for a given sample at one concentration. The measurements were then fit to the Szyskowski–Langmuir equation [45]:

$$\sigma = \sigma_w - \alpha T \ln(1 + \beta c) \tag{1}$$

where α, β are optimally fitted constants, T is temperature, σ_w is the surface tension of pure water and c is the dissolved carbon concentration (mg·C·L^{-1}). Each pendant drop was suspended 60 s before surface tension was measured, in order to allow organics in the bulk to equilibrate with the droplet surface layer [46]. Table 1 provides a summary of the α and β parameters for all samples considered. Bulk measurements are known to not represent the droplet activation regime. Thus parameters derived from bulk measurements (Table 1) are applied to inferred droplet concentrations at activation [33–35,47].

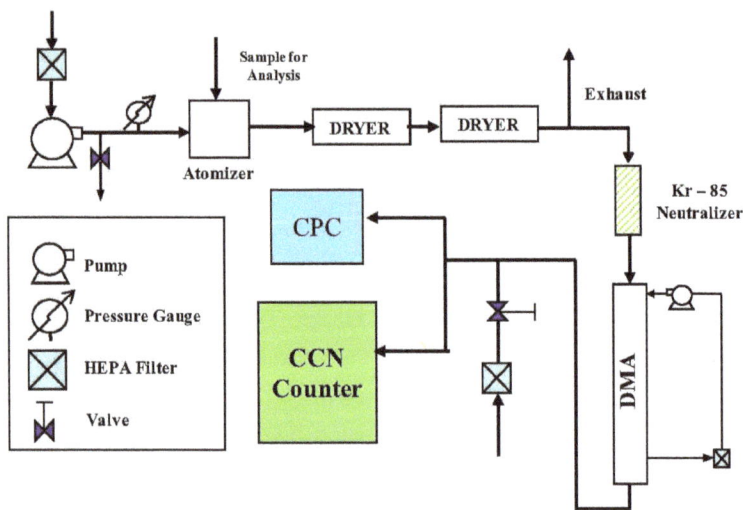

Figure 2. Experimental setup for characterizing size-resolved Cloud Condensation Nuclei activity of samples analyzed in this study.

2.4. Droplet Size Measurements of Activated CCN

The optical particle counter used for detection of CCN concentrations also provides the size of the activated droplets; thus the SMCA procedure can determine the size of activated CCN with known dry diameter at the exit of the CFSTGC. We adopt the method of Threshold Droplet Growth Analysis (TDGA) (e.g., but not limited to [39,48,49]) to detect the possible presence of droplets growing more slowly than calibration $(NH_4)_2SO_4$ aerosol [50]. If present then, a coupled measurement modeling approach can be used to infer the effective water vapor uptake coefficient [14,51]. TDGA compares the D_p of CCN with s_c equal to the instrument saturation (*i.e.*, CCN with a dry diameter equal to the cutoff diameter, d) against the wet diameter, $D_{p,(NH4)2SO4}$ of CCN with identical s_c, but composed of $(NH_4)_2SO_4$. If the presence of organics does not delay droplet growth, $D_p \sim D_{p,(NH4)2SO4}$ (e.g., [34,52]). Supersaturation depression is observed at high CCN concentration and can affect droplet size [53]. Thus both CCN measurement and analysis should account for CCN concentrations.

3. Analytical Theory

3.1. Köhler Theory

Köhler Theory Analysis (KTA) [33] can be used to infer the average molar volume (molecular weight, M_j, over density ρ_j) of the organic fraction, j, of CCN. It has been shown to work well for low molecular weight species, such as those presented here. Using measurements of s_c *versus* d to determine the Fitted CCN Activity parameter (FCA), $\omega = s_c d^{-1.5}$, KTA infers the average molar volume of the WSOC, $\frac{M_j}{\rho_j}$ [33,43],

$$\frac{M_j}{\rho_j} = \frac{\varepsilon_j v_j}{\frac{256}{27} \left(\frac{M_w}{\rho_w}\right)^2 \left(\frac{1}{RT}\right)^3 \sigma^3 \omega^{-2} - \sum_{i \neq j} \frac{\rho_i}{M_i} \varepsilon_i v_i} \tag{2}$$

where M_w, ρ_w are the molecular weight and density of water, respectively, R is the universal gas constant, T is the ambient temperature, σ is the droplet surface tension at the point of activation, ε is the volume fraction, and v is the effective van't Hoff factor. Subscripts I and j refer to the inorganic and organic components, respectively. ε_k is related to the mass fraction, m_k, of solute k (k being either of i or j) as:

$$\varepsilon_k = \frac{m_k/\rho_k}{\sum_{i \neq j} m_i/\rho_i + m_j/\rho_j} \tag{3}$$

Two measures of molar volume uncertainty are used: (*i*) the standard deviation of all the $\frac{M_j}{\rho_j}$ inferences (Equation (2)), and, (*ii*) estimates determined from, $\Delta\frac{M_j}{\rho_j} = \sqrt{\sum_x (\Phi_x \Delta x)^2}$, where Δx is the uncertainty of each of the measured parameters x, (*i.e.*, any of σ, ω, ε_i, ε_j, v_i, and v_j) and Φ_x is the sensitivity of molar volume to x, $\Phi_x = \frac{\partial}{\partial x}\left(\frac{M_o}{\rho_o}\right)$, using the formulas of [33,34,43]. The maximum of both estimates is the reported uncertainty of $\frac{M_j}{\rho_j}$. KTA has been shown to constrain molecular weight estimates for laboratory aerosol (having organic mass fraction between 20 and 50%) with an average error of 20% [33], 40% for complex biomass burning aerosol [43], 30% for secondary organic matter [35], and to within 25% for primary marine organic matter [34].

3.2. Inferring Surface Tension

The low concentration of WSOC in the PILS samples limits the determination (using direct measurements) of their surface tension depression for CCN-relevant concentrations. However, if CCN activity data are available for mixtures of WSOC and a salt (e.g., $(NH_4)_2SO_4$), KTA can be used to concurrently infer $\frac{M_j}{\rho_j}$ and σ using

185

an iterative procedure [34]. If enough salt is present in the sample, the contribution of organic solute to s_c is small, and an iterative procedure is not required; the effect of the organic on CCN activity amounts to its impact on surface tension, and can then be inferred as [35],

$$\sigma = \sigma_w \left(\frac{s_c}{s_c^*} \right)^{\frac{2}{3}}$$

(4)

where s_c is the measured critical supersaturation, and s_c^* is the predicted value (from Köhler theory). Assuming $\sigma = \sigma_w$, the surface tension of pure water computed at the average CFSTGC column temperature [35], then

$$s_c^* = \frac{2}{3} \left(\frac{4 M_w \sigma_w}{RT\rho_w} \right)^{\frac{3}{2}} \left(\frac{M_w d^3}{\rho_w} \sum_i \frac{\varepsilon_i \nu_i \rho_i}{M_i} \right)^{\frac{-1}{2}}$$

(5)

where "i" denotes all inorganic solutes present in the aerosol. Each surface tension inference is then related to the WSOC concentration at the critical diameter (Equation (6) of [34], and fit to the Szyskowski–Langmuir adsorption isotherm (Equation (1)). The partitioning of the surfactant from the bulk to the droplet monolayer should be considered. This method of inferring surface tension depression has been shown to work well for dissolved organic matter isolated from seawater [34]. Thus in the case of marine POMA, where the surfactant may partition mostly to the surface, partitioning effects are less important than bulk properties and why Moore *et al.* [34] were able to achieve good closure.

4. Results and Discussion

4.1. Surface Tension

For the low WSOC concentrations measured in the PILS samples, organics have minimal effect on surface tension (Figure 3). Hence, Equation (4) is used to infer σ at carbon concentrations relevant for CCN activation (roughly 1000 mg·C·L^{-1}). WSOC from biomass burning aerosol and marine aerosol have been shown to contain strong surfactants that depress surface tension by 25%–42% [43,47,54,55] at similar concentration. The inferred σ values for both SC and CR aerosol (Figure 3) exhibit weak surface tension depression ($-\Delta\sigma/\sigma \simeq 5\%$) at concentrations >1000 mg·C·L^{-1}. The surface tension depression results are similar to results from dissolved organic marine matter [34]. Both CR and SC samples were influenced by ship emissions, SOMA, and POMA and contained soluble organics. The solubility of the organics is unknown, however the potential effects of limited solubility on CCN activity for SC samples may be outweighed by significant depression in surface tension at the droplet surface (Figure 3). Finding the average organic molar mass from KTA

186

will help constrain if the SC and CR samples indeed have different organic aerosol compositions that affect aerosol solubility and surfactant properties.

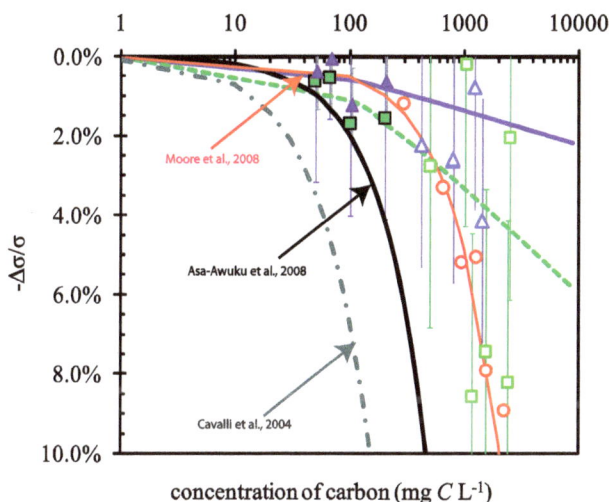

Figure 3. Surface tension depression as a function of dissolved carbon concentration. Measurements (closed symbols) and Inferred values (open symbols) of the CR (blue triangles) and SC aerosol (green squares) are shown as data points. For comparison, curves are added with data from freshly emitted biomass burning aerosol (solid black; [43], marine organic aerosol (grey dash-dot; [54]) and dissolved marine organic matter (solid red; [34]).

4.2. CCN Activity

Figures 4 and 5 show the critical supersaturation, s_c, as a function of dry diameter, d, for aerosol atomized from both CR and SC samples. The data for $(NH_4)_2SO_4$ aerosol have been added for comparison. Data for $(NH_4)_2SO_4$ are consistent with the expectation that the aerosol is highly CCN active, lies to the left of both CR and SC data sets, and behaves as classical Kohler CCN without surface tension depression. The activation slope of particles composed of soluble and insoluble material should exhibit a s_c that scales with $d^{-1.5}$ (e.g., $(NH_4)_2SO_4$ data in Figure 4). At low supersaturations ($s_c \leq 0.6\%$), the WSOC concentration at the point of activation was low (<1000 mg$\cdot$C$\cdot$L^{-1}) and surface tension depression was small (Figure 3), hence $s_c \sim d^{-1.5}$ for both samples, and KTA is applied for this region of the CCN spectrum. The CR sample contained material that was less hygroscopic than sulfate, but more hygroscopic than in the SC sample (*i.e.*, for a given s, the d of the CR is greater than SC, Figure 4). SC and CR activation curves converge at high s, likely because the WSOC concentration is high enough to notably decrease surface depression, more for organics in SC than for CR (Figure 3). The difference in CCN

187

activity is consistent with studies to date (e.g., [16,56]), showing that hygroscopic components tend to be incorporated in cloud droplets where their less hygroscopic counterparts prefer to remain in the interstitial air. It is also noted that the CR samples excluded droplets <5 μm diameter thus the excluded smaller droplets may potentially contain less hygroscopic materials.

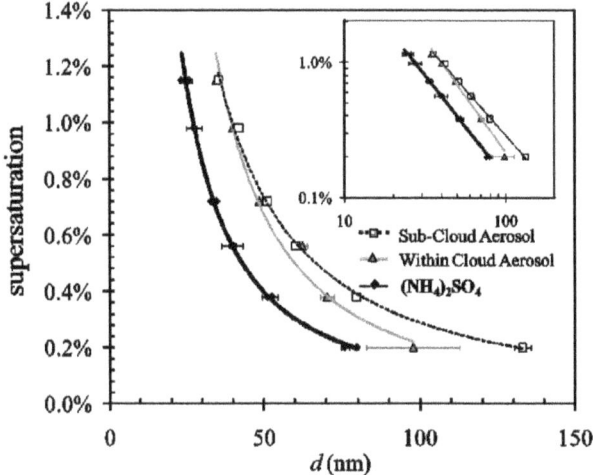

Figure 4. CCN activity of SC and CR samples. $(NH_4)_2SO_4$ (solid black line) is added for comparison. The CCN activity curve exponents for $(NH_4)_2SO_4$, CR, and SC are -1.51, -1.66, and -1.35, respectively.

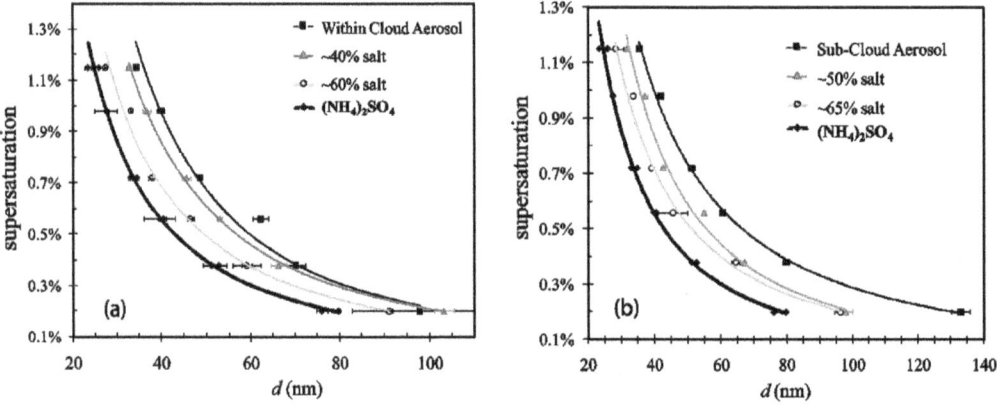

Figure 5. CCN activity of MASE samples with the addition of $(NH_4)_2SO_4$. Salt contents are expressed in terms of mass fraction. **(a)** Cloud Residuals and **(b)** Sub-Cloud samples.

4.3. Inferred Molar Volumes and Uncertainties

Application of KTA requires the input of the organic mass concentration, $m_{organic}$, which is obtained by dividing the WSOC carbon concentration (Table 1) by the carbon-to-organic mass ratio, C/OC. In this study, $C/OC \approx 0.27$ is applied, the value for oxalic acid ($C_2H_6O_4$, 90 g·moL^{-1}), the most abundant organic acid measured in our samples (Table 1). $C/OC \approx 0.27$ is very close to 0.29, the value for dissolved primary organic marine matter and marine coarse mode aerosol [34,54]. Here SC expect it to apply to the whole sample. The van't Hoff factor for the organic fraction is assumed to be unity. The presence of organics from ship emissions are assumed to introduce little variability to the organic mass fractions estimated. The ionic composition taken from the chemical analysis of the PILS vials is then converted into a mixture of inorganic salts and organics with the ISORROPIA II aerosol thermodynamic equilibrium model [57]. Lu *et al.* [36] find the composition of aerosol sampled to be bimodal, composed of ammonium bisulfate and sea-salt; when samples are mixed in the PILS, chloride depletion is both expected and observed in the samples (Table 2). The results of the KTA analysis for each sample are summarized in Tables 3 and 4. Assuming an organic density of 1.4 g·cm^{-3} [35,58], the SC sample has a high inferred molecular weight (M_j = 2413 ± 536 g·mol^{-1}; Table 4), possibly long-chained aliphatic compounds within primary organic matter transferred to the aerosol from bubble bursting at the seawater surface [16,34,54,59,60]. The marine nature of the SC sample is further supported by its inferred surface tension depression (Figure 3, which is remarkably consistent with dissolved organic matter [34]. The average molecular weight inferred for CR aerosol is substantially less (143 ± 25 g·mol^{-1}; Table 3), consistent with the presence of low molecular weight carboxylic acids (*i.e.*, C_2–C_9 mono and dicarboxylic acids) measured in cloud-processed marine aerosol [15]. Water-soluble oxidation products from ship VOC emissions can also contribute to the WSOC, although the information at hand is not sufficient to conclusively show this. As in previous KTA studies (e.g., [33]), inferred average molar volume is subject to an estimated 30% error; the greatest source of uncertainty stems from the FCA parameter (Tables 3 and 4). The low depression in surface tension (Figure 3) suggests that the molecular weight distribution in the WSOC may not contain compounds characteristic of HULIS (300–750 g·mol^{-1}) [43,47], but instead can be described by the superposition of two "modes", one from SOMA (low molecular weight compounds), and one from POMA (higher molecular weight compounds). For the SC sample, we assume that the SOMA is primarily composed of oxalic acid (90 g·mol^{-1}, the predominant compound identified in the PILS samples and constitutes 0.6% of the organic mass, Table 1), and the remaining mass (99.4%) is POMA characteristic of Redfield-type molecules, $(CHO_2)_{106}(NH_3)_{16}H_3PO_4$, 3444 g·mol^{-1} [61,62]. With this assumed two-component model composition, the average M_j in SC is equal to 2852 g·mol^{-1}, and consistent (to within uncertainty) with the inferred KTA value

(Table 4). Similarly, if most of the organic in the CR aerosol is a mixture of ship emissions and oxalate, and that the POMA mode is consistent with phenathrene (178 g·mol^{-1} and 99.6% of the organic mass), and SOMA with oxalic acid (90 g·mol^{-1} and 0.4% of the organic mass, Table 1), the average molar mass for the organic distribution yields 177 g·mol^{-1}, a value within 25% of our KTA estimate.

Table 2. Mass fraction (%) of constituents in the aerosol samples considered in this study. Composition of the inorganic fraction was obtained by an aerosol thermodynamic equilibrium model (ISORROPIA-II), using the ionic composition of Table 1.

	Cloud Residuals (CR)	Sub-Cloud (SC)
Organic	88.3	82
NH_4	3.4	3.6
NaCl	0.6	0
$(NH_4)_2SO_4$	0.0	5.1
$NaNO_3$	2.3	0.0
NH_4NO_3	0.0	2.4
Na_2SO_4	4.1	7.0
$CaSO_4$	1.1	0.0

Table 3. Köhler Theory Analysis for Cloud-Residual (CR) samples.

Property x (units)	Average Value of x	Uncertainty Δx	Sensitivity, Φ_x $(m^3 \cdot mol^{-1} \cdot x^{-1})$	$\frac{M_j}{\rho_j}$ Contribution (%)
σ (N·m^{-1})	6.89×10^{-2}	1.38×10^{-3}	3.83×10^{-3}	5.72
ω (m$^{1.5}$)	6.80×10^{-2}	5.53×10^{-15}	2.59×10^{9}	12.42
ν_{NH_4Cl}	2	0.5	6.43×10^{-6}	3.49
ν_{NaCl}	2	0.5	1.54×10^{-6}	0.83
ν_{NaNO_3}	2	0.5	4.18×10^{-6}	2.27
$\nu_{Na_2SO_4}$	3	-	-	-
$\nu_{organic}$	1	0.20	4.39×10^{-5}	9.52
ε_{NaNO_3}	0.05	-	-	-
$\varepsilon_{Na_2SO_4}$	0.07	-	-	-
ε_{NH_4Cl}	0.10	1.95×10^{-3}	8.50×10^{-4}	0.80
$\varepsilon_{organic}$	0.88	6.36×10^{-3}	4.10×10^{-3}	2.82
ρ_{NH_4Cl}	1.53	-	-	-
ρ_{NaCl}	2.16	-	-	-
ρ_{NaNO_3}	2.3	-	-	-
$\rho_{Na_2SO_4}$	2.68	-	-	-
$\rho_{organic}$	1.4	-	-	-
$\rho_{aerosol}$	1.59	-	-	-
M_J/ρ_J (cm^3·mol^{-1})	1.60×10^{-4}	17.5%	-	-
M_J (g·mol^{-1})	143	25	-	-

Table 4. Köhler Theory Analysis for Sub-Cloud (SC) samples.

Property x (units)	Average Value of x	Uncertainty Δx	Sensitivity, Φ_x $(m^3 \cdot mol^{-1} \cdot x^{-1})$	$\frac{M_j}{\rho_j}$ Contribution (%)
σ (N·m^{-1})	6.58×10^{-2}	1.32×10^{-3}	1.97×10^{-3}	8.37
ω (m$^{1.5}$)	8.75×10^{-14}	4.43×10^{-15}	9.89×10^9	14.13
ν_{NH_4Cl}	2	0.5	6.35×10^{-5}	10.24
$\nu_{(NH_4)_2SO_4}$	2.5	0.5	3.74×10^{-6}	0.60
$\nu_{NH_4NO_3}$	2	0.5	2.82×10^{-6}	0.45
$\nu_{Na_2SO_4}$	3	-	-	-
$\nu_{organic}$	1	0.20	7.21×10^{-5}	4.65
$\varepsilon_{NH_4NO_3}$	0.06	-	-	-
$\varepsilon_{Na_2SO_4}$	0.10	-	-	-
ε_{NH_4Cl}	0.09	1.95×10^{-3}	4.21×10^{-3}	2.64
$\varepsilon_{(NH_4)_2SO_4}$	0.12	2.03×10^{-3}	4.10×10^{-3}	2.69
$\varepsilon_{organic}$	0.82	6.36×10^{-3}	4.63×10^{-3}	9.50
ρ_{NH_4Cl}	1.53	-	-	-
$\rho_{(NH_4)_2SO_4}$	1.77	-	-	-
$\rho_{NH_4NO_3}$	1.73	-	-	-
$\rho_{Na_2SO_4}$	2.68	-	-	-
M_J/ρ_J (cm^3·mol^{-1})	1.72×10^{-3}	22.2%	-	-
M_J (g·mol^{-1})	2413	536	-	-

4.4. The Effect of Organics on Droplet Growth Kinetics

Figure 6 illustrates the OPC droplet size measurements for all supersaturations and samples considered. The flow rate within the instrument was maintained at 0.5 L·min^{-1} and the sheath to aerosol flow rate ratio was 10:1 to so that particles have the same residence time in the CFSTGC. Similar to the WSOC droplet data presented in [34,35,52,63], almost all of the growth droplet data lie within the measurement uncertainty and are in agreement with (NH$_4$)$_2$SO$_4$ calibration aerosol. According to TDGA, the water uptake coefficient is similar to that of the water uptake coefficient of (NH$_4$)$_2$SO$_4$ (α_c ~$_{(NH4)2SO4}$). Given that the water uptake coefficient, $_{(NH4)2SO4}$ ~0.2 [50,51], this suggest that the WSOC would not slow down activation kinetics of CCN compared to (NH$_4$)$_2$SO$_4$. This is consistent with the analysis of CCN data collected from a wide range of environments [51]. The WSOC does not appreciably impact droplet growth kinetics (or the effective water vapor mass transfer coefficient) as aerosol particles produced from the SC and CR grow like (NH$_4$)$_2$SO$_4$.

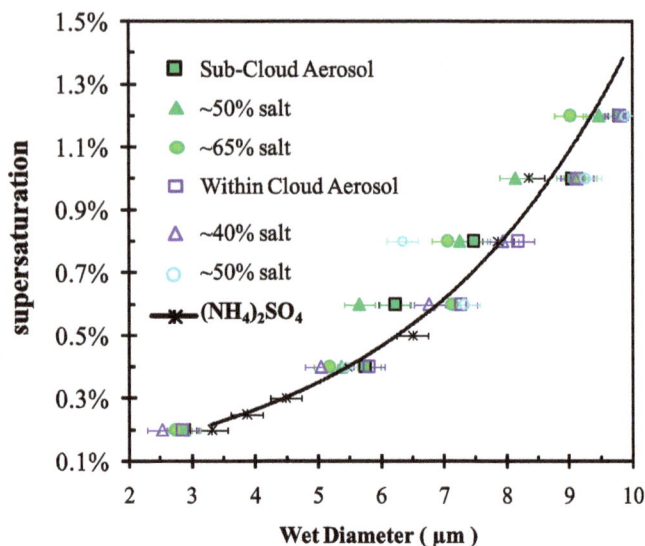

Figure 6. Size of activated CCN generated from the cloud residual samples (CR, solid symbols), sub-cloud samples (SC, open symbols) and $(NH_4)_2SO_4$ (solid black line). Droplet sizes are presented for CCN with s_c equal to the instrument supersaturation.

5. Summary and Implications

CCN activity, chemical composition and droplet growth measurements coupled with Köhler Theory Analysis are used to characterize the cloud droplet formation potential of water-soluble organic matter collected from cloud droplet (CR) residuals and sub-cloud (SC) aerosol during the MASE 2005 campaign over the eastern Pacific Ocean off the coast of California. The organics within CR samples were found to be more hygroscopic than in the SC sample, most likely from the enhanced levels of soluble organic acids (e.g., oxalic) formed during cloud processing. Both the direct and inferred surface tension measurements show neither sample contained strong surfactants; in fact, surface tension depression is consistent with the effect from primary marine organic matter.

Inferred average molecular weights of both CR and SC samples are consistent with a bimodal molecular weight distribution; one composed of oxalic acid (produced through glyoxylic oxidation or the oxidative decay of larger diacids in cloud droplets [15]) with primary organic matter (SC sample), or, organics from ship emissions (CR sample). These results are consistent with recently published work [18,64] that shows SOMA as a significant contributor to CCN activity in marine regions. Finally, all samples display similar droplet growth kinetics to CCN composed of $(NH_4)_2SO_4$, suggesting that water-soluble organics do not significantly impact the growth rate of CCN. Hence, CCN activity of the organics affects CCN activity through their contribution of solute from SOMA and possibly a slight surface tension depression from POMA. In both cases, increasing the size of CCN will also have a major impact, which is not shown here.

Acknowledgments: The work in this study was supported by an NSF CAREER Award. In addition, we would like to thank Dr. Rodney Weber and his students at the Georgia Institute of Technology for the use of the Total Organic Carbon (TOC) Turbo Siever Analyzer and Dr. Christos Fountoukis for his help in running the ISORROPIA II code.

Author Contributions: Akua Asa-Awuku is co-corresponding author and conducted laboratory experiments, collected and analyzed data sets, and played a significant role in the writing of the manuscript. Armin Sorooshian collected samples in the field, aided in analysis and writing. Richard C. Flagan and John H. Seinfeld devised flight experiments. Athanasios Nenes is co-corresponding author and conceived and devised experiments, analysis, and significantly contributed to writing.

Conflicts of Interest: The authors declare no conflict of interest.

References

1. Saxena, P.; Hildemann, L.M.; McMurry, P.H.; Seinfeld, J.H. Organics alter hygroscopic behavior of atmospheric particles. *J. Geophys. Res.* **1995**, *100*, 18755–18770.

2. Facchini, M.C.; Fuzzi, S.; Zappoli, S.; Andracchio, A.; Gelencsér, A.; Kiss, G.; Krivácsy, Z.; Mészáros, E.; Hansson, H.-C.; Alsberg, T.; *et al.* Partitioning of the organic aerosol component between fog droplets and interstitial air. *J. Geophys. Res.* **1999**, *104*, 26821–26832.

3. Nenes, A.; Charlson, R.J.; Facchini, M.C.; Kulmala, M.; Laaksonen, A.; Seinfeld, J.H. Can chemical effects on cloud droplet number rival the first indirect effect? *Geophys. Res. Lett.* **2002**, *29*, 29–21.

4. Decesari, S.; Facchini, M.C.; Mircea, M.; Cavalli, F.; Fuzzi, S. Solubility properties of surfactants in atmospheric aerosol and cloud/fog water samples. *J. Geophys. Res.* **2003**, *108*, 1984–2012.

5. Ervens, B.; Feingold, G.; Clegg, S.L.; Kreidenweis, S.M. A modeling study of aqueous production of dicarboxylic acids: 2. Implications for cloud microphysics. *J. Geophys. Res.* **2004**, *109*, D15206.

6. Saxena, P.; Hildemann, L.M. Water-soluble organics in atmospheric particles: A critical review of the literature and application of thermodynamics to identify candidate compounds. *J. Atmos. Chem.* **1996**, *24*, 57–109.

7. Mircea, M.; Facchini, M.C.; Decesari, S.; Fuzzi, S.; Charlson, R.J. The influence of the organic aerosol component on CCN supersaturation spectra for different aerosol types. *Tellus. B Chem. Phys. Meteorol.* **2002**, *54*, 74–81.

8. Alfonso, L.; Raga, G.B. The influence of organic compounds on the development of precipitation acidity in maritime clouds. *Atmos. Chem. Phys.* **2004**, *4*, 1097–1111.

9. O'Dowd, C.D.; Facchini, M.C.; Cavalli, F.; Ceburnis, D.; Mircea, M.; Decesari, S.; Fuzzi, S.; Yoon, Y.J.; Putaud, J.-P. Biogenically driven organic contribution to marine aerosol. *Nature* **2004**, *431*, 676–680.

10. Zhao, C.; Ishizaka, Y.; Peng, D. Numerical Study on Impacts of Multi-Component Aerosols on Marine Cloud Microphysical Properties. *J. Meteorol. Soc. Jpn. Ser. II* **2005**, *83*, 977–986.

11. Kanakidou, M.; Seinfeld, J.H.; Pandis, S.N.; Barnes, I.; Dentener, F.J.; Facchini, M.C.; Van Dingenen, R.; Ervens, B.; Nenes, A.; Nielsen, C.J.; *et al.* Organic aerosol and global climate modelling: A review. *Atmos. Chem. Phys.* **2005**, *5*, 1053–1123.

12. Yoon, Y.J.; Ceburnis, D.; Cavalli, F.; Jourdan, O.; Putaud, J.P.; Facchini, M.C.; Decesari, S.; Fuzzi, S.; Sellegri, K.; Jennings, S.G.; *et al.* Seasonal characteristics of the physicochemical properties of North Atlantic marine atmospheric aerosols. *J. Geophys. Res.* **2007**, *112*, D04206.

13. Ovadnevaite, J.; Ceburnis, D.; Martucci, G.; Bialek, J.; Monahan, C.; Rinaldi, M.; Facchini, M.C.; Berresheim, H.; Worsnop, D.R.; O'Dowd, C. Primary marine organic aerosol: A dichotomy of low hygroscopicity and high CCN activity. *Geophys. Res. Lett.* **2011**, *38*, L21806.

14. Moore, R.H.; Raatikainen, T.; Langridge, J.M.; Bahreini, R.; Brock, C.A.; Holloway, J.S.; Lack, D.A.; Middlebrook, A.M.; Perring, A.E.; Schwarz, J.P.; *et al.* CCN spectra, hygroscopicity, and droplet activation kinetics of secondary organic aerosol resulting from the 2010 Deepwater Horizon oil spill. *Environ. Sci. Technol.* **2012**, *46*, 3093–3100.

15. Sorooshian, A.; Lu, M.-L.; Brechtel, F.J.; Jonsson, H.; Feingold, G.; Flagan, R.C.; Seinfeld, J.H. On the source of organic acid aerosol layers above clouds. *Environ. Sci. Technol.* **2007**, *41*, 4647–4654.

16. Ceburnis, D.; O'Dowd, C.D.; Jennings, G.S.; Facchini, M.C.; Emblico, L.; Decesari, S.; Fuzzi, S.; Sakalys, J. Marine aerosol chemistry gradients: Elucidating primary and secondary processes and fluxes. *Geophys. Res. Lett.* **2008**, *35*, L07804.

17. Rinaldi, M.; Decesari, S.; Finessi, E.; Giulianelli, L.; Carbone, C.; Fuzzi, S.; O'Dowd, C.D.; Ceburnis, D.; Facchini, M.C. Primary and secondary organic marine aerosol and oceanic biological activity: Recent results and new perspectives for future studies. *Adv. Meteorol.* **2010**, *2010*, 1–10.

18. Coggon, M.M.; Sorooshian, A.; Wang, Z.; Metcalf, A.R.; Frossard, A.A.; Lin, J.J.; Craven, J.S.; Nenes, A.; Jonsson, H.H.; Russell, L.M.; *et al.* Ship impacts on the marine atmosphere: Insights into the contribution of shipping emissions to the properties of marine aerosol and clouds. *Atmos. Chem. Phys.* **2012**, *12*, 8439–8458.

19. Sorooshian, A.; Wang, Z.; Coggon, M.M.; Jonsson, H.H.; Ervens, B. Observations of sharp oxalate reductions in stratocumulus clouds at variable altitudes: Organic acid and metal measurements during the 2011 E-PEACE campaign. *Environ. Sci. Technol.* **2013**, *47*, 7747–7756.

20. Sorooshian, A.; Varutbangkul, V.; Brechtel, F.J.; Ervens, B.; Feingold, G.; Bahreini, R.; Murphy, S.M.; Holloway, J.S.; Atlas, E.L.; Buzorius, G.; *et al.* Oxalic acid in clear and cloudy atmospheres: Analysis of data from International Consortium for Atmospheric Research on Transport and Transformation 2004. *J. Geophys. Res. D: Atmos.* **2006**, *111*, D10S27.

21. Ervens, B.; Feingold, G.; Frost, G.J.; Kreidenweis, S.M. A modeling study of aqueous production of dicarboxylic acids: 1. Chemical pathways and speciated organic mass production. *J. Geophys. Res.* **2004**, *109*, D15205.

22. Lim, H.; Carlton, A.G.; Turpin, B.J. Isoprene forms secondary organic aerosol in Atlanta: Results from time-resolved measurements during the Atlanta supersite experiment. *Environ. Sci. Technol.* **2005**, *39*, 4441–4446.

23. Carlton, A.G.; Turpin, B.J.; Lim, H.-J.; Altieri, K.E.; Seitzinger, S. Link between isoprene and secondary organic aerosol (SOA): Pyruvic acid oxidation yields low volatility organic acids in clouds. *Geophys. Res. Lett.* **2006**, *33*, L06822.

24. Coggon, M.M.; Sorooshian, A.; Wang, Z.; Craven, J.S.; Metcalf, A.R.; Lin, J.J.; Nenes, A.; Jonsson, H.H.; Flagan, R.C.; Seinfeld, J.H. Observations of continental biogenic impacts on marine aerosol and clouds off the coast of California. *J. Geophys. Res. D: Atmos.* **2014**, *119*, 6724–6748.

25. Eyring, V.; Köhler, H.W.; van Aardenne, J.; Lauer, A. Emissions from international shipping: 1. The last 50 years. *J. Geophys. Res.* **2005**, *110*, D17305.

26. Zheng, Z.; Tang, X.; Asa-Awuku, A.; Jung, H.S. Characterization of a method for aerosol generation from heavy fuel oil (HFO) as an alternative to emissions from ship diesel engines. *J. Aerosol Sci.* **2010**, *41*, 1143–1151.

27. Russell, L.M.; Noone, K.J.; Ferek, R.J.; Pockalny, R.A.; Flagan, R.C.; Seinfeld, J.H. Combustion Organic Aerosol as Cloud Condensation Nuclei in Ship Tracks. *J. Atmos. Sci.* **2000**, *57*, 2591–2606.

28. Twomey, S. Pollution and the planetary albedo. *Atmos. Environ.* **1974**, *8*, 1251–1256.

29. Coakley, J.A.; Bernstein, R.L.; Durkee, P.A. Effect of ship-stack effluents on cloud reflectivity. *Science* **1987**, *237*, 1020–1022.

30. Albrecht, B.A. Aerosols, cloud microphysics, and fractional cloudiness. *Science* **1989**, *245*, 1227–1230.

31. Ackerman, A.S.; Toon, O.B.; Taylor, J.P.; Johnson, D.W.; Hobbs, P.V.; Ferek, R.J. Effects of aerosols on cloud albedo: Evaluation of Twomey's parameterization of cloud susceptibility using measurements of ship tracks. *J. Atmos. Sci.* **2000**, *57*, 2684–2695.

32. Moore, R.H.; Cerully, K.; Bahreini, R.; Brock, C.A.; Middlebrook, A.M.; Nenes, A. Hygroscopicity and composition of California CCN during summer 2010. *J. Geophys. Res.* **2012**, *117*, D00V12.

33. Padró, L.T.; Asa-Awuku, A.; Morrison, R.; Nenes, A. Inferring thermodynamic properties from CCN activation experiments: Single-component and binary aerosols. *Atmos. Chem. Phys.* **2007**, *7*, 5263–5274.

34. Moore, R.H.; Ingall, E.D.; Sorooshian, A.; Nenes, A. Molar mass, surface tension, and droplet growth kinetics of marine organics from measurements of CCN activity. *Geophys. Res. Lett.* **2008**, *35*, L07801.

35. Asa-Awuku, A.; Nenes, A.; Gao, S.; Flagan, R.C.; Seinfeld, J.H. Water-soluble SOA from Alkene ozonolysis: Composition and droplet activation kinetics inferences from analysis of CCN activity. *Atmos. Chem. Phys.* **2010**, *10*, 1585–1597.

36. Lu, M.-L.; Conant, W.C.; Jonsson, H.H.; Varutbangkul, V.; Flagan, R.C.; Seinfeld, J.H. The Marine Stratus/Stratocumulus Experiment (MASE): Aerosol-cloud relationships in marine stratocumulus. *J. Geophys. Res. D: Atmos.* **2007**, *112*, D10209.

37. Marple, V.A.; Chien, C.M. Virtual impactors: A theoretical study. *Environ. Sci. Technol.* **1980**, *14*, 976–985.

38. Noone, K.J.; Ogren, J.A.; Heintzenberg, J.; Charlson, R.J.; Covert, D.S. Design and calibration of a counterflow virtual impactor for sampling of atmospheric fog and cloud droplets. *Aerosol Sci. Technol.* **1988**, *8*, 235–244.

39. Roberts, G.C.; Nenes, A. A continuous-flow streamwise thermal-gradient CCN chamber for atmospheric measurements. *Aerosol Sci. Technol.* **2005**, *39*, 206–221.

40. Lance, S.; Nenes, A.; Medina, J.; Smith, J.N. Mapping the operation of the DMT continuous flow CCN counter. *Aerosol Sci. Technol.* **2006**, *40*, 242–254.

41. Rose, D.; Gunthe, S.S.; Mikhailov, E.; Frank, G.P.; Dusek, U.; Andreae, M.O.; Pöschl, U. Calibration and measurement uncertainties of a continuous-flow cloud condensation nuclei counter (DMT-CCNC): CCN activation of ammonium sulfate and sodium chloride aerosol particles in theory and experiment. *Atmos. Chem. Phys.* **2008**, *8*, 1153–1179.

42. Moore, R.H.; Nenes, A.; Medina, J. Scanning mobility CCN analysis—A method for fast measurements of size-resolved CCN distributions and activation kinetics. *Aerosol Sci. Technol.* **2010**, *44*, 861–871.

43. Asa-Awuku, A.; Sullivan, A.P.; Hennigan, C.J.; Weber, R.J.; Nenes, A. Investigation of molar volume and surfactant characteristics of water-soluble organic compounds in biomass burning aerosol. *Atmos. Chem. Phys.* **2008**, *8*, 799–812.

44. Brechtel, F.J.; Kreidenweis, S.M. Predicting particle critical supersaturation from hygroscopic growth measurements in the humidified TDMA. Part I: Theory and sensitivity studies. *J. Atmos. Sci.* **2000**, *57*, 1854–1871.

45. Langmuir, I. The constitution and fundamental properties of solids and liquids. II. liquids.1. *J. Am. Chem. Soc.* **1917**, *39*, 1848–1906.

46. Taraniuk, I.; Graber, E.R.; Kostinski, A.; Rudich, Y. Surfactant properties of atmospheric and model humic-like substances (HULIS). *Geophys. Res. Lett.* **2007**, *34*, L16807.

47. Giordano, M.R.; Short, D.Z.; Hosseini, S.; Lichtenberg, W.; Asa-Awuku, A.A. Changes in droplet surface tension affect the observed hygroscopicity of photochemically aged biomass burning aerosol. *Environ. Sci. Technol.* **2013**, *47*, 10980–10986.

48. Lance, S.; Nenes, A.; Mazzoleni, C.; Dubey, M.K.; Gates, H.; Varutbangkul, V.; Rissman, T.A.; Murphy, S.M.; Sorooshian, A.; Flagan, R.C.; *et al.* Cloud condensation nuclei activity, closure, and droplet growth kinetics of Houston aerosol during the Gulf of Mexico Atmospheric Composition and Climate Study (GoMACCS). *J. Geophys. Res. D: Atmos.* **2009**, *114*, D00F15.

49. Asa-Awuku, A.; Moore, R.H.; Nenes, A.; Bahreini, R.; Holloway, J.S.; Brock, C.A.; Middlebrook, A.M.; Ryerson, T.B.; Jimenez, J.L.; DeCarlo, P.F.; *et al.* Airborne cloud condensation nuclei measurements during the 2006 Texas Air Quality Study. *J. Geophys. Res. D: Atmos.* **2011**, *116*, D11201.

50. Raatikainen, T.; Moore, R.H.; Lathem, T.L.; Nenes, A. A coupled observation—modeling approach for studying activation kinetics from measurements of CCN activity. *Atmos. Chem. Phys.* **2012**, *12*, 4227–4243.

51. Raatikainen, T.; Nenes, A.; Seinfeld, J.H.; Morales, R.; Moore, R.H.; Lathem, T.L.; Lance, S.; Padró, L.T.; Lin, J.J.; Cerully, K.M.; *et al.* Worldwide data sets constrain the water vapor uptake coefficient in cloud formation. *Proc. Natl. Acad. Sci. USA* **2013**, *110*, 3760–3764.

52. Engelhart, G.J.; Asa-Awuku, A.; Nenes, A.; Pandis, S.N. CCN activity and droplet growth kinetics of fresh and aged monoterpene secondary organic aerosol. *Atmos. Chem. Phys.* **2008**, *8*, 3937–3949.

53. Lathem, T.L.; Nenes, A. Water vapor depletion in the DMT continuous-flow CCN chamber: Effects on supersaturation and droplet growth. *Aerosol Sci. Technol.* **2011**, *45*, 604–615.

54. Cavalli, F.; Facchini, M.C.; Decesari, S.; Mircea, M.; Emblico, L.; Fuzzi, S.; Ceburnis, D.; Yoon, Y.J.; O'Dowd, C.D.; Putaud, J.-P.; *et al.* Advances in characterization of size-resolved organic matter in marine aerosol over the North Atlantic. *J. Geophys. Res. D: Atmos.* **2004**, *109*, D24215.

55. Kiss, G.; Tombácz, E.; Hansson, H.-C. Surface tension effects of humic-like substances in the aqueous extract of tropospheric fine aerosol. *J. Atmos. Chem.* **2005**, *50*, 279–294.

56. Hallberg, A.; Ogren, J.A.; Noone, K.J.; Okada, K.; Heintzenberg, J.; Svenningsson, I.B. The Influence of Aerosol Particle Composition on Cloud Droplet Formation. In *The Kleiner Feldberg Cloud Experiment 1990*; Springer Netherlands: Houten, the Netherlands, 1994; pp. 153–171.

57. Fountoukis, C.; Nenes, A. ISORROPIA II: A computationally efficient thermodynamic equilibrium model for K^+-Ca^{2+}-Mg^{2+}-NH_4^+-Na^+-SO_4^{2-}-NO_3^--Cl^--H_2O aerosols. *Atmos. Chem. Phys.* **2007**, *7*, 4639–4659.

58. Turpin, B.J.; Lim, H.-J. Species contributions to $PM_{2.5}$ mass concentrations: Revisiting common assumptions for estimating organic MASS. *Aerosol Sci. Technol.* **2001**, *35*, 602–610.

59. Oppo, C.; Bellandi, S.; Degli Innocenti, N.; Stortini, A.M.; Loglio, G.; Schiavuta, E.; Cini, R. Surfactant components of marine organic matter as agents for biogeochemical fractionation and pollutant transport via marine aerosols. *Mar. Chem.* **1999**, *63*, 235–253.

60. Hansell, D.A.; Carlson, C.A. *Biogeochemistry of Marine Dissolved Organic Matter*; Academic Press: London, UK, 2002.

61. Redfield, A.C.; Ketchum, B.H.; Richards, F.A. The influence of organisms on the composition of sea-water. *Sea* **1963**, *2*, 26–77.
62. Schulz, H.D.; Zabel, M. *Marine Geochemistry*; Springer Science & Business Media: Berlin, Germany, 2013.
63. Asa-Awuku, A.; Engelhart, G.J.; Lee, B.H.; Pandis, S.N.; Nenes, A. Relating CCN activity, volatility, and droplet growth kinetics of β-caryophyllene secondary organic aerosol. *Atmos. Chem. Phys. Disc.* **2008**, *8*, 10105–10151.
64. Wonaschütz, A.; Coggon, M.; Sorooshian, A.; Modini, R.; Frossard, A.A.; Ahlm, L.; Mülmenstädt, J.; Roberts, G.C.; Russell, L.M.; Dey, S.; *et al.* Hygroscopic properties of smoke-generated organic aerosol particles emitted in the marine atmosphere. *Atmos. Chem. Phys.* **2013**, *13*, 9819–9835.

The Influence of Sandstorms and Long-Range Transport on Polycyclic Aromatic Hydrocarbons (PAHs) in PM$_{2.5}$ in the High-Altitude Atmosphere of Southern China

Minmin Yang, Yan Wang, Qiang Liu, Aijun Ding and Yuhua Li

Abstract: PM$_{2.5}$ (Particulate Matter 2.5) samples were collected at Mount Heng and analyzed for polycyclic aromatic hydrocarbons (PAHs). During sampling, a sandstorm from northern China struck Mount Heng and resulted in a mean PM$_{2.5}$ concentration of 150.61 μg/m^3, which greatly exceeded the concentration measured under normal conditions (no sandstorm: 58.50 μg/m^3). The average mass of PAHs in PM$_{2.5}$ was 30.70 μg/g, which was much lower than in the non-sandstorm samples (80.80 μg/g). Therefore, the sandstorm increased particle levels but decreased PAH concentrations due to dilution and turbulence. During the sandstorm, the concentrations of 4- and 5-ring PAHs were below their detection limits, and 6-ring PAHs were the most abundant. Under normal conditions, the concentrations of 2-, 3- and 6-ring PAHs were higher, and 4- and 5-ring PAHs were lower relative to the other sampling sites. In general, the PAH contamination was low to medium at Mount Heng. Higher LMW (low molecular weight) concentrations were primarily linked to meteorological conditions, and higher HMW (high molecular weight) concentrations primarily resulted from long-range transport. Analysis of diagnostic ratios indicated that PM$_{2.5}$ PAHs had been emitted during the combustion of coal, wood or petroleum. The transport characteristics and origins of the PAHs were investigated using backwards Lagrangian particle dispersion modeling. Under normal conditions, the "footprint" retroplumes and potential source contributions of PAHs for the highest and lowest concentrations indicated that local sources had little effect. In contrast, long-range transport played a vital role in the levels of PM$_{2.5}$ and PAHs in the high-altitude atmosphere.

Reprinted from *Atmosphere*. Cite as: Yang, M.; Wang, Y.; Liu, Q.; Ding, A.; Li, Y. The Influence of Sandstorms and Long-Range Transport on Polycyclic Aromatic Hydrocarbons (PAHs) in PM$_{2.5}$ in the High-Altitude Atmosphere of Southern China. *Atmosphere* **2015**, *6*, 1633–1651.

1. Introduction

Haze pollution in China has occurred frequently in recent decades, with a vital contribution from $PM_{2.5}$ (Particulate Matter 2.5). Sandstorms also frequently occur in China, resulting in serious air pollution. $PM_{2.5}$ and PAHs (polycyclic aromatic hydrocarbons) have become the focus of governments and researchers. Many studies have been conducted on PAHs and $PM_{2.5}$ in the suburban and urban atmosphere; however, very few studies have evaluated the PAH concentrations associated with $PM_{2.5}$ at high altitudes in China. Further studies of the PAH concentration distributions and sources in $PM_{2.5}$ were conducted in this research. Furthermore, the influence of long-range transport on PAH concentrations in the atmosphere of a heavily polluted area is also discussed.

Regarding $PM_{2.5}$ pollution, the standard established by the U.S. Environmental Protection Agency (US EPA) has been progressively tightened and is currently set at $35\ \mu g/m^3$ for a 24-h period, with an annual average of $12\ \mu g/m^3$. $PM_{2.5}$ originates from both natural and anthropogenic sources, which can directly emit particles or emit precursor gases that subsequently form particles in the atmosphere [1]. $PM_{2.5}$ can reduce visibility because its particle diameter is near the wavelength of light, in addition to exerting climate and human health effects, particularly in the form of respiratory system disease due to its diverse chemical composition. Furthermore, $PM_{2.5}$ is a carrier for toxic pollutants, such as PAHs or polychlorinated biphenyls (PCBs). $PM_{2.5}$ particles have a lifespan of days to weeks and can be transported over distances of thousands of kilometers due to their small size [2]. Therefore, it is important to study the distribution of $PM_{2.5}$ concentrations and their emission sources.

PAHs, especially those associated with fine particles ($PM_{2.5}$), are mutagenic and carcinogenic [3]. During incomplete combustion or the pyrolysis of organic materials such as wood, fossil fuels and coal, PAHs, a group of semi-volatile compounds, are formed. In terms of environmental effects, PAHs can also inhibit growth of diatoms and even affect the global carbon cycle [4,5]. In terms of human health, PAHs, especially those with high molecular weight, are associated with several pathologies and may cause acute health effects, adverse birth outcomes or higher levels of risk for lung cancer [3,6,7]. In China, the limit for the daily average concentration of BaP in ambient air is $0.01\ \mu g/m^3$. In the 1970s, the US EPA designated 16 PAHs as priority pollutants. In particular, high molecular weight (HMW) PAHs (MW ≥ 202) with 4–6 aromatic rings frequently result from combustion [8]. In addition, PAHs can disperse regionally and travel to remote places through long-range atmospheric transport; the influence of this process is analyzed in this study. PAHs are primarily emitted from anthropogenic sources, including petrogenic and pyrogenic sources [9]. Petrogenic sources are introduced to the atmosphere through oil spills of crude and refined petroleum, and pyrogenic sources are released through the combustion of coal, petroleum or biomass [9–11]. The major sources in China are biofuel burning, domestic coal combustion, and industrial emissions [12]. Therefore, it is important to analyze the concentrations, distributions and potential emission sources of PAHs associated with $PM_{2.5}$, particularly in the atmosphere, to effectively control air pollution caused by $PM_{2.5}$.

Most studies on PAHs have been conducted in low-lying urban or rural areas around the world [13–15]. However, few studies have described PAH concentrations in $PM_{2.5}$, especially concerning sandstorms and long-range transport at high mountain sites far from ground-level pollution sources. Areas such as Mount Heng (the sampling site in this study), particularly for those located in the free troposphere of heavily polluted areas, are unique environments for investigating $PM_{2.5}$ PAH concentrations. Although high mountain sites are usually considered the most pristine continental areas [16], they may receive significant amounts of PAHs from the deposition of particulates after long-range atmospheric transport [17,18]. The particulate ($PM_{2.5}$) samples were collected at the summit of Mount Heng (1269 m asl), located in the transition region between the boundary layer and free troposphere. The concentrations and distributions of chemical compounds at high mountain sites can provide valuable information regarding the long-range transport and accumulation of such pollutants. The objectives of this study were to determine the $PM_{2.5}$ and particulate PAHs levels and to identify potential emission sources of PAHs and $PM_{2.5}$ at Mount Heng during the spring of 2009.

2. Materials and Methods

2.1. Sampling

$PM_{2.5}$ samples were collected at the top of Mount Heng (27.3°N, 112.7°E) continuously every day from March to May of 2009, except on rainy or very foggy days. Mount Heng is located in the Hunan province of southern China and is approximately 500 km from the East China Sea and South China Sea. Sampling was conducted at the Mount Heng Meteorological Station, which is located at an elevation of 1269 m asl near Zhurong Peak, located within the free troposphere as shown in Figure 1. Measurements were performed in 2009 on the roof of a small house that is part of the meteorological observatory. The local climate is characterized by distinct seasons and influenced by air masses that originate from various directions. Although Mount Heng is less developed than coastal cities in China, it has experienced rapid industrialization and urbanization in the last two decades [19]. Furthermore, its high elevation and geographic location are suitable for studying the long-range transport of pollutants.

A medium-volume PM air sampler (TH150A, Tianhong, Wuhan) with a $PM_{2.5}$-selective inlet was used for ambient sample collection at a calibrated flow rate of 100 mL/min. $PM_{2.5}$ particles were collected on quartz fiber filters (90 mm in diameter) that had been pre-baked in a muffle furnace for 24 h at 600 °C to minimize the concentration in the blank. Then, the quartz fiber filters were stored in a constant temperature and humidity incubator (WS 150 III) at 25 °C and a relative humidity of 50% for 48 h before weighing.

A sandstorm in northern China struck Mount Heng on 25 April 2009 (25-1: 8:30–20:15 and 25-2: 21:00–8:05 the next day) and 26 April 2009 (26-1: 8:30–12:30, 26-2:12:50–16:30, 26-3: 16:50–20:40 and 26-4: 20:50–8:40 the next day). During this time, six of the particulate samples were collected. As shown in Figure 1b, when the sandstorm occurred on 25 and 26 April 2009, the UV (ultraviolet) aerosol index increased. However, this index was lower when no sandstorm occurred on 19 and 20 May 2009 (Figure 1a).

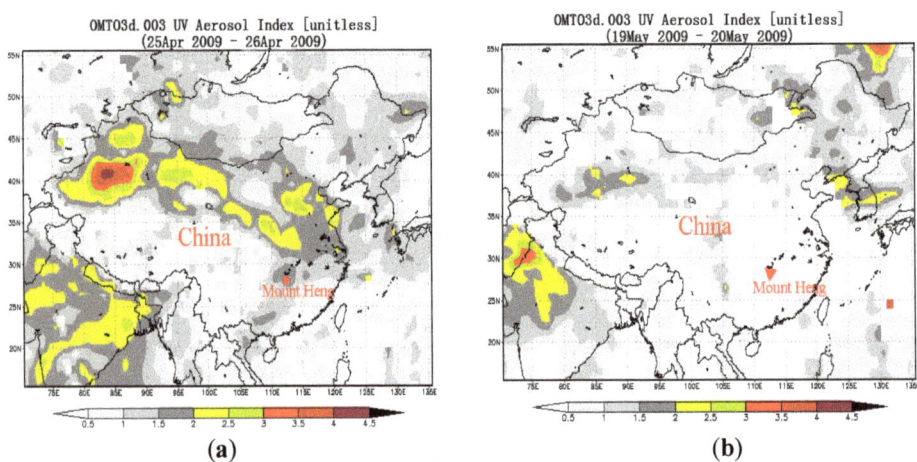

Figure 1. Sampling locations and UV (ultraviolet) aerosol index under normal conditions (**a**) and during the sandstorm (**b**).

2.2. Analytical Techniques

After field sampling, all filters were immediately sealed in aluminum foil and stored in a freezer at −20 °C before analysis. Next, the compounds collected on the quartz fiber filters were extracted in an accelerated solvent extractor (ASE 3000, Dionex) using a solvent mixture of acetone and n-hexane (v/v = 1:1) at an extraction temperature of 100 °C for 5 min. Perylene-D12 was added to the samples as a surrogate before extraction to estimate the extraction efficiency and losses during the sample extraction and concentration steps. The extracts were rotary evaporated to a volume of approximately 10 mL before concentrating in a concentrator (N-EVAP 112) to approximately 1 mL. Finally, each extract was diluted with 1 mL of n-hexane and spiked with phenanthrene-d10 as an internal standard for analysis by gas chromatography/mass spectrometry (GC/MS).

Sixteen priority PAH species listed by the US EPA were quantified using GC/MS (QP2010, SHIMADZU) with a capillary GC column (30 m × 0.25 mm × 0.25 μm, DB-5 ms). The ion source, injector and interface temperatures were 200.0, 300.0 and 250.0 °C, respectively. The column oven temperature was initially 45.0 °C for 1.0 min before increasing to 130.0 °C at a rate of 45.0 °C /min, to 240.0 °C at a rate of 7.0 °C /min and to 320.0 °C at a rate of 12.0 °C /min, which was then held for 8.0 min. The analytical method was based on US EPA Method 8270 [20]. Sixteen PAHs were measured and quantified. Although naphthalene is on this list, its levels are affected by many factors, and it was not investigated in this study due to its high volatility in particles and its presence in the ambient air and carrier gas. Therefore, only the sums of the other 15 PAHs were analyzed, as in many previous studies [21,22]. The fifteen priority PAH species are acenaphthene (Ace), acenaphthylene (Acy), fluorene (Flu), phenanthrene (PhA), anthracene (AnT), fluoranthene (FluA), pyrene (Pyr), benz (*a*) anthracene (BaA), chrysene (Chr), benzo (*b*) fluoranthene (BbF), benzo (*k*) fluoranthene (BkF), benzo (*a*) pyrene (BaP), indeno (1,2,3-*cd*) pyrene (InP), dibenz (*a*,*h*) anthracene (DbA) and benzo (*g*,*h*,*i*) perylene (BP).

2.3. Quality Assurance and Control

The analyses were performed in accordance with the technical specifications of the US EPA. Data were acquired in single-ion monitoring mode (SIM) and quantified using the internal standard method. For qualitative analysis, compounds were matched using the retention times and ion mass fragments of standard mixtures of PAHs from the National Institute of Standards and Technology (NIST). Calibration curves were created based on the response factors of certain PAH species in the standard solution *versus* phenanthrene-d10, the internal standard used to quantify the individual PAH species. A standard mixture of 16 PAHs was diluted to six concentrations and analyzed by GC/MS using the same procedure. The mean extraction recoveries for the $PM_{2.5}$ surrogates varied from 83.3 to 100.3%. The field and laboratory blanks were treated in parallel with the particulate matter samples during sampling, storage and chemical analysis. The blank levels were subtracted to obtain the reported PAH concentrations in the $PM_{2.5}$.

2.4. Modeling Tools and Methodology

In this study, a Lagrangian dispersion model, the hybrid single-particle Lagrangian integrated trajectory (HYSPLIT) model, was used to simulate transport and dispersion [23]. The positions of the particles were calculated based on the mean wind and turbulence transport components after their release at a receptor for backward runs [24]. Detailed information regarding this model was presented by Draxler and Hess [23] and Draxler [25]. This model was used to conduct hourly backward simulations of particle dispersion to calculate the potential source contributions (PSCs) of $PM_{2.5}$ (*i.e.*, the sources that contribute to the simulated mixing ratios at the receptor) based on the methods used by various authors [24,26–28]. In addition, $PM_{2.5}$ was the carrier of the PAHs detected in this study and was emitted from the same potential sources. Here, the word "potential" indicates that this calculation was based on transport alone [29] and that the generation of secondary organic aerosols was not considered as a factor during transport. We calculated the residence time at an altitude of 0–100 m to obtain a "footprint" of the retroplume of the released air particles, which represented the probability distribution or residence time of a simulated air mass based on the method developed by Stohl *et al.* [29,30]. The gridded contributions of $PM_{2.5}$ (*i.e.*, the PSCs) to an observed air mass were calculated by multiplying the emission rate by the footprint retroplume with an emissions inventory [24] at a spatial resolution of 0.1° latitude and 0.1° longitude. The LPDM (Lagrangian Particle Dispersion Model) was developed on a horizontal grid of 0.1° × 0.1° (latitude and longitude) and was fully capable of representing the regional transport of air pollutants to the Mount Heng region. An updated $PM_{2.5}$ emissions inventory (INTEX_B; 2006) was used to calculate the PSCs of $PM_{2.5}$ [31] at a horizontal resolution of 0.2° × 0.2° (latitude and longitude) over China. Emissions may not significantly vary from year to year, and the same inventory was applied in this study. Details regarding the application of the LPDM were presented by Ding *et al.* [24].

3. Results and Discussion

3.1. PM$_{2.5}$ and PAH Concentrations and Distributions at Mount Heng

Figure 2 shows the daily $PM_{2.5}$ levels measured at Mount Heng during the spring of 2009. When the sandstorm struck Mount Heng on 25 and 26 April 2009, the $PM_{2.5}$ concentration ranged from 86.65 to 212.66 $\mu g/m^3$ with a mean concentration of 156.61 $\mu g/m^3$. In the absence of a sandstorm, the $PM_{2.5}$ concentration ranged from 32.69 $\mu g/m^3$ to 92.50 $\mu g/m^3$ with a mean of 58.50 $\mu g/m^3$. The average mass concentrations measured at Mount Heng were greater than those measured at Mount Tai, with a mean concentration of 42.24 $\mu g/m^3$ in the autumn-winter of 2008 [22]. When the sandstorm struck Mount Heng, the $PM_{2.5}$ concentration rapidly increased, indicating that the sandstorm largely affected the air quality.

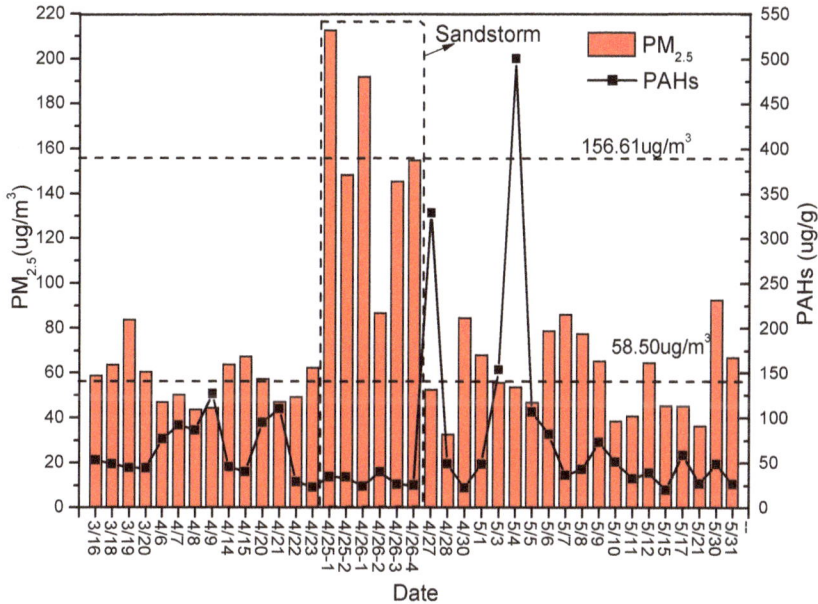

Figure 2. PM$_{2.5}$ and total daily PAH levels measured at Mount Heng.

Table 1. Concentrations of the individual polycyclic aromatic hydrocarbons (PAHs) during the sandstorm (25 and 26 April 2009) and under normal conditions (no sandstorm) ($\mu g/g$).

PAH	Sandstorm				No Sandstorm			
	Max	Min	Average	Stdev	Max	Min	Average	Stdev
Ace	1.9	1.14	1.52	0.3	2.39	0.3	1.14	0.58
Acy	6.46	2.18	3.99	1.49	34.91	1.32	5.2	5.65
Flu	4.82	1.49	2.72	1.26	30.04	0.67	4.58	4.81
PhA	9.85	3.42	5.35	2.33	31.69	1.52	9.59	6.46
AnT	3.16	0.82	2.05	1	8.12	0.82	2.24	1.48
FluA	5.87	1.68	3.54	1.47	71.21	ND	9.74	14.44
Pyr	5.06	1.22	2.65	1.35	61.91	ND	8.17	11.85
BaA	ND	ND	ND	ND	31.08	ND	3.48	6.95
Chr	ND	ND	ND	ND	57.07	ND	5.42	10.34
BbF	3.6	ND	1.19	1.84	82.76	ND	9.14	15.66
BkF	3.14	ND	0.79	1.32	22.67	ND	4.64	5.21
BaP	3.89	ND	1.21	1.88	27.6	ND	4.76	6.21
InP	2.47	ND	0.68	1.09	37.65	ND	4.71	7.55
DbA	ND	ND	ND	ND	12.96	ND	1.06	2.3
BP	6.6	3.7	5.02	0.99	54.01	ND	6.92	10.2
SUM	56.83	15.64	30.7	16.31	566.08	4.63	80.8	109.7

ND: not detected.

205

The mass concentrations of the individual PAH species under the influence of the sandstorm at Mount Heng are shown in Table 1. When the sandstorm struck Mount Heng, the total PAH mass concentrations in $PM_{2.5}$ ranged from 15.64 to 56.83 µg/g with a mean concentration of 30.70 µg/g. In addition, PhA was the predominant compound (5.35 µg/g), followed by BP, Acy and Pyr, which contributed nearly 60% to the total concentration during the sandstorm. Under normal conditions, the total PAH mass concentration ranged from 4.63 to 566.08 µg/g with a mean of approximately 80.15 µg/g; this variation is large and resulted from the varying meteorological conditions and emissions sources during the sampling campaign. The most abundant PAH species observed were FluA, PhA, BbF, and Pyr, which accounted for nearly 50% of the total.

Figure 3 compares the published data on PAH concentrations in $PM_{2.5}$ with the values observed at Mount Heng. The volume concentrations of the total PAHs in $PM_{2.5}$ were comparable under sandstorm (4.70 ng/m^3) and normal (4.50 ng/m^3) conditions, as shown in the top left corner of Figure 3. The total PAH concentrations at Mount Heng were much greater than those reported in southwestern urban Atlanta [3] and were generally lower than those reported at Mount Tai [22] in northern China and in the city of Beijing [15]. However, these values were comparable to the mean concentrations reported in Hong Kong [32]. In this study, PAHs with 2–6 rings were measured. Figure 3 shows that the concentrations of the 2- and 3-ring PAHs were generally higher at Mount Heng than at Mount Tai [22], except for FluA. The 6-ring PAH concentrations were comparable to the values reported at Mount Tai [22] and were much higher than those in Hong Kong [32] and Atlanta [3] under normal (no sandstorm) conditions. The concentrations of 4- and 5-ring PAHs were lower than those at the other sites, except Atlanta. In this study, 2–6 ring PAHs displayed the same distributions during the sandstorm as the 2-, 3- and 6-ring PAHs, which had greater concentrations. However, the 4- and 5-ring PAHs had lower concentrations than at the other sampling sites, and the overall PAH contamination at Mount Heng was low to medium.

In terms of mass concentrations, PAHs with three and six rings were the predominant compounds, accounting for 51.5% of the total under normal conditions and 66.8% during the sandstorm. In contrast, the mass concentrations of PAHs were generally lower during the sandstorm, as shown in Figure 2. The potentially carcinogenic PAHs, particularly BaA, BbF, BkF, BaP, InP and DbA (four and five rings) (according to the International Agency for Research on Cancer (IARC), were present at high concentrations and accounted for 12.0% of the total at Mount Heng.

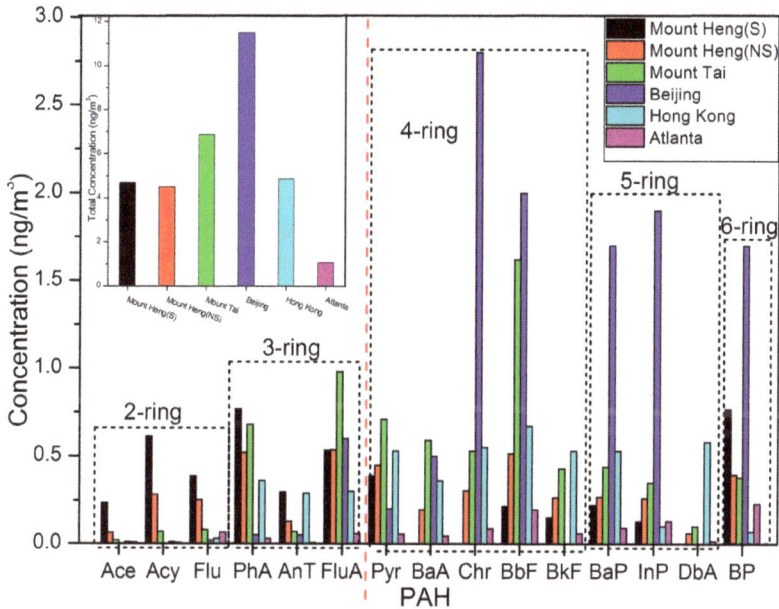

Figure 3. Average concentrations of PAHs in PM$_{2.5}$ at Mount Heng and other sites (S: sandstorm, NS: no sandstorm).

As shown in Figure 2, the mass concentrations of PAHs were generally lower during the sandstorm. When the sandstorm struck the site on 25 and 26 April 2009, the PM$_{2.5}$ volume concentrations were much greater, whereas the total PAHs were present at lower mass concentrations. Six samples were collected at various times, and the concentration profiles of the 15 PAH species and the corresponding PM$_{2.5}$ levels and ring distributions are shown in Figure 3. The average daily concentrations of the individual PAH species were all higher on 26 April than on 25 April 2009, except for BaA, Chr, BbF, BkF, BaP, InP and DbA, which were below the detection limit. The concentration of Acy was greater during the night than during the day, whereas the total PAHs were present at comparable levels. Acy is generally emitted from cement [33], diesel vehicle emissions [34,35] and combustion sources [36]. As shown in Figure 3, the concentrations of the 2-ring PAHs were greater during the night than during the day. On 26 April 2009, four particle samples were collected, and the concentrations of PhA and Flu were higher in the particulate sample collected between 12:50 and 16:34. PhA primarily results from incomplete combustion and the pyrolysis of fuels [34], and Flu results from diesel and gasoline vehicle emissions, general combustion, industrial oil burning and coal combustion [33–35]. These sources are consistent with the findings presented in the third section. During nighttime, the 2-ring PAH concentrations were higher and equaled those on

207

25 April 2009. Therefore, we can conclude that the 2-ring PAH concentrations increased due to their volatility when the temperature fell at night. Furthermore, PAHs with five aromatic rings were not detected on 26 April 2009, but contributed 16.08% and 12.97% of the total on 25 April 2009.

During the sandstorm, BaA, Chr and DbA were not detected in the samples, as shown in Table 1. In general, BaA originates from the steel industry [33], and Chr primarily originates from gasoline vehicle emissions [34,35] and industrial oil burning [33,35]. Thus, the steel industry and gasoline-powered vehicles potentially did not contribute to the PAH concentrations under sandstorm conditions, which is supported by the diagnostic ratio analysis of PAHs in the third section.

3.2. Diagnostic Ratio Analysis of PAHs

As indicated above, emission sources were estimated based on specific pollutants, and diagnostic ratios were analyzed to clarify their dominant sources. Due to high volatility of 2–3-ring PAHs, only 4–6-ring PAHs were used for source identification in this study. Linear regression analyses were conducted between various aromatic PAH concentrations to estimate their emission sources, as shown in Figure 4. Based on Figure 4a,b, the concentrations of 4-, 5- and 6-ring PAHs were strongly correlated with each other (R^2 = 0.97, 0.97 and 0.93, respectively). These good correlations indicate that HMW (4-, 5- and 6-ring) PAHs could be emitted from sources that are similar and distinct from those of the 2-ring PAHs. Portions of the PAHs with three and four aromatic rings resulted from similar emissions sources with a similar transport pathway.

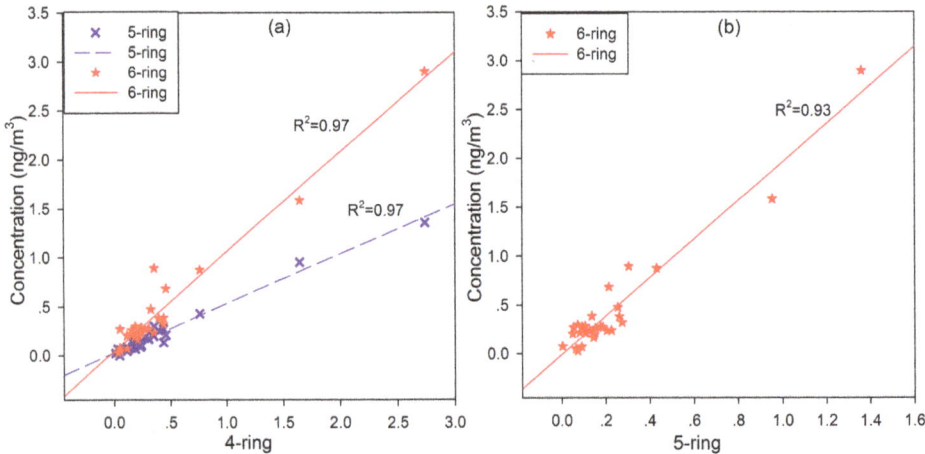

Figure 4. (a) Relationships between the 4-ring PAH concentrations and those of 5- and 6-ring PAHs; (b) relationship between the concentrations of the 5- and 6-ring PAHs.

The concentrations of certain individual marker PAHs and their ratios have been widely used as indicators to identify PAH sources [8,22,37–41]. Common diagnostic ratios of LMW (low molecular weight, 2-3-ring PAHs)/HMW (4-6-ring PAHs), BaA/(BaA + Chr) and InP/(InP + BP) were calculated in this study for source evaluation. LMW/HMW ratios of less than 1.0 indicate pyrogenic sources that include incomplete combustion of fossil fuels or wood [42–44], and LMW/HMW ratios greater than 1.0 (indicating petrogenic sources) include refined oil or petroleum products [44]. In this study, the LMW/HMW ratios ranged from 0.45 to 8.93, which implies that pyrogenic and petrogenic sources contributed to the contamination at the site. PAH isomer pairs with similar molecular weights are further investigated to distinguish between petrogenic and pyrogenic sources. The LMW/HMW ratios were plotted against the BaA/(BaA + Chr) and InP/(InP + BP) ratios to identify the emission sources affecting Mount Heng, as shown in Figure 5. The BaA/(BaA + Chr) ratio has been used to indicate dominant fossil fuel emissions sources [15]. Petroleum combustion dominates PAH profiles when the BaA/(BaA + Chr) ratio is less than 0.2, a ratio between 0.20 and 0.35 results from a mixture of petroleum and coal combustions, and a ratio higher than 0.35 primarily result from coal combustion [8,44]. In this study, according to Figure 5a, most of the ratios are concentrated near 0.35, indicating that the dominant source was petroleum or coal combustion. The BaA and Chr concentrations were below the detection limit; thus, their ratios were not analyzed during the sandstorm. Furthermore, InP/(InP + BP) ratios smaller than 0.2 suggest a petroleum source, and this ratio ranged from 0.2 to 0.5, indicative of petroleum combustion; if this ratio is greater than 0.5, combustion of biomass, including grass, wood and coal, is inferred [8,45]. At Mount Heng, the main sources were petroleum combustion (according to Figure 5b).

The PAH compositional ratios presented above indicate that the dominant PAH sources affecting Mount Heng were coal, wood and petroleum combustion. As indicated in the correlation analysis, PAHs with 4, 5 and 6 rings were emitted from the same sources based on the correlation analysis, and the ratios of (BaA + Chr) and InP/(InP + BP) confirmed that these PAHs primarily resulted from the combustion of petroleum. No large differences were attributed to the sandstorm.

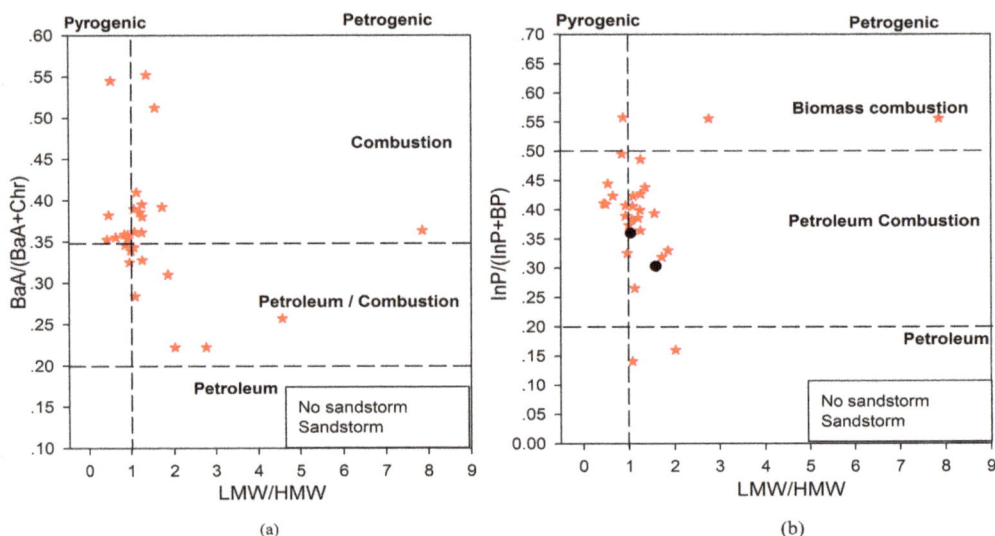

Figure 5. The ratios of LMW/HMW PAHs *versus* the ratios of BaA/(BaA + Chr) (**a**) and InP/(InP + BP) (**b**) in PM$_{2.5}$ for source identification.

3.3. Implications of Long-Range Transport for Air Pollution

A ratio analysis was used to estimate the ages of the air masses. Generally, BaA is degraded more easily and rapidly than its isomer Chr during transport due to the higher reactivity of the former species [21,46–48]. The ratios of more-reactive to less-reactive PAHs can be used to estimate whether an air mass is fresh or aged [21]. A higher ratio (>0.7) indicates that the air mass underwent relatively little photochemical processing and that the major pollutants were from local emissions, whereas a lower ratio suggests that the PAHs were primarily from long-range transport. The ratios of BaA/Chr were calculated in this study. During the sandstorm, the BaA and Chr concentrations were below the detection limit, which suggests that the air masses were relatively aged. Under normal (no sandstorm) conditions, most of the ratios were below 0.70. Therefore, the sampled air masses primarily resulted from remote emissions sources.

Pollutants emitted from distant regions in the free troposphere can travel long distances and strongly influence variations in pollutant concentrations [49]. Under normal conditions, the PAH concentrations varied greatly, and several special samples were analyzed to determine their LMW ($\sum$2–3 ring) and HMW ($\sum$4–6 ring) PAH concentrations. The backward trajectories of the air masses were calculated using an online version of the HYSPLIT (Hybrid Single Particle Lagrangian Integrated Trajectory) model from the NOAA (National Oceanic and Atmospheric Administration) Resources Laboratory, as shown in Figure A1. The

210

total concentrations of the LMW and HMW PAHs in the $PM_{2.5}$ were calculated. The concentrations of the LMW PAHs were 4–8 times greater than those of the HMW PAHs on 21, 31, 15 and 11 May and 9 April 2009. In particular, the 2-ring PAH concentrations were higher and accounted for 52.6% of the total PAHs on 9 April 2009. The concentrations of the HMW PAHs were approximately equal to the concentrations of the LMW PAHs on 19 and 20 March 2009, 20 and 22 April 2009 and 5 May 2009. The concentrations of the LMW PAHs were approximately 50% of the HMW PAHs on 27 April and 3, 4 and 9 May 2009, These variations were potentially caused by differences in sources and air mass transport. On 9 April 2009, the particle samples were collected before a rainstorm, and the air masses were traveling from the sea to the east. The samples collected between 11 May and 15 May 2009 were collected before rainy or foggy weather. Therefore, the higher LMW PAH concentrations on these days were potentially linked to the meteorological conditions. The air masses associated with higher HMW concentrations primarily traveled from eastern coastal cities (e.g., Shanghai, Shandong, and Guangdong Province), including two of China's main industrial regions in southern China, one centered in Shanghai and the other centered near Guangzhou. Thus, the higher HMW concentrations resulted from the transport of pollutants to the site, and the air mass that traveled from the south on 9 May 2009 had been impacted by a non-ferrous metal smelting plant in Hunan Province, which also contributed to the higher concentrations. The LMW concentrations were approximately equal to (or slightly higher than) the HMW concentrations and originated from three sources near coastal areas. On 4 May 2009, the total PAH concentrations were the highest, with the corresponding air masses originating in Shanghai three days earlier. In addition, many nearby factories also contributed to the PAH concentrations. Most of the PAH species collected on 4 May 2009 had undergone little photochemical processing, which explains their higher total concentrations. The LMW PAH concentrations were all higher than the HMW PAH concentrations during the sandstorm.

Based on the observed ratios (BaA/Chr), the PAH concentrations were primarily affected by emissions located far from the site and by some local sources. Figure 6a–h show the footprint retroplumes (residence times) and PSCs of the $PM_{2.5}$ samples collected on 25 and 26 April 2009. In the simulation, particle transport was simulated backwards in time over a 72-hour period.

The residence times at an altitude of 100 m were calculated to obtain the "footprint" retroplumes of released air particles in Figure 6a–f during the sandstorm, which represent the distribution of probability or the residence time of a simulated air mass [29,30]. The longer residence time (red and yellow colors in the figure) indicates a higher probability of influence on the sampling site. In addition, $PM_{2.5}$ was the primary vehicle for PAHs and was assumed to originate from the same sources discussed in this study. The air masses all originated in northern China and carried

211

the pollutants to the site during the sandstorm. The nighttime samples collected on 26 April 2009 (Figure 6h) displayed the lowest probability influence from the surroundings. The PM$_{2.5}$ concentrations in the samples on 26 April 2009 (12:50–16:30) were lower than those in the other samples collected during the sandstorm (as shown in Figure 2), which is consistent with the results of the residence time.

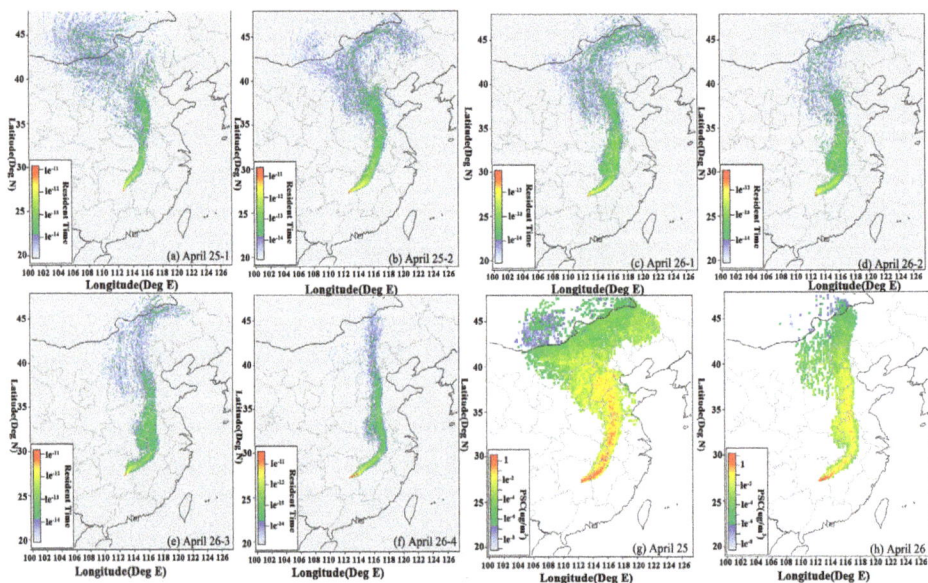

Figure 6. Illustrations of three-day backward retroplumes and potential source contributions (PSCs) during the sandstorm. (**a–f**) Map of the "footprint" (*i.e.*, 100 m) retroplume (residence time) and (**g–h**) maps of the PSCs of PM$_{2.5}$ (PAHs) at Mount Heng.

In addition, the PSCs of the PAHs were also calculated (Figure 6g–h) in this study. Air pollutants from northern China, including eastern Mongolia, could be transported to Mount Heng within three days, and significant source contributions in several regions were present along the route on 26 April 2009, including central eastern Hunan Province, northwestern Jiangxi Province and the nearby Shandong and Shanxi Provinces in northern China (Figure 6g). In addition, high emission source contributions on 26 April 2009 (Figure 6h) occurred along a similar route with a smaller area and lower concentrations in these regions. The emission sources were primarily in northeastern Hunan Province. Therefore, the PM$_{2.5}$ concentrations were higher on 25 April than on 26 April 2009 based on the PSC of the PM$_{2.5}$. The average measured concentration of PM$_{2.5}$ was 180.49 µg/m^3 on 25 April 2009, which was greater than the concentration of 144.67 µg/m^3 on 26 April 2009. This finding is consistent with the PSC results. In addition, the total PAH concentrations were higher on 25 April 2009, and most of the individual PAH concentrations were higher on 25 April than on 26 April 2009.

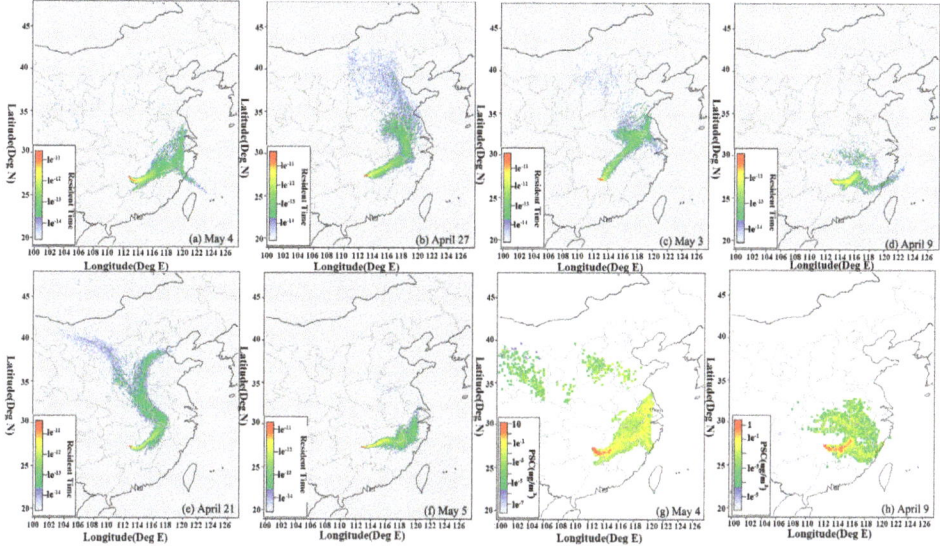

Figure 7. Illustrations of three-day backward retroplumes and the potential source contributions (PSCs) of higher PAH levels. (**a–f**) Map of the "footprint" (*i.e.*, 100 m) retroplume (residence time) and (**g–h**) maps of the PSCs of $PM_{2.5}$ (PAHs) at Mount Heng.

The highest total PAH concentrations were measured on 4 May, 27 April, 3 May, 9 April, and 21 April 2009 and the lowest on 15 May, 30 April, 23 April, 31 May and 21 May 2009 in the absence of sandstorm conditions, as shown in Figure 2. Their "footprint" retroplumes (residence times) of released air particles and the PSCs of the $PM_{2.5}$ (PAHs) are shown in Figures 7 and 8. The air masses associated with higher concentrations evidently all originated east of the site and passed over urban areas, particularly the air mass of 4 May 2009, which is shown as a large yellow area and extends across Jiangxi Province from west to east. Therefore, the total concentration was the highest on 4 May 2009 (as shown in Figure 2), which is consistent with the simulation results. The influences of emissions from the coastal cities of eastern China were able to reach the site within three days, and the dominant pollutants traveled from eastern Hunan Province and western Jiangxi Province, as indicated by Figure 7a–h. The large amounts of pollution emitted from the coastal cities of eastern China were carried to the site via atmospheric transport and resulted in higher $PM_{2.5}$ (PAH) concentrations. In contrast, to demonstrate the influences of the transport of lower concentrations, their "footprint" retroplumes (residence times) of released air particles and the PSCs of $PM_{2.5}$ (PAHs) are shown in Figure 8a–h. The air masses

213

associated with the two lowest concentrations all originated from the ocean to the south, and their corresponding areas depicted in red and yellow are clearly smaller, which indicates that almost no pollutants were carried to this site. The paths of the last three air masses (31 and 21 May, 22 April 2009) originated from near the site, which indicates that the pollutants were primarily from local emissions sources. Their concentrations were lower than those of the other samples. Therefore, the air masses originating from the ocean and carrying few pollutants resulted in lower $PM_{2.5}$ (PAHs) concentrations, and the local emission sources contributed little to the pollutant concentrations. The pollutants from long-range transport contributed more to the site.

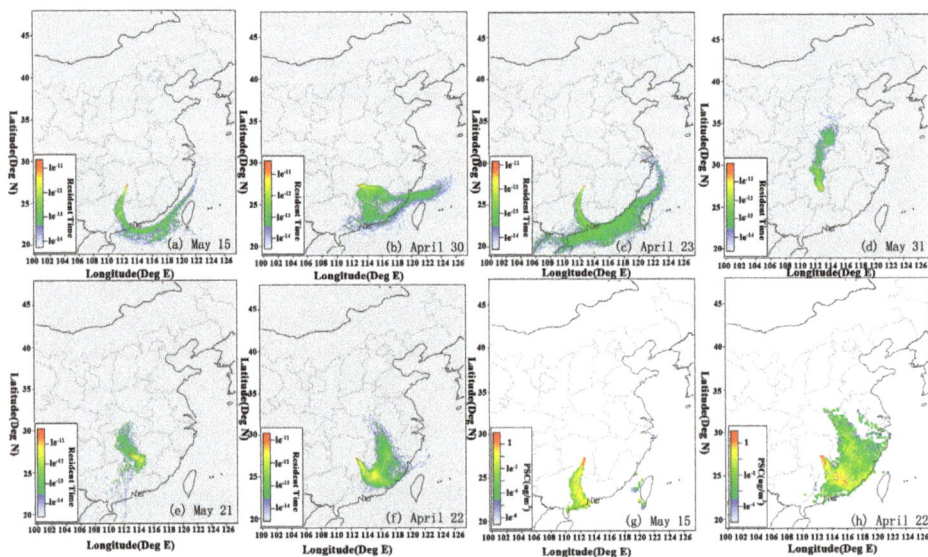

Figure 8. Illustrations of three-day backward retroplumes and potential source contributions (PSCs) of lower PAH levels. (**a–f**) Map of the "footprint" (*i.e.*, 100 m) retroplume (residence time) and (**g–h**) maps of the PSCs of $PM_{2.5}$ (PAHs) at Mount Heng.

Regardless of the presence of the sandstorm, the combination of backward simulation dispersion via the PSCs of the $PM_{2.5}$ (PAHs) revealed the importance of transport for the pollutant concentrations. During the sandstorm, the air masses all originated from northern China, and the distribution of the PSCs affected the $PM_{2.5}$ (PAHs) concentrations. Under normal (no sandstorm) conditions, the air masses associated with higher or lower concentrations originated from the same respective directions, and their variations in route resulted in higher or lower concentrations. The higher and lower concentrations demonstrate that the atmospheric transport significantly influenced the pollutant concentrations at Mount Heng.

4. Conclusions

A study of PAHs in $PM_{2.5}$ was conducted in northern China between March and May in 2009. A sandstorm from northern China struck Mount Heng on 25 and 26 April 2009 during the sampling campaign, and the sandstorm greatly affected the $PM_{2.5}$ and particulate PAH concentrations. During the sandstorm, the $PM_{2.5}$ levels rapidly increased to 156.61 $\mu g/m^3$, whereas before or after the sandstorm, the normal value was 58.50 $\mu g/m^3$. The mean mass concentration of the PAHs in $PM_{2.5}$ was 30.70 $\mu g/g$, and those of BaA, Chr and DbA were below the detection limits during the sandstorm. Overall, PhA, BP, Acy and Pyr were the predominant compounds, which together accounted for approximately 60% of the total PAH concentration. The mass PAH concentration at Mount Heng varied greatly around a mean of 80.15 $\mu g/g$, and the most abundant compounds were FluA, PhA, BbF and Pyr, which accounted for nearly 50% of the total under normal (no sandstorm) conditions. The individual PAH species were present at low to medium concentrations at Mount Heng. The 4- and 5-ring PAH concentrations were lower at Mount Heng than at the other sites during the sandstorm. In addition, the 3- and 6-ring PAHs were the most abundant, accounting for 66.8% of the total PAH concentration during the sandstorm and 51.5% under normal conditions. Under normal conditions, special samples were selected for analysis of the LMW and HMW PAH concentrations. Higher LMW PAH concentrations were linked to meteorological conditions, and the higher HMW concentrations were primarily affected by the transport of pollutants.

Linear regression analyses were performed between the total PAH concentrations and the corresponding $PM_{2.5}$ levels, and no clear relationship existed between these values. The relationships among the various-ring PAH concentrations suggested that 4-, 5- and 6-ring PAHs were potentially emitted from a similar source. The ratios of LMW/HMW, BaA/(BaA + Chr) and InP/(InP + BP) were used to identify the emissions sources, and indicated that the PAHs had primarily been emitted from the combustion of wood, coal or petroleum. The isomer pair of BaA/Chr suggested that most of the air masses collected at Mount Heng resulted from long-range transport. The combination of "footprint" retroplumes of released air particles (residence time) and the PSCs of the $PM_{2.5}$ (PAHs) with higher and lower concentrations indicated that local emission sources had little effect on the PAH concentrations (*i.e.*, the concentrations in the air masses that were primarily collected from local sources were much lower) and that long-range transport played a vital role in the pollutant concentrations at Mount Heng under normal conditions.

Acknowledgments: This study is based on work supported by funding from the National Natural Science Foundation of China (grants 21177073) and the China National Basic Research Program (2005CB422203). We are grateful to the Mount Heng Meteorological Station for providing access to the experimental site, to Jinan Academy of Environmental sciences (Zaifeng

Wang and Houyong Zhang) for providing experimental instruments, and to AJE (American Journal Experts) for editing the paper.

Author Contributions: Minmin Yang and Yan Wang wrote the article and design the study, Qiang Liu and Aijun Ding provided the figures of resident time and PSCs. Yuhua Li helped perform the statistical analysis.

Conflicts of Interest: The authors declare no conflict of interest.

Appendix A.

In this appendix, concentrations of HMW (4-6 ring) and LMW (2-6 ring) PAHs were compared and the backward trajectories of the air masses collected were presented in Figure A1. The HYSPLIT model was used to calculate backward trajectories for each day. The air masses were categorized into three categories based on concentrations. Figure A1a presented the backward trajectories that the concentrations of LMW PAHs were much higher than those of HMW and the backward trajectories that concentrations of HMW PAHs were approximately equal to the concentrations of LMW PAHs were presented in Figure A1b. Backward trajectories of HMW PAH concentrations smaller than the LMW PAH concentrations were shown in Figure A1c. The aim of this section is to indicate that the potential influence on concentrations of HMW and LMW PAHs respectively, including meteorological conditions and long range transport.

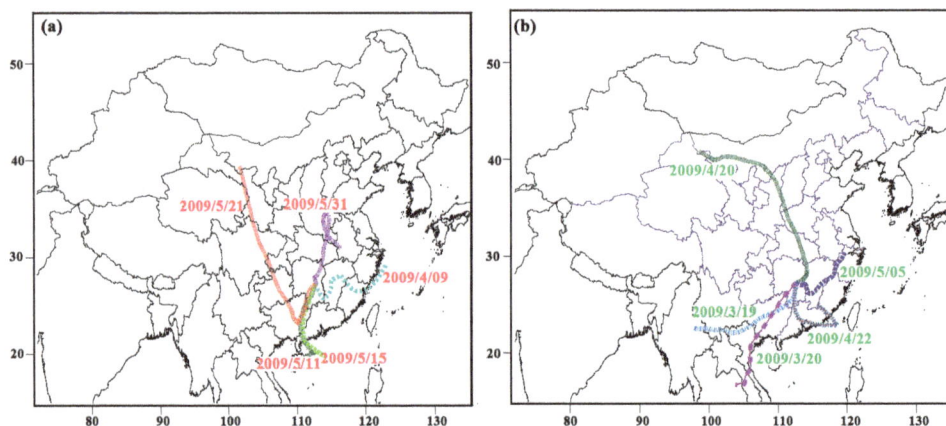

Figure A1. *Cont.*

216

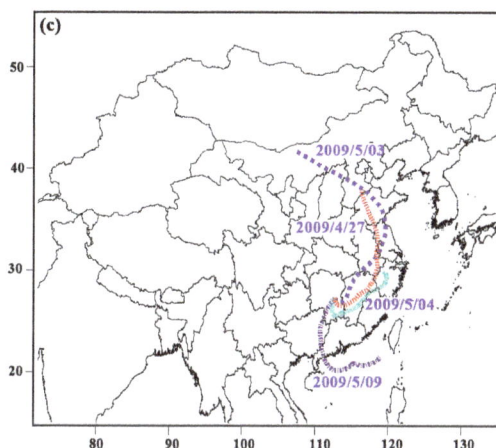

Figure A1. The backward trajectories of the selected air masses based on the LMW and HMW PAH concentrations, (**a**) LMW » HMW, (**b**) LMW ≈ HMW, (**c**) LMW « HMW.

References

1. Wang, Y.; Hopke, P.K.; Xia, X.; Rattigan, O.V.; Chalupa, D.C.; Utell, J.E. Source apportionment of airborne particulate matter using inorganic and organic species as tracers. *Atmos. Environ.* **2012**, *55*, 525–532.
2. Pui, D.Y.H.; Chen, S.-C.; Zuo, Z. PM$_{2.5}$ in China: Measurements, sources, visibility and health effects and mitigation. *Particuology* **2013**, *13*, 1–26.
3. Li, Z.; Porter, E.N.; Sjödin, A.; Needham, L.L.; Lee, S.; Russell, A.G.; Mulholland, J.A. Characterization of PM$_{2.5}$ bound polycyclic aromatic hydrocarbons in Atlanta—Seasonal variations at urban, suburban, and rural ambient air monitoring sites. *Atmos. Environ.* **2009**, *43*, 4187–4193.
4. Neff, J. Polycyclic aromatic hydrocarbons in the aquatic environment and cancer risk to aquatic organisms and man. *Am. Environ. Prot. Agency Rep.* **1982**, *600*, 9–82.
5. Bopp, S.K.; Lettieri, T. Gene regulation in the marine diatom *thalassiosira pseudonana* upon exposure to polycyclic aromatic hydrocarbons (PAHs). *Gene* **2007**, *396*, 293–302.
6. Wilhelm, M.; Ghosh, J.K.; Su, J.; Cockburn, M.; Jerrett, M.; Ritz, B. Traffic-related air toxics and preterm birth: A population-based case-control study in Los Angeles County, California. *Environ. Health* **2011**, *10*, 89–89.
7. Straif, K.; Baan, R.; Grosse, Y.; Secretan, B.; El Ghissassi, F.; Cogliano, V. Carcinogenicity of household solid fuel combustion and of high-temperature frying. *Lancet Oncol.* **2006**, *7*, 977–978.
8. Yunker, M.B.; Macdonald, R.W.; Vingarzan, R.; Mitchell, R.H.; Goyette, D.; Sylvestre, S.; Mitchelld, D. PAHs in the Fraser River basin: A critical appraisal of PAH ratios as indicators of PAH source and composition. *Org. Geochem.* **2002**, *33*, 489–515.

9.   Zakaria, M.P.; Takada, H.; Tsutsumi, S.; Ohno, K.; Yamada, J.; Kouno, E.; Kumata, H. Distribution of polycyclic aromatic hydrocarbons (PAHs) in rivers and estuaries in Malaysia: A widespread input of petrogenic PAHs. *Environ. Sci. Technol.* **2002**, *36*, 1907–1918.

10. Zhang, Y.; Tao, S. Global atmospheric emission inventory of polycyclic aromatic hydrocarbons (PAHs) for 2004. *Atmos. Environ.* **2009**, *43*, 812–819.

11. Chang, K.-F.; Fang, G.-C.; Chen, J.-C.; Wu, Y.-S. Atmospheric polycyclic aromatic hydrocarbons (PAHs) in Asia: A review from 1999 to 2004. *Environ. Pollut.* **2006**, *142*, 388–396.

12. Wang, Y.; Li, P.H.; Li, H.L.; Liu, X.H.; Wang, W.X. PAHs distribution in precipitation at Mount Taishan China. Identification of sources and meteorological influences. *Atmos. Res.* **2010**, *95*, 1–7.

13. Liu, F.; Xu, Y.; Liu, J.; Liu, D.; Li, J.; Zhang, G.; Li, X.; Zou, S.; Lai, S. Atmospheric deposition of polycyclic aromatic hydrocarbons (PAHs) to a coastal site of Hong Kong, South China. *Atmos. Environ.* **2013**, *69*, 265–272.

14. Tian, F.; Chen, J.; Qiao, X.; Wang, Z.; Yang, P.; Wang, D.; Ge, L. Sources and seasonal variation of atmospheric polycyclic aromatic hydrocarbons in Dalian, China: Factor analysis with non-negative constraints combined with local source fingerprints. *Atmos. Environ.* **2009**, *43*, 2747–2753.

15. Wu, Y.; Yang, L.; Zheng, X. Characterization and source apportionment of particulate PAHs in the roadside environment in Beijing. *Sci. Total. Environ.* **2014**, *470*, 76–83.

16. Arellano, L.; Fernández, P.; Tatosova, J.; Stuchlik, E.; Grimalt, J.O. Long-range transported atmospheric pollutants in snowpacks accumulated at different altitudes in the Tatra Mountains (Slovakia). *Environ. Sci. Technol.* **2011**, *45*, 9268–9275.

17. Carrera, G.; Fernández, P.; Grimalt, J.O.; Ventura, M.; Camarero, L.; Catalan, J.; Nickus, U.; Thies, H.; Psenner, R. Atmospheric deposition of organochlorine compounds to remote high mountain lakes of Europe. *Environ. Sci. Technol.* **2002**, *36*, 2581–2588.

18. Fernández, P.; Carrera, G.; Grimalt, J.O.; Ventura, M.; Camarero, L.; Catalan, J.; Nickus, U.; Thies, H.; Psenner, R. Factors governing the atmospheric deposition of polycyclic aromatic hydrocarbons to remote areas. *Environ. Sci. Technol.* **2003**, *37*, 3261–3267.

19. Sun, M.; Wang, Y.; Wang, T.; Fan, S.; Wang, W.; Li, P.; Guo, J.; Li, Y. Cloud and the corresponding precipitation chemistry in south China: Water-soluble components and pollution transport. *J. Geophys. Res.: Atmos.* **2010**, *115*, D22.

20. Techniquesb, A.P. Method 8270c Semivolatile Organic Compounds by Gas Chromatography/Mass Spectrometry (Gc/Ms). 1996; Available online: http://mx.gemc.gov.cn/upload/20120201035748.pdf (accessed on 22 October 2015).

21. Ding, X.; Wang, X.-M.; Xie, Z.-Q. Atmospheric polycyclic aromatic hydrocarbons observed over the North Pacific Ocean and the Arctic area: Partial distribution and source identification. *Atmos. Environ.* **2007**, *41*, 2061–2072.

22. Li, P.H.; Wang, Y.; Li, Y.H.; Wang, Z.F.; Zhang, H.Y. Characterization of polycyclic aromatic hydrocarbons deposition in $PM_{2.5}$ and cloud/fog water at Mount Taishan (China). *Atmos. Environ.* **2010**, *44*, 1996–2003.

23. Draxler, R.R.; Hess, G.D. An overview of the HYSPLIT_4 modelling system for trajectories dispersion, and deposition. *Aus. Meteorol. Maga.* **1998**, *47*, 295–308.

24. Ding, A.; Wang, T.; Fu, C. Transport characteristics and origins of carbon monoxide and ozone in Hong Kong, South China. *J. Geophys. Res.: Atmos.* **2013**, *118*, 9475–9488.

25. Draxler, R.R. Demonstration of a global modeling methodology to determine the relative importance of local and long-distance sources. *Atmos. Environ.* **2007**, *41*, 776–789.

26. Custódio, D.; Ferreira, C.; Alves, C.; Duarte, M.; Nunes, T.; Cerqueira, M.; Pio, C.; Frosini, D.; Colombi, C.; Gianelle, V. Urban aerosol in Oporto, Portugal: Chemical characterization of PM_{10} and $PM_{2.5}$. Available online: http://adsabs.harvard.edu/abs/2014EGUGA.16.2454C (accessed on 22 October 2015).

27. Ding, A.; Fu, C.; Yang, X.; Sun, J.; Zheng, L.; Xie, Y.; Herrmann, E.; Nie, W.; Petäjä, T.; Kerminen, V.-M. Ozone and fine particle in the western Yangtze River Delta: An overview of 1 year data at the SORPES station. *Atmos. Chem. Phys.* **2013**, *13*, 5813–5830.

28. Nie, W.; Ding, A.; Wang, T.; Kerminen, V.-M.; George, C.; Xue, L.; Wang, W.; Zhang, Q.; Petäjä, T.; Qi, X. Polluted dust promotes new particle formation and growth. *Sci. Rep.* **2014**, *4*.

29. Ding, A.; Wang, T.; Xue, L.; Gao, J.; Stohl, A.; Lei, H.; Jin, D.; Ren, Y.; Wang, X.; Wei, X. Transport of north China air pollution by midlatitude cyclones: Case study of aircraft measurements in summer 2007. *J. Geophys. Res.: Atmos.* **2009**, *114*, D8.

30. Stohl, A.; Forster, C.; Eckhardt, S.; Spichtinger, N.; Huntrieser, H.; Heland, J.; Schlager, H.; Wilhelm, S.; Arnold, F.; Cooper, O. A backward modeling study of intercontinental pollution transport using aircraft measurements. *J. Geophys. Res.: Atmos.* **2003**, *108*, 4370–4370.

31. Zhang, Q.; Streets, D.G.; Carmichael, G.R.; He, K.; Huo, H.; Kannari, A.; Klimont, Z.; Park, I.; Reddy, S.; Fu, J. Asian emissions in 2006 for the NASA INTEX-B mission. *Atmos. Chem. Phys.* **2009**, *9*, 5131–5153.

32. Guo, H.; Lee, S.C.; Ho, K.F.; Wang, X.M.; Zou, S.C. Particle-associated polycyclic aromatic hydrocarbons in urban air of Hong Kong. *Atmos. Environ.* **2003**, *37*, 5307–5317.

33. Yang, H.-H.; Lee, W.-J.; Chen, S.-J.; Lai, S.-O. PAH emission from various industrial stacks. *J. Hazard. Mater.* **1998**, *60*, 159–174.

34. Ho, K.F.; Lee, S.C.; Chiu, G.M. Characterization of selected volatile organic compounds, polycyclic aromatic hydrocarbons and carbonyl compounds at a roadside monitoring station. *Atmos. Environ.* **2002**, *36*, 57–65.

35. Kulkarni, P.; Venkataraman, C. Atmospheric polycyclic aromatic hydrocarbons in Mumbai, India. *Atmos. Environ.* **2000**, *34*, 2785–2790.

36. Park, S.S.; Kim, Y.J.; Kang, C.H. Atmospheric polycyclic aromatic hydrocarbons in Seoul, Korea. *Atmos. Environ.* **2002**, *36*, 2917–2924.

37. Brown, J.N.; Peake, B.M. Sources of heavy metals and polycyclic aromatic hydrocarbons in urban stormwater runoff. *Sci. Total. Environ.* **2006**, *359*, 145–155.

38. Budzinski, H.; Jones, I.; Bellocq, J.; Pierard, C.; Garrigues, P.H. Evaluation of sediment contamination by polycyclic aromatic hydrocarbons in the Gironde estuary. *Mar. Chem.* **1997**, *58*, 85–97.

39. Dunbar, J.C.; Lin, C.-I.; Vergucht, I.; Wong, J.; Durant, J.L. Estimating the contributions of mobile sources of PAH to urban air using real-time PAH monitoring. *Sci. Total. Environ.* **2001**, *279*, 1–19.

40. Feng, J.; Hu, M.; Chan, C.K.; Lau, P.S.; Fang, M.; He, L.; Tang, X. A comparative study of the organic matter in PM$_{2.5}$ from three Chinese megacities in three different climatic zones. *Atmos. Environ.* **2006**, *40*, 3983–3994.

41. Zhang, W.; Zhang, S.; Wan, C.; Yue, D.; Ye, Y.; Wang, X. Source diagnostics of polycyclic aromatic hydrocarbons in urban road runoff, dust, rain and canopy throughfall. *Environ. Pollut.* **2008**, *153*, 594–601.

42. Mai, B.; Qi, S.; Zeng, E.Y.; Yang, Q.; Zhang, G.; Fu, J.; Sheng, G.; Peng, P.; Wang, Z. Distribution of polycyclic aromatic hydrocarbons in the coastal region of Macao, China: Assessment of input sources and transport pathways using compositional analysis. *Environ. Sci. Technol.* **2003**, *37*, 4855–4863.

43. Mai, B.-X.; Fu, J.-M.; Sheng, G.-Y.; Kang, Y.-H.; Lin, Z.; Zhang, G.; Min, Y.-S.; Zeng, E.Y. Chlorinated and polycyclic aromatic hydrocarbons in riverine and estuarine sediments from Pearl River Delta, China. *Environ. Pollut.* **2002**, *117*, 457–474.

44. Soclo, H.H.; Garrigues, P.H.; Ewald, M. Origin of polycyclic aromatic hydrocarbons (PAHs) in coastal marine sediments: Case studies in Cotonou (Benin) and Aquitaine (France) areas. *Mar. Pollut. Bull.* **2000**, *40*, 387–396.

45. Ravindra, K.; Sokhi, R.; van Grieken, R. Atmospheric polycyclic aromatic hydrocarbons: Source attribution, emission factors and regulation. *Atmos. Environ.* **2008**, *42*, 2895–2921.

46. Butler, J.D.; Crossley, P. Reactivity of polycyclic aromatic hydrocarbons adsorbed on soot particles. *Atmos. Environ.* **1981**, *15*, 91–94.

47. Kamens, R.M.; Guo, Z.; Fulcher, J.N.; Bell, D.A. The influence of humidity, sunlight, and temperature on the daytime decay of polyaromatic hydrocarbons on atmospheric soot particles. *Environ. Sci. Technol.* **1988**, *22*, 103–108.

48. Schauer, J.J.; Rogge, W.F.; Hildemann, L.M.; Mazurek, M.A.; Cass, G.R.; Simoneit, B.R. Source apportionment of airborne particulate matter using organic compounds as tracers. *Atmos. Environ.* **1996**, *30*, 3837–3855.

49. Gao, J.; Wang, T.; Ding, A.; Liu, C. Observational study of ozone and carbon monoxide at the summit of mount Tai (1534m asl) in central-eastern China. *Atmos. Environ.* **2005**, *39*, 4779–4791.

Effect of Nearby Forest Fires on Ground Level Ozone Concentrations in Santiago, Chile

María A. Rubio, Eduardo Lissi, Ernesto Gramsch and René D. Garreaud

Abstract: On 4 and 8 January 2014, at the height of the austral summer, intense wildfires in forests and dry pastures occurred in the Melipilla sector, located about 70 km to the southwest of Santiago, the Chilean capital, affecting more than 6 million inhabitants. Low level winds transported the forest fire plume towards Santiago causing a striking decrease in visibility and a marked increase in the concentration of both primary (PM_{10} and CO) and secondary (Ozone) pollutants in the urban atmosphere. In particular, ozone maximum concentrations in the Santiago basin reached hourly averages well above 80 ppb, the national air quality standard. This ozone increase took place at the three sampling sites considered in the present study. These large values can be explained in terms of high NOx concentrations and NO_2/NO ratios in biomass burning emissions.

Reprinted from *Atmosphere*. Cite as: Rubio, M.A.; Lissi, E.; Gramsch, E.; Garreaud, R.D. Effect of Nearby Forest Fires on Ground Level Ozone Concentrations in Santiago, Chile. *Atmosphere* **2015**, *6*, 1926–1938.

1. Introduction

Air plumes originating in forest fires are rich in primary pollutants such as particles, carbon monoxide (CO), non-methane volatile organic compounds (VOCs), and nitrogen oxides (NOx) [1–3] Photochemical transformations of those air masses can produce secondary pollutants, including ozone [4], that are transported over large distances [1,3,5–9]. These plumes can also affect nearby cities leading to ground level concentrations that largely surpass air quality standards [8,10,11]. Nevertheless, there are few evaluations regarding ozone behavior in large cities exposed to polluted masses transported from nearby forest fires [3,8,11,12] Results range from moderate decreases at Arizona and Central Texas, US, and western Mexico, when the cities are located less than 400 km from fire [12], to large increases as in the city of Edmonton, Canada, when the city is located 300 km from the fire [11]. On the other hand, measurements in Mexico City (MCMA) did not demonstrate significant differences in ground level ozone concentrations in periods with active fires [13]. Evaluation of the effect of nearby wild fires on the concentration of pollutants in the air at ground level of large cities is particularly complex due to the mixing of local emissions with the fire derived polluted plumes [14]. Interestingly, it has been reported that in western

U.S, the ozone was significantly correlated with forest fires in the surrounding $5° \times 5°$ and $10° \times 10°$ grids, but not with wild fires in the nearest $1° \times 1°$ region (110×110 km, approximately), reflecting a subtle balance between ozone production and destruction in NOx rich environments [15].

Santiago, the Chilean capital, is a large city (with just over 6 million inhabitants, accounting for 40% of the national population with a density of 393 inhabitants per km^2) located at a subtropical latitude ($33°S$), about 100 km from the Pacific coast and just to the east of the Andes cordillera (Figure 1). It features a semi-arid climate with annual mean precipitation of 310 mm, almost exclusively concentrated in winter months [16]. During the dry summer months (November to March), the city is exposed to relatively high concentrations of secondary oxidants [17,18], particularly in its east side where 35% of the summer days surpassed the ozone national safety standard of 60 ppbv for an 8 h mobile running average or 80 ppbv for one hour [19,20].

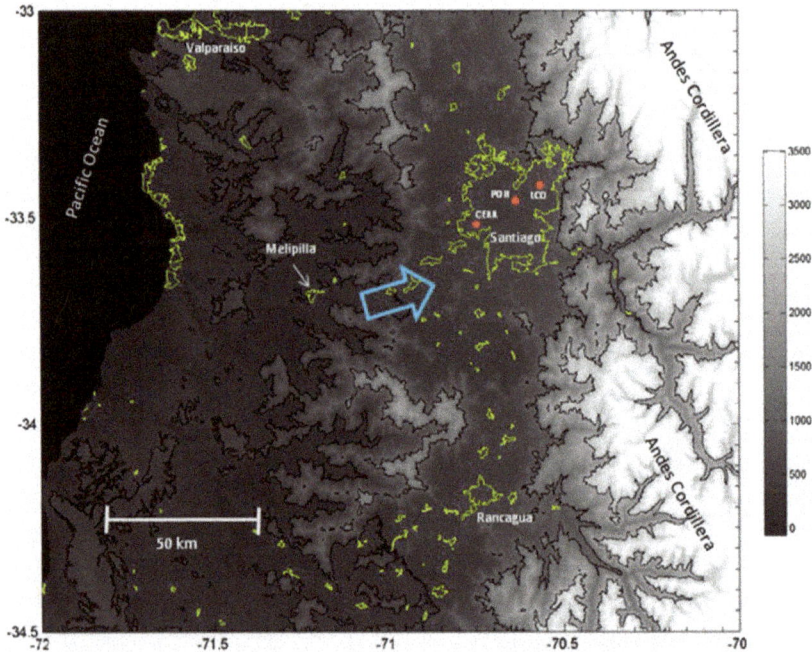

Figure 1. Topographic map of central Chile. Gray shading in m ASL; also indicated the 300, 1200 and 2400 m above sea level (ASL) topographic contours (black lines). Yellow lines outline main cities, red dots indicate the location of the MACAM (Air Quality Monitoring Program in Santiago Metropolitan Area) stations Cerrillos, Parque O'Higgins and Las Condes. The blue thick arrow indicates the predominant low-level winds from the SW during summer afternoon (adapted from [21], see also Figure 3).

During 4 and 8 January 2014, two sizable forest fires took place in the Melipilla sector located about 70 km from downtown Santiago. These fires covered approximately 15 km^2 of a Mediterranean landscape characterized by a mixture of pasture, open woodland and shrubland, including some *Eucalyptus* and *Acacia caven* trees [22,23]. The dominant vegetation and the changes introduced in recent years have been discussed by Schulz *et al.* [23]. The pre-Columbian vegetation of Central Chile is dominated by pastures and *Acacia caven* shrubland that have been replaced by vineyard and exotic trees such *Pinus radiata* and *Eucalyptus globulus*. These fires produced a noticeable increase in visible particles in the urban atmosphere of Santiago. In this short contribution, we present a critical discussion of ozone and other pollutants' behavior during those days. Given the possible effect of the large amounts of oxidants in the urban air on Santiago's inhabitants [24–26], we analyze the ozone concentrations and discuss the possible reasons for the extremely high values found in different areas of Santiago's basin.

2. Observations and Methods

Meteorological data (air temperature, relative humidity, incoming solar radiation and wind direction and speed) in the central part of Santiago (near downtown) was obtained from two automatic weather stations (USACH and DGF-UCh) recording 15 min averages. Station USACH (33.45°S, 70.68°W, 528 m. above sea level (ASL)) is a Novalynx Corp. Station, model 110-WS-16. Station DGF-UCh (33.46°S, 70.66°W, 542 m. ASL) includes a standard Campbell Scientific AWS and a laser ceilometer (model VAISALA CL-31) retrieving backscatter reflectivity profiles from the surface up to 6 km above ground level every 30 s. The laser reflectivity has been related with the aerosol loading in the mixed layer [27]. One-minute average wind speed and wind direction at 10 m above the ground were also available from station El Paico (33.72°S, 71.02°W, 312 m. ASL), operated by the National Weather Service (DMC). Station El Paico is located in the Maipo valley connecting the Melipilla sector (where the fires took place) with Santiago.

Ozone, NO_2, NO, PM_{10}, $PM_{2.5}$ and CO concentrations at ground level were obtained from the Air Quality Monitoring Program in Santiago Metropolitan Area (MACAM-2) operated by the Environmental Ministry (SINCA, 2014) [28]. Specifically, hourly average values were gathered from three sampling stations located in Santiago (Figure 1): Cerrillos (CERR) in the west side of the city, (33.49°S; 70.71°W; 528 m. ASL); Parque O'Higgins (POH) in downtown (33.46°S, 70.66°W, 562 m. ASL) and Las Condes (LC) at the North East side of the city (33.37°S, 70.52°W; 811 m. ASL). The map in Figure 1 also shows the location of Melipilla, the small town around which the forest fires took place in January 2014.

In all the stations, ozone was measured using a Thermo UV-Photometric ozone analyzer, model 49i (measuring range: 0–0.5 ppm). NO-NO_2-NO_X were measured using a Thermo chemiluminsescent gas analyzer, model 42i (measuring range 0.05–100 ppm with a detection limit of 0.40 ppb). CO was analyzed using a Thermo CO Analyzer, model 40i (measuring range 0–50 mg/m^3 with a detection limit of 4.0 ppm). PM$_{10}$ was measured using a PM$_{10}$ Monitor TEOM 1405 (measuring range of 0–1,000,000 µg/m^3. PM$_{2.5}$ was measured using a MET-ONE BAM-1020 monitor (measuring range of 0.1–10 µg/m^3, with a detection limit ⩽1.0 µg/m^3.

3. Results and Discussion

3.1. Meteorological Aspects

Ground level concentrations of primary and secondary pollutants in the proximity of wildfires strongly depend on the direction and intensity of the low-level flow. As evident in the satellite imageries given in Figure 2, the forest fires in the Melipilla sector released a considerable amount of pollutants (ash, particulate matter, gasses) that were transported to Santiago by the southwesterly low-level flow that regularly develops during the afternoon in summer months (Figure 3) [21]. On both 4 and 8 January 2014, the near-surface wind speed at El Paico reached about 10 m/s by noon (not shown). Thus, the fire plume that originated during the morning hours took less than 3 h to reach the city of Santiago, located about 70 km upwind of the Melipilla sector. The arrival of the polluted plume produced a striking reduction in visibility, readily evident for the general population (especially for the case of 4 January), as illustrated by the time-height diagram of the backscatter reflectivity from the laser ceilometer at DGF-UCh (Figure 4). The reflectivity, indicative of the aerosol loading in the mixed layer [27], exhibits a sharp increase around 2 PM, 4 January, encompassing the first 500 m above ground level. The marked increase in aerosol loading was accompanied by a simultaneous stalling of the air temperature (that often maximize around 4 PM) and a reduction of the global solar radiation relative to the previous day (when very similar synoptic conditions prevailed), which integrated over the course of the afternoon represents a deficit of about 15% of the insolation for a typical clear sky day in Santiago.

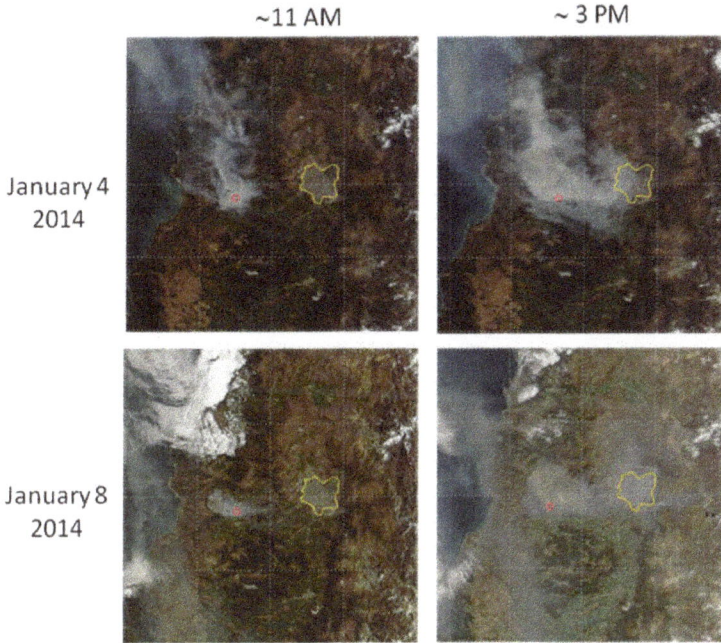

Figure 2. Corrected reflectance (True Color) scenes from MODIS for 4 and 8 January 2014, over central Chile (same region as in Figure 1). The plume of the forest fires in the Melipilla sector (red circle) reaching Santiago (yellow polygon) is evident in all the images. The left column shows images from the AQUA satellite at about 11 a.m., the right column shows images from the TERRA satellite at about 3 p.m.

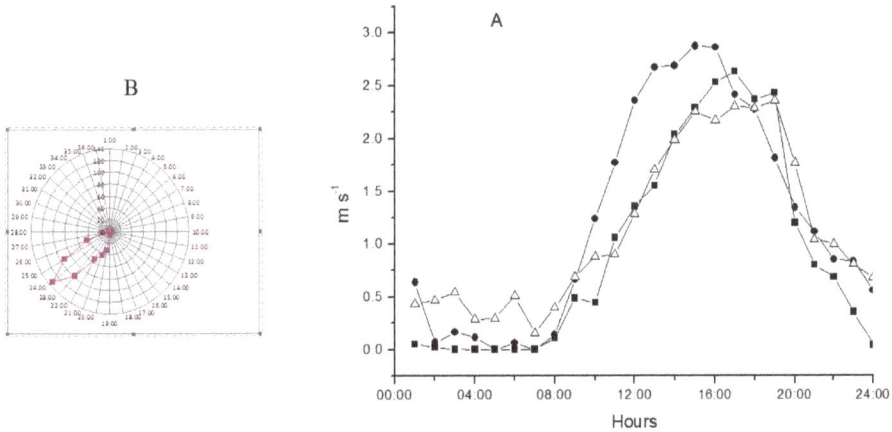

Figure 3. (a) Wind speed and wind in days with active fires (■ January 8; ● January 4; and days without fires Δ; (b) Wind rose for 4 January, 14:00–24:00.

225

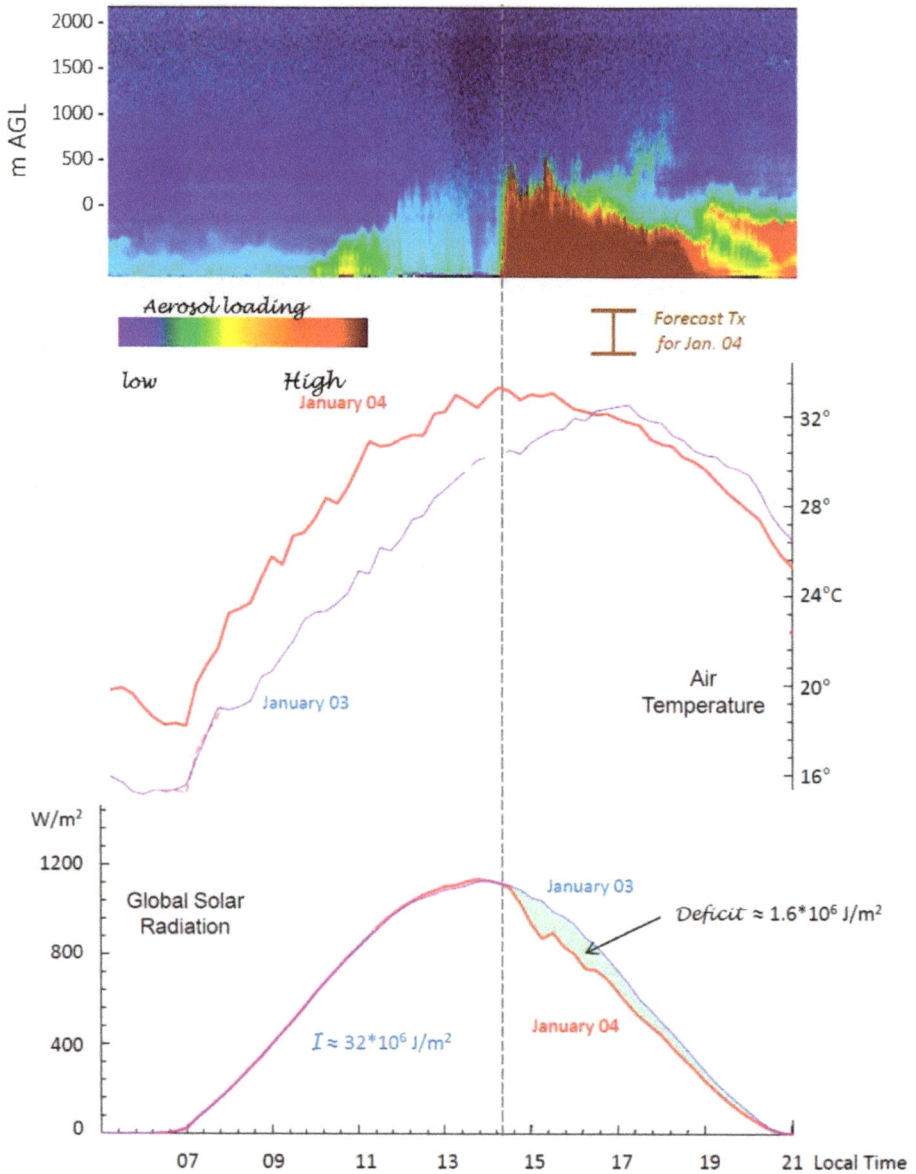

Figure 4. Upper panel: Time-height diagram of the backscatter reflectivity from a laser ceilometer (VAISALA CL-31) located at Santiago downtown during 4 January 2014 (active fire). The reflectivity has been related with the aerosol loading in the mixed layer [27] which is qualitatively indicated here from low to high. Middle panel: Air temperature 4 January (solid red line) and the previous day (3 January, blue line). Lower panel: Global solar radiation reaching the surface at Santiago for 3 January (blue line) and 4 January (red line).

3.2. Impacts on Primary Pollutants

The arrival of the forest fire plumes in Santiago markedly increased the concentration of pollutants measured at ground level relative to non-fire days, as illustrated by the time series of MP_{10} in CERR and LC (Figure 5). A similar behavior is observed for different pollutants in the three sampling sites summarized in Table 1. This table shows the maximum daily concentrations of CO, PM_{10} and O_3 during days with active fires in the Melipilla sector and the average daily maximum for non-fire days during summer (days from 2011, 2012 and 2013). Note that the presence of active fires in Melipilla increased nearly three times the maximum CO concentration in Santiago relative to the background reference values. Similar differences were obtained in fires near to Cordoba city (Argentina) in a scenario with characteristics similar to those of Santiago [29].

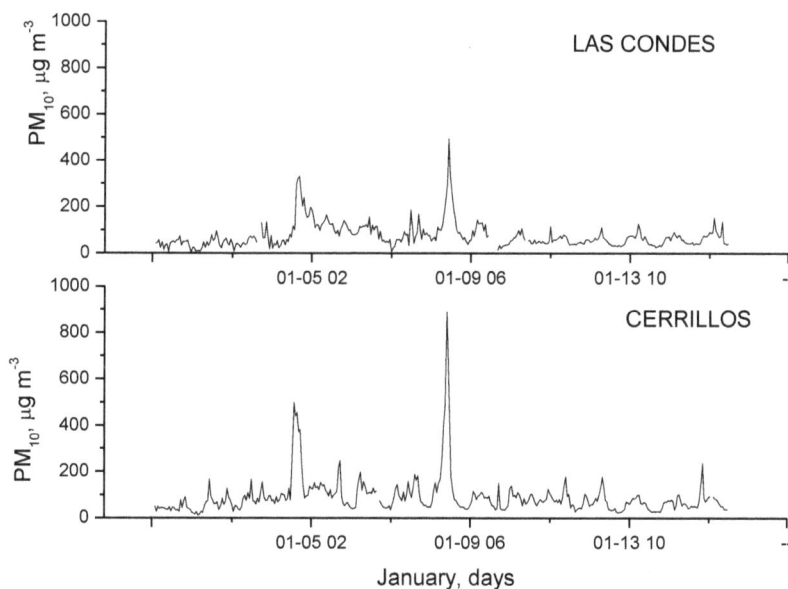

Figure 5. PM_{10} time profile measured in Cerrillos and las Condes from 1 to 14 January 2014.

The occurrence of nearby wildfires also changed the diurnal cycle of the pollutants. In absence of fires, the highest CO and PM_{10} concentrations in POH and CERR occur during the morning rush hours (7:00–8:00 a.m.) consistent with the elevated emissions from mobile sources in that period, while the maximum concentrations at LC occur around 11:00 a.m., as expected for air masses transported from downtown (Figure 6). In contrast, during days with active wildfires, the maximum CO and PM_{10} levels were attained between 15:00 and 16:00, irrespective

227

of the sampling location, suggesting the dominant role of the transport processes (Figure 6).

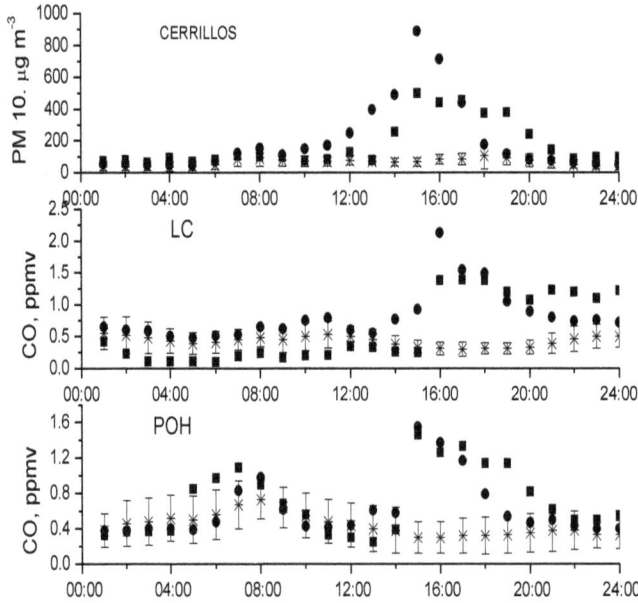

Figure 6. Comparison of data obtained in days with active fires (4 January 2014 and 8 January 2014) (■,●) and the average of background data for days without fires during January 2014 (∗).

Table 1. Historical average maximum values (January days from years 2011, 2012 and 2013) and maximum values in days with active fires (4 January 2014 and 8 January 2014).

Site	Day	CO (ppmv)	PM_{10} ($\mu g \cdot m^{-3}$)	O_3 (ppbv)
Cerrillos	Historical Average	0.61 ± 0.09	116 ± 19	45.1 ± 4.7
	4 January 2014	1.84	493	127
	8 January 2014	2.26	888	92
Parque O'higgins	Historical Average	0.50 ± 0.24	83.8 ± 22.4	43.4 ± 20.0
	4 January 2014	1.46	433	133
	8 January 2014	1.55	493	91
Las Condes	Historical Average	0.57 ± 0.18	109 ± 32	76 ± 17
	4 January 2014	1.39	331	142
	8 January 2014	2.13	493	105

3.3. Ozone in Wildfire Days

Ozone concentrations were also considerably enhanced during the upwind forest fires. Since ozone concentrations depend upon the sampling place and the maximum daily temperature [30–32] the data are plotted as a function of the air temperature in Santiago at each location (Figure 7). This figure shows that nearby fires significantly contributed to the high ozone concentration all over the Santiago basin. The values obtained when fires were active are well outside the 95% confidence limit of the data obtained using the historical background reference days. The contribution of nearby fires to ozone urban levels can be estimated from the difference between measured values and those expected from the maximum daily temperature [33] included in Table 2. The extra urban ozone associated to Melipilla's wildfires went up to 65 ppb, amounting to a 100% increase. This is a value considerably larger than those reported elsewhere [3]. Furthermore, ozone concentration during wildfire days consistently surpassed the 80 ppb limit in the three locations considered, a level considered harmful for humans, animals and vegetation [34,35] even during photochemically inactive seasons [8].

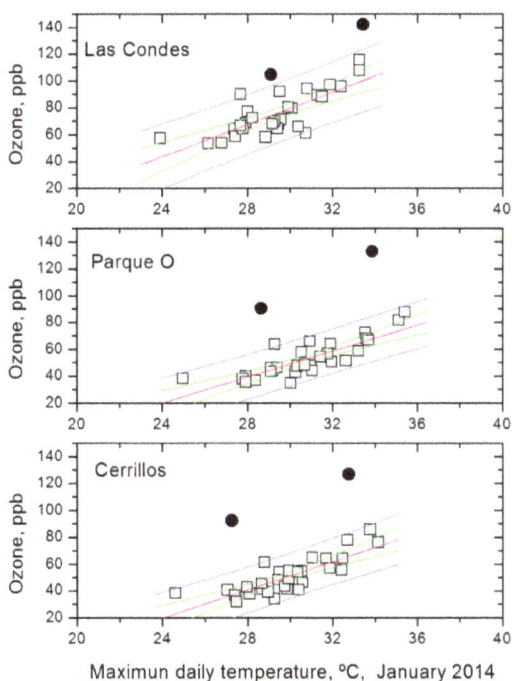

Figure 7. Lineal relationship between Ozone maximum one hour average and daily maximum temperature. (□) with Confidence band and Prediction band (95%). Measurements in days with nearby fires (■).

Similar to the evolution of other pollutants, ozone maximum concentrations took place during the late afternoon of the days with wildfire in Melipilla, approximately two hours later than in background reference days (Figure 8). Notably, at these times, there is a noticeable increase in NO_2 concentrations in the aging plume (Figure 9) that can push up the ground level ozone concentration as discussed below.

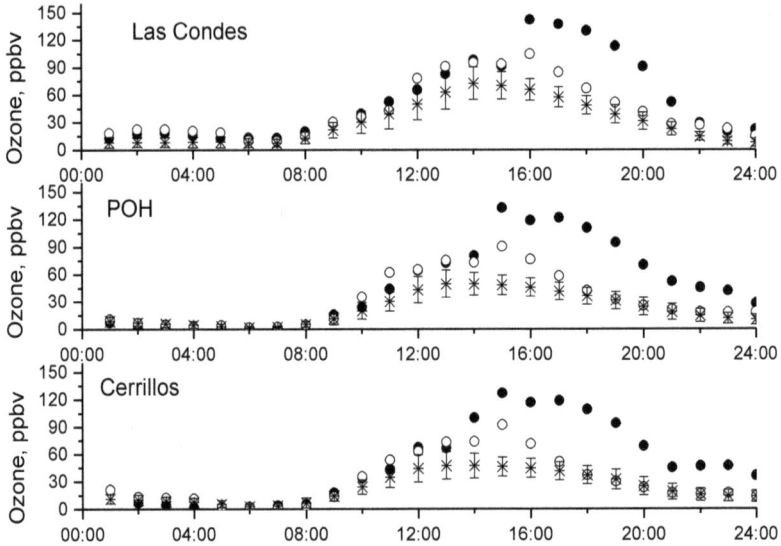

Figure 8. Ozone time profiles in background reference days (∗) and days with active fires in Melipilla region (○) (•).

Table 2. Contribution of Melipilla's fires to ozone maximum concentration in Santiago's atmosphere (ground level).

Sites	Fire Day	Calculated (ppbv)	Measured (ppbv)	Difference (ppbv)
Las Condes	4	100.19	142.15	41.96
	8	74.20	104.7	30.5
ParqueO'higgins	4	67.66	132.8	65.14
	8	42.40	90.55	48.15
Cerrillos	4	66.10	126.83	60.73
	8	36.95	92.32	55.37

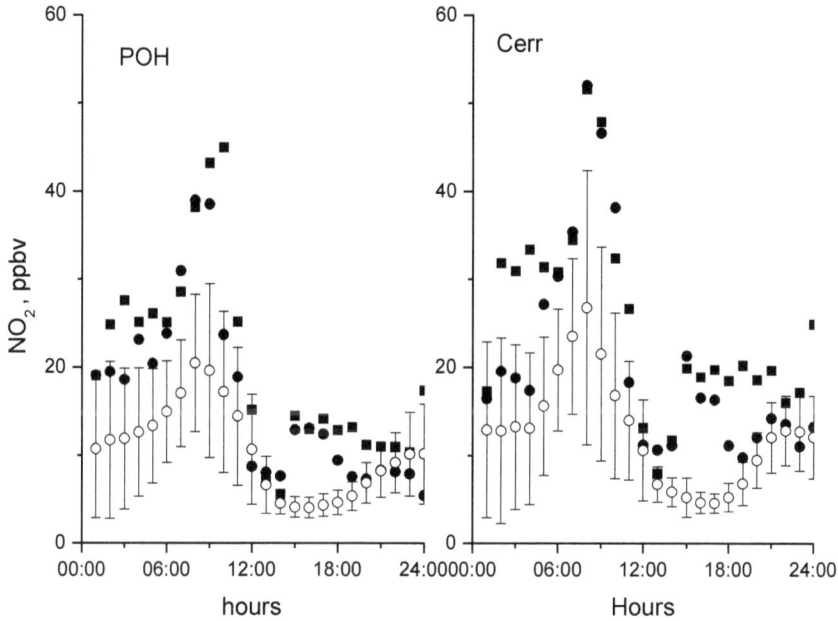

Figure 9. NO$_2$ daily profile, reference background days (□) and days with active wildfires in Melipilla region. (■), (●).

4. Discussion

Ozone concentration in the traveling plume increases with ageing but decreases with dilution. The interplay of these two factors can increase or decrease the concentrations of the oxidants with the fire to receptor distance [15], and the ozone concentration in the plumes is depleted as often as it is enhanced [36]. Furthermore, it could be expected that, due to back reflection of solar radiation by the increased amounts of particles (Figure 4), formation of ozone in the plume originating from wildfires could be reduced, particularly at ground level. However, the data obtained in this work and in previous studies [8] show a clear increase in ozone concentrations in plumes from wildfires and at ground level in nearby cities [11].This increase can be related to:

(i)   An increase in the rate of the chemical ozone production [3].
(ii) High emissions of ozone precursors, such as VOCs and NOx [4,37–39].
(iii) A reduced rate of ozone capture due to closure of plants stomata [2].
(iv) PAN decomposition as a late source of NO$_2$ [4]. PAN is rapidly formed in smoke plumes, with *ca.* 40% if the initial emitted NO$_X$ being converted to PAN in the first few hours after emission, contributing to downwind ozone formation [36].

231

(v) Significant emissions of formaldehyde [37] and HONO [40–42] relevant sources of hydroxyl radicals in Santiago´s atmosphere [18].

Point (i) and point (iii) should not be enough to explain the large differences observed. Since, Melipilla fires involve dry pastures, *acacia caven* and some *eucalyptus globulus* with very low ozone capture rates.

NO_2 arising from PAN decomposition can be an important promoter of high ozone levels in aged plumes [4,36] but its contribution to ozone levels in cities nearby the wildfire is unlikely, given the high NOx emission rates. In fact, the "direct" emission of NO_2 from wildfires has been estimated to have a NO_2/CO ratio of 0.005 ± 0.002 mol/mol [37]. NOx emissions are determined by the amount of nitrogen present in the fuel [43] and is emitted both at the flaming and smoldering stages [43,44].

Regarding the role of precursors (NOx, VOCS, formaldehyde and HONO), it has to be considered that daytime ozone steady state concentrations near to wildfires can be considered to be approximately given by Leighton's relationship:

$$[O_3]_{ss} = (J_{NO2}/k_{NO}) ([NO_2]/[NO]) \tag{1}$$

where J_{NO2} is the NO_2 photolysis constant and k_{NO} is specific rate constant of NO plus O_3 reaction. The role of the precursors in the Los Angeles type photochemical smog is to transform NO (the main primary NOx emitted by mobile sources) into NO_2 allowing an increase in ozone concentrations. Emission NO_2/NOx ratios in biomass burning are, for a variety of fuels, in the 0.1–0.3 range [40,45]. These values are larger than those arising from vehicles, the most important NOx source in large cities [46]. During the year 2006 at the monitoring station in Wuppertal (Germany), an annual average NO_2/NOx emission ratio of 0.12 ± 0.01 was reported by Kurtenbach *et al.*, [47]. The same group reported, from a traffic tunnel study in 1997, a much smaller emission ratio (0.04 ± 0.01). Similarly, in mainly ozone free road tunnels in Hong Kong, an NO_2/NOx ratio smaller than 0.02 was reported [48].

The relatively high NO_2/NOx emitted in biomass burning implies that high ozone concentrations could be achieved independently of precursor emissions since they are no longer necessary to photochemically generate large $[NO_2]/[NO]$ ratios and, hence, high ozone concentrations. In agreement with these considerations, at the morning rush hour of days with active fires, the ratio NO_2/NO in Cerrillos was 35 and 7.5, values much higher than those measured in background reference days (3.5 and 1.4, respectively). NOx directly measured upon the flames contains approximately 10% of NO_2 [43] suggesting that NO emission is the dominant process. However, in the fire and in the plume, local NO concentrations are very high and, at the current high temperatures, oxidation of NO to NO_2 can be a fast process that increases the NO_2/NOx ratio during the plume travel from the fire locus to the city (about 4–6 h), Figure 9.

The effect of the fire-derived plume upon urban ozone is complex due to incorporation of local emissions to contaminated air masses. This incorporation is not straightforward and sometimes is difficult to establish [29] and modeling has been minimally successful [14,49–51]. However, this incorporation of contaminants in the travelling plume does not seem to be relevant in the present scenario. The reported data show that ozone values measured in Cerrillos are very similar to those found at POH. If the location of Cerrillos at the entrance of the plume to Santiago's basin is considered, it can be concluded that the urban ozone is mostly explained in terms of the chemistry taking place during the transfer from the fire to the city and is minimally influenced by local emissions. In this regard, it is interesting to note that on normal days the maximum ozone in the city occurs about six hours after the rush hours. This delay time is of the same order of magnitude as that necessary to bring to the city the air masses charged with wildfire emissions. Interestingly, the presence of the plume decreases the difference between LC and the other stations (Table 2) that generally present smaller ozone concentrations.

5. Conclusions

Major forest fires in the Melipilla sector in central Chile increased ozone concentrations in the urban atmosphere of Santiago, reaching values higher than 80 ppb (the national one hour average limit). This increase took place at the three sampling sites considered and is explained in terms of a large NO_2/NO ratio in biomass burning emissions, the age of the plume reaching Santiago, and the high initial concentrations of ozone precursors, such as NOx, VOCs and OH• radical sources, such as formaldehyde and HONO, present in the urban atmosphere.

Acknowledgments: This work has been supported by DICYT—USACH 021541RC and CEDENNA-USACH. RG is partially supported by FONDAP Grant 15110009.

Author Contributions: Rubio, Lissi and Gramsch performed the air-chemistry analysis. Garreaud provided the meteorological context.

Conflicts of Interest: The authors declare no conflict of interest.

References

1.   Real, E.; Law, K.S.; Weinzierl, B.; Fiebig, M.; Petzold, A.; Wild, O.; Methven, J.; Arnold, S.; Stohl, A.; Huntrieser, H.; *et al.* Processes influencing ozone levels in Alaskan forest fire plumes during long-range transport over the North Atlantic. *J. Geophys. Res.* **2007**, *112*, 1–19.

2.   Hodnebrog, O.; Solberg, S.; Stordal, F.; Svendby, T.M.; Simpson, D.; Gauss, M.; Hilboll, A.; Pfister, G.G.; Turquety, S.; Richter, A.; et al. Impact of forest fires, biogenic emissions and high temperatures on the elevated Eastern Mediterranean ozone levels during the hot summer of 2007. *Atmos. Chem. Phys.* **2012**, *12*, 8727–8750.

3.   Martins, V.; Miranda, A.I.; Carvalho, A.; Schaap, M.; Borrego, C.; Sá, E. Impact of forest fires on particulate matter and ozone levels during the 2003, 2004 and 2005 fire seasons in Portugal. *Sci. Total Environ.* **2012**, *414*, 53–62.

4.   Jaffe, D.A.; Wigder, N.L. Ozone production from wildfires: A critical review. *Atmos. Environ.* **2012**, *51*, 1–10.

5.   Val Martin, M.; Honrath, R.E.; Owen, R.C.; Pfister, G.; Fialho, P.; Barata, F. Significant enhancements of nitrogen oxides, black carbon, and ozone in the North Atlantic lower free troposphere resulting from North American boreal wildfire. *J. Geophys. Res.* **2006**, *111*, D23S60.

6.   Suarez, L.; Castillo, L.; Marín, M.; Carrillo, G.; Rímac, L.; Pomalaya, J.; Penacho, R. Study of the seasonal variation of tropospheric ozone and aerosols related to the biomass burning in Amazonian. *Mosaico. Cient.* **2006**, *3*, 1–5.

7.   Morris, G.A.; Hersey, S.; Thompson, A.M.; Pawson, S.; Nielsen, J.E.; Colarco, P.R.; Wallace, M.S.; McMillan, W.W.; Stohl, A.; Turquety, S.; et al. Alaskan and Canadian forest fires exacerbate ozone pollution over Houston, Texas, on 19 and 20 July 2004. *J. Geophys. Res.* **2006**, *111*.

8.   Pfister, G.G.; Wiedinmyer, C.; Emmons, L.K. Impacts of the fall 2007 California wildfires on surface ozone: Integrating local observations with global model simulations. *Geophys. Res. Lett.* **2008**, *35*.

9.   Mebust, A.K.; Cohen, R.C. Space-based observations of fire NOx emission coefficients: A global biome-scale comparison. *Atmos. Chem. Phys.* **2014**, *14*, 2509–2524.

10. Evans, L.F.; King, N.K.; Packham, D.R.; Stephens, E.T. Ozone measurements in smoke from forest fire. *Environ. Sci. Technol.* **1974**, *8*, 75–76.

11. Cheng, L.; McDonald, K.M.; Angle, R.P.; Sandhu, H.S. Forest fire enhanced photochemical air pollution. A case study. *Atmos. Environ.* **1998**, *32*, 673–681.

12. Chalbot, M.C.; Kavouras, I.C.; Dubois, D.W. Assessment of the contribution of wildfires in ozone concentrations in the Central U: Mexico Border Region. *Aerosol Air Qual. Res.* **2013**, *13*, 838–848.

13. Lei, W.; Molina, L.T. Modelling the impact of biomass burning on air quality in and around Mexico City. *Atmos. Chem. Phys.* **2013**, *13*, 2299–2319.

14. Crounse, J.D.; DeCarlo, P.F.; Blake, D.R.; Emmons, L.K.; Campos, E.C.; Apel, E.C.; Clarke, A.D.; Weinheimer, A.J.; McCabe, D.C.; Yokelson, R.J.; et al. Biomass burning and urban air pollution over the Central Mexican Plateau. *Atmos. Chem. Phys.* **2009**, *9*, 4929–4944.

15. Jaffe, D.A.; Chand, D.; Hafner, W.; Westerling, A.; Spracklen, D. Influence of fires on O_3 concentrations in the Western U.S. *Environ. Sci. Technol.* **2008**, *42*, 5885–5891.

16. Garreaud, R. The Andes climate and weather. *Adv. Geosci.* **2009**, *7*, 1–9.

17. Rappengluek, B.; Schmitz, R.; Bauerfeind, M.; Cereceda-Balic, F.; von Baer, D.; Jorquera, H.; Silva, Y.; Oyola, P. An urban photochemistry study in Santiago de Chile. *Atmos. Environ.* **2005**, *39*, 2913–2931.

18. Elshorbany, Y.F.; Kleffmann, J.; Kurtenbach, R.; Lissi, E.; Rubio, M.A.; Villena, G.; Gramsch, E.; Rickard, A.R.; Pilling, M.J.; Wiesen, P. Summertime photochemical ozone formation in Santiago, Chile. *Atmos. Environ.* **2009**, *43*, 6398–6407.

19. Romero, H.; Ihl, M.; Rivera, A.; Zalazar, P.; Azocar, P. Rapid urban growth, land-use changes and air pollution in Santiago, Chile. *Atmos. Environ.* **1999**, *33*, 4039–4047.

20. Rubio, M.A.; Lissi, E.; Villena, G.; Caroca, V.; Gramsch, E.; Ruiz, A. Estimation of hydroxyl and hydroperoxyl radicals concentrations in the urban atmosphere of Santiago. *J. Chil. Chem. Soc.* **2005**, *50*, 375–379.

21. Rutllant, J.; Garreaud, R. Episodes of strong flow down the western slope of the subtropical Andes. *Mon. Weather Rev.* **2004**, *132*, 611–622.

22. Armesto, J.; Arroyo, M.T.K.; Hinojosa, L.F. The Mediterranean enviroment of central Chile. In *The Physical Geography of South America*; Veblen, T., Young, K., Orme, A., Eds.; Oxford University Press: Oxford, UK, 2007.

23. Schulz, J.J.; Cayuela, L.; Echeverria, C.; Salas, J. Monitoring land cover change of the dryland landscape of Central Chile (1975–2008). *Appl. Geogr.* **2010**, *30*, 436–447.

24. Krishna, M.T.; Chauhan, A.J.; Freww, A.J.; Holgate, S.T. Toxicological mechanisms underling oxidant pollutants-induced airway injury. *Rev. Environ. Health* **1998**, *13*, 59–61.

25. Yang, W.; Omaye, S.T. Air Pollutants oxidative stress and human health. *Mutat. Res. Genet. Toxicol. Environ. Mutagen.* **2009**, *674*, 45–54.

26. Shindell, D.; Kuylenstierna, J.C.I.; Vignati, E.; van Dingenen, R.; Amann, M.; Klimont, Z.; Anenberg, S.C.; Muller, N.; Janssens-Maenhout, G.; Raes, F.; *et al.* Simultaneous mitigating near-term climate change and improving human health and food security. *Science* **2012**, *335*, 183–189.

27. Muñoz, R.C.; Alcafuz, R. Variability of urban aerosols over Santiago, Chile: Comparison of surface PM10 concentrations and remote sensing with ceilometer and lidar. *Aerosol Air Qual. Res.* **2012**, *12*, 8–19.

28. SINCA 2014. National Service of Air Quality Information. Available online: http:// SINCA.mma.gob.cl (accessed on 1 November 2015).

29. Olcese, L.E.; Toselli, B.M. Unexpected high levels of ozone measured in Cordoba, Argentina. *J. Atmos. Chem.* **1998**, *31*, 269–279.

30. Rubio, M.A.; Oyola, P.; Gramsch, E.; Lissi, E.; Pizarro, J.; Villena, G. Ozone and peroxyacetylnitrate in downtown Santiago, Chile. *Atmos. Environ.* **2004**, *38*, 4931–4939.

31. Khoder, M.I. Diurnal seasonal and weekdays-weekend variations of ground level ozone concentrations in an urban area in greater Cairo. *Environ. Monit. Assess.* **2009**, *149*, 349–362.

32. Im, U.; Markakis, K.; Poupkou, A.; Melas, D.; Unal, A.; Gerasopoulos, E.; Daskalakis, N.; Kindap, T.; Kanakidou, M. The impact of temperature changes on summer time ozone and its precursors in the Eastern Mediterranean. *Atmos. Chem. Phys.* **2011**, *11*, 3847–3864.

33. Rubio, M.A.; Lissi, E.A. Temperature as thumb rule predictor of ozone levels in Santiago de Chile ground air. *J. Chil. Chem. Soc.* **2014**, *59*, 2427–2431.

34. Krupa, S.V.; Kickert, R.V. The greenhouse effect: Impact of ultraviolet B radiation, carbon dioxide and ozone on vegetation. *Environ. Pollut.* **1989**, *61*, 263–393.

35. Bell, M.L.; McDermontt, A.; Zeger, S.L.; Samet, J.M.; Dominici, F. Ozone and short term mortality in 95 U.S. urban communities. *JAMA* **2004**, *292*, 2372–2378.

36. Alvarado, M.J.; Logan, J.A.; Mao, J.; Apel, E.; Riemer, D.; Blake, D.; Cohen, R.C.; Min, K.E.; Perrin, A.E.; Browne, E.C.; *et al.* Nitrogen oxides and PAN in plumes from boreal fires during ARCTAS-B and their impact on ozone: An integrated analysis of aircraft and satellite observations. *Atmos. Chem. Phys.* **2010**, *10*, 9739–9760.

37. Young, E.; Paton-Walsh, C. Emission ratios of the tropospheric Ozone precursors nitrogen dioxide and formaldehyde from Australia's black Saturday fires. *Atmosphere* **2011**, *2*, 617–632.

38. Konovalov, I.B.; Beeckmann, M.; Kuznetsova, I.N.; Yurova, A.; Zvyagintsev, A.M. Atmospheric impacts of the 2010 Russian wildfires: Integrating modeling and measurements of an extreme air pollution episode in the Moscow region. *Atmos. Chem. Phys.* **2011**, *11*, 10031–10056.

39. Schreier, S.F.; Richter, A.; Kaiser, J.W.; Burrows, J.P. The empirical relationship between satellite-derived tropospheric NO_2 and fire radiative power and possible implications for fire emissions rates of NOx. *Atmos. Chem. Phys.* **2014**, *14*, 2447–2466.

40. Burling, I.R.; Yokelson, R.J.; Griffith, D.W.T.; Johnson, T.J.; Veres, P.; Roberts, J.M.; Warneke, C.; Urbanski, S.P.; Reardon, J.; Weise, D.R.; *et al.* Laboratory measurements of trace gas emissions from biomass burning of fuel types from the south eastern and south western United States. *Atmos. Chem. Phys.* **2010**, *10*, 11115–11130.

41. Roberts, J.M.; Veres, P.; Warneke, C.; Neuman, J.A.; Washenfelder, R.A.; Brown, S.S.; Baasandorj, M.; Burkholder, J.B.; Burling, I.R.; Johnson, T.J.; *et al.* Measurement of HONO, HNCO, and other inorganic acids by negative ion proton-transfer chemical-ionization mass spectrometry (NI-PT-CIMS): Application to biomass burning emissions. *Atmos. Meas. Tech.* **2010**, *3*, 981–990.

42. Nie, W.; Ding, A.J.; Xie, Y.N.; Xu, Z.; Mao, H.; Kerminen, V.; Zheng, L.F.; Qi, X.M.; Yang, X.Q.; Sun, J.N.; *et al.* Influence of biomass burning plumes on HONO chemistry in eastern China. *Atmos. Chem. Phys. Discuss.* **2014**, *14*, 7859–7887.

43. Andreae, M.O.; Merlet, P. Emissions of trace gases and aerosols from biomass burning. *Glob. Biogeochem. Cycles.* **2001**, *15*, 955–966.

44. Contreras-Moctezuma, J.; Rodriguez-Trejo, D.A.; Retama-Hernandez, A.; Sanchez-Rodriguez, J.J.M. Smoke gases of wildfires in *pinus hartwegii* Forest. *Agrociencia* **2003**, *37*, 309–316.

45. Yokelson, R.J.; Griffith, D.W.T.; Ward, D.E. Open-path Fourier transforms infrared studies of large scale laboratory biomass fires. *J. Geophys. Res.* **1996**, *101*, 21067–21080.

46. Wu, Y.; Zhang, S.J.; Li, L.M.; Ge, Y.S.; Shu, J.W.; Zhou, Y.; Xu, Y.Y.; Hu, J.N.; Liu, H.; Fu, L.X.; *et al.* The challenge to NOx emission control for heavy-duty diesel vehicles in China. *Atmos. Chem. Phys.* **2012**, *12*, 9365–9379.

47. Kurtenbach, R.; Kleffmann, J.; Niedojadlo, A.; Wiesen, P. Primary NO$_2$ emissions and their impact on air quality in traffic environments in Germany. *Environ. Sci. Eur.* **2012**, *24*, 2–8.

48. Yao, X.; Lau, N.T.; Chan, C.K.; Fang, M. The use of tunnel concentration profile data to determine ratio of NO2/NOx directly emitted from vehicles. *Atmos. Chem. Phys. Discuss.* **2005**, *5*, 12723–12740.

49. Bein, K.J.; Zhao, Y.; Johnston, M.V.; Wexler, A.S. Interactions between boreal wildfire and urban emissions. *J. Geophys. Res.* **2008**, *113*, DO 7304.

50. Singh, H.B.; Cai, C.; Kaduwela, A.; Weinheimer, A.; Wisthaler, A. Interaction of fire emissions and urban pollution over California: Ozone formation and air quality simulations. *Atmos. Environ.* **2012**, *56*, 45–51.

51. Wigder, N.L.; Jaffe, D.A.; Saketa, F.A. Ozone and particulate matter enhancements from regional wildfires observed at Mount Bachelor during 2004–2011. *Atmos. Environ.* **2013**, *75*, 24–31.

Frequency and Character of Extreme Aerosol Events in the Southwestern United States: A Case Study Analysis in Arizona

David H. Lopez, Michael R. Rabbani, Ewan Crosbie, Aishwarya Raman, Avelino F. Arellano Jr. and Armin Sorooshian

Abstract: This study uses more than a decade's worth of data across Arizona to characterize the spatiotemporal distribution, frequency, and source of extreme aerosol events, defined as when the concentration of a species on a particular day exceeds that of the average plus two standard deviations for that given month. Depending on which of eight sites studied, between 5% and 7% of the total days exhibited an extreme aerosol event due to either extreme levels of PM_{10}, $PM_{2.5}$, and/or fine soil. Grand Canyon exhibited the most extreme event days (120, *i.e.*, 7% of its total days). Fine soil is the pollutant type that most frequently impacted multiple sites at once at an extreme level. PM_{10}, $PM_{2.5}$, fine soil, non-Asian dust, and Elemental Carbon extreme events occurred most frequently in August. Nearly all Asian dust extreme events occurred between March and June. Extreme Elemental Carbon events have decreased as a function of time with statistical significance, while other pollutant categories did not show any significant change. Extreme events were most frequent for the various pollutant categories on either Wednesday or Thursday, but there was no statistically significant difference in the number of events on any particular day or on weekends *versus* weekdays.

Reprinted from *Atmosphere*. Cite as: Lopez, D.H.; Rabbani, M.R.; Crosbie, E.; Raman, A.; Arellano, A.F., Jr.; Sorooshian, A. Frequency and Character of Extreme Aerosol Events in the Southwestern United States: A Case Study Analysis in Arizona. *Atmosphere* **2016**, 7, 1.

1. Introduction

Severe aerosol pollution events pose a major threat to society due to significant reductions in visibility and air quality, in addition to adverse impacts on public health and daily operations. Fine particulate matter ($PM_{2.5}$) is linked to various health impacts regardless of whether there is chronic or acute exposure, with effects ranging from lung cancer to cardiovascular disease [1,2]. Extreme pollution events are thought to be especially important with regard to annual acute mortality [3], in addition to leading to temporary shutdown of daily activities such as school and work in parts of the world [4]. Semi-arid and arid regions are particularly vulnerable to such events due to dust emissions, and this is especially dramatic during haboob events [5–7]. In recent decades, the Southwestern United States (Southwest) has experienced significant population growth, land use change, and is moving towards a more arid regime with higher temperatures, less precipitation, and lower soil moisture [8]. These changes promote increased dust emissions [9,10] and wildfires [11,12], with a rapidly growing population left vulnerable to the effects of the emissions. These issues coupled to the impact of dust and wildfire emissions on the hydrologic cycle and snowpack behavior at higher altitudes in the Southwest [9,13] warrants an examination of extreme aerosol events.

An ideal location to study extreme aerosol events is Arizona, which represents a state in the Southwest that is impacted by both dust and wildfires, in addition to having one of the fastest growing populations in the United States that is prone to the effects of poor air quality. The absolute population growth between 2000 and 2009, in Tucson and Phoenix, the two largest cities in Arizona, rank as the 33rd and 4th largest in the United States, respectively (U.S. Census Bureau, 2009). Sources of wind-blown dust impacting this and other Southwest states include naturally un-vegetated or anthropogenically disturbed soil surfaces, such as dry lakes ("playas"), dry washes, gravel pits, construction sites, oil and gas development sites, fields (after harvest), and long-range transport of Asian dust [14–20].

Aside from the ubiquity of dust in the Southwest, the greater Western United States is becoming increasingly vulnerable to the effects of wildfires owing to both a warmer climate and fire-control strategies over past decades resulting in conditions that promote larger and more frequent fires [11,21]. Depending on the fuel type and burning conditions, biomass burning leads to extensive emissions of various gaseous (e.g., nitrogen oxides (NO_x), ozone (O_3), carbon monoxide (CO), Volatile Organic Compounds (VOCs)) and particulate species (e.g., Elemental Carbon (EC), Organic Carbon (OC), inorganics), but also soil emissions due to lofting of soil in areas of turbulent mixing surrounding flames [22–24].

The goal of this study is to examine long-term data (2001—2014) from the EPA IMPROVE network across Arizona to characterize the frequency, spatial range, and origin of extreme aerosol events. The following questions are addressed: (i) what is the frequency of extreme aerosol events across Arizona and how many are due to EC-enriched air masses, Asian dust, non-Asian dust, or some other source? (ii) how frequently do these events occur at all or subsets of the study sites on the same day? and (iii) how are these events distributed between months of the year, days of the week, and inter-annually?

2. Experimental Methods

2.1. EPA IMPROVE

This study utilizes aerosol composition data from the Interagency Monitoring of Protected Visual Environments (IMPROVE) network [25]. IMPROVE aerosol monitoring stations are located primarily in National Parks and Wilderness Areas and collect ambient aerosol on filters over a period of 24 h every third day. Samples are analyzed for ions, metals, Organic Carbon (OC) and Elemental Carbon (EC). Among the elemental measurements, X-Ray Fluorescence (XRF) is used for Fe and heavier elements while Particle-Induced X-Ray Emission (PIXE) is used for elements Na to Mn. Fine soil concentrations reported in this study are calculated using the following equation [25]:

$$\text{Fine Soil} \, (\mu g \cdot m^{-3}) = 2.2[Al] + 2.49[Si] + 1.63[Ca] + 2.42[Fe] + 1.94[Ti] \quad (1)$$

with regard to this equation, the components and their contributions were previously confirmed in comparisons of local re-suspended soils and ambient particles in the Western United States [25]. As this study is concerned with extreme concentrations of fine soil, it is expected that this equation can successfully capture all soil-rich air masses regardless of whether minor variations exist in the factors used in Equation (1). Species mass concentrations discussed in this study are from the fine fraction of aerosols ($PM_{2.5}$). Sampling protocols and additional details are provided elsewhere [26].

In this study, we use data from eight sites in Arizona (see map in Figure 1) and over time spans ranging from as early as data were possible starting in January 2001 until August 2014 (Table 1). Three of the eight sites are impacted more significantly by urban emissions owing to their closer proximity to populated cities; the Phoenix site is centrally located in the metropolitan area of the most populated city in Arizona, while Saguaro National Monument and Saguaro West are separated by ~50 km and are on the east and west sides, respectively, of Tucson, which is the second largest city. Chiricahua National Monument is a high-altitude site that is in a remote vegetated area with the nearest major urban area being Tucson (~150 km to the west). Nearby aerosol sources include the Willcox Playa and the Apache Power Plant, which are ~45 km to the west. This site is the closest to the Chihuahuan Desert. Tonto is ~90 km to the east/northeast of Phoenix and Queen Valley is ~60 km to the east of Phoenix, and thus these two sites are vulnerable to emissions transported from Phoenix. Organ Pipe is near the border of the United States and Mexico and is vulnerable to dust emissions and anthropogenic emissions from the nearby town of Sonoita, Mexico (~10,000 inhabitants). Grand Canyon is in a remote site in Northern Arizona and is removed from anthropogenic emissions.

Table 1. Summary of IMPROVE sites and date ranges over which data are analyzed.

Site Name	Latitude (°)	Longitude (°)	Altitude (m)	Date Range
Chiricahua (Chi)	32.0994	−109.389	1554	January 2001–August 2014
Grand Canyon (GC)	35.9731	−111.9841	2267	January 2001–August 2014
Organ Pipe (OP)	31.9506	−112.8016	504	December 2002–August 2014
Phoenix (Ph)	33.5038	−112.096	342	April 2001–August 2014
Queen Valley (QV)	33.2939	−111.2858	661	April 2001–August 2014
Saguaro NM (SNM)	32.1746	−110.737	941	April 2001–August 2014
Saguaro West (SW)	32.2486	−111.2178	714	October 2001–August 2014
Tonto (Ton)	33.6548	−111.1068	775	January 2001–August 2014

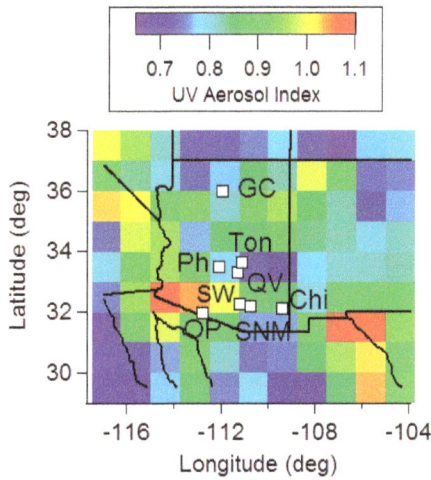

Figure 1. Spatial map of the eight EPA IMPROVE stations examined in Arizona overlaid on a four year average (2005–2008) of OMI ultraviolet aerosol index data, which includes influence from light-absorbing aerosol constituents, such as dust and smoke.

2.2. NAAPS Aerosol Model

Simulation data providing information about long-range dust transport are obtained from the Navy Aerosol Analysis and Prediction system (NAAPS; http://www.nrlmry.navy.mil/aerosol_web/). NAAPS relies on global meteorological fields from the Navy Operational Global Atmospheric Prediction System (NOGAPS) [27,28] analyses and provides output at a spatial resolution of $1° \times 1°$, at six hour intervals, and with 24 vertical levels reaching 100 mb [29]. NAAPS has been used extensively to study intercontinental transport of dust to North America e.g., [18,29–32]. Sources of dust are defined in NAAPS using the USGS Land Cover Characteristics Database, which was created with Advanced Very High Resolution Radiometer (AVHRR) data. TOMS aerosol index data was used to further refine dust source regions. Dust emission occurs when the friction velocity exceeds a threshold value (value depending on land type) and when the surface moisture and snow depth are lower than a critical value (0.3 and 0.4 cm, respectively). The model operationally assimilates remotely-sensed aerosol optical depth (AOD) data from MODIS [33].

2.3. Satellite Data

Ultraviolet aerosol index (UV AI) data are obtained from the Ozone Monitoring Instrument (OMI). Data were obtained at a resolution $1° × 1.25°$ using a minimum threshold value of 0.5 [34]. The UV AI parameter serves as a proxy for absorbing aerosol particles [35], which are predominantly comprised of smoke and dust. Figure 1 shows a spatial map of a four-year average of OMI data to provide a backdrop of where light-absorbing aerosol particles (primarily dust in study region) are most abundant relative to where the eight IMPROVE stations are located.

2.4. Criteria for Events

In past work, criteria to define an extreme aerosol event have included the use of a cutoff threshold of a parameter value (i.e., average $± i ×$ standard deviation, with i starting at 1 and increasing; e.g., [36]) or when parameter values were below and above specific quantile values (e.g., 3). Here we take a similar approach to define an extreme event for specific aerosol parameters (e.g., $PM_{2.5}$, PM_{10}, and fine soil) as when the measured concentration on a given day at any of the eight sites exceeds the average concentration plus two times the standard deviation for the month in which the event occurred over the time range of data used for that particular site. This criterion leads to concentrations that exceed the 90th percentile of mass concentrations in each category. The choice of this criterion reflects a balance between removing sensitivity to month-dependent factors and being sufficiently strict to isolate only a few cases that were the most polluted. The conclusions of this study, especially the number of extreme days in the various categories presented, are sensitive to the criteria definition. The numerical threshold criteria values (i.e., average $+ 2 ×$ standard deviation) for each site and month are shown in Table S1 (Supplementary Material).

Those events with extreme fine soil concentrations are referred to hereafter as extreme dust events. However, it is noted that extreme PM_{10} events that did not reach extreme fine soil levels could have also been due to dust that was concentrated in coarse aerosol ($D_p \geqslant 2.5$ µm). Extreme fine soil events are further classified as having influence from Asia or not using output from the NAAPS model to validate long-range transport to the study region. The criteria for Asian dust was to observe a clear aerosol plume being advected from Asia to Arizona with multiple repeated NAAPS output plots as depicted in Figure S1 of the Supplement. It is cautioned that this classification scheme using NAAPS has limitations in that (i) the dust transport results are driven by a model rather than fully by observations, and (ii) the relative influence of Asian dust *versus* local sources is uncertain. Thus, although the term "Asian dust" is used subsequently, this is not meant to indicate that the fine soil measurement is fully due to long-range transport of dust from Asia. A suite of previous studies discussing source attribution of aerosol to long-range transport from Asia to North America have relied on NAAPS. For example, Cottle *et al.* [31] used NAAPS with HYSPLIT back-trajectories, and sunphotometer and lidar data to show that springtime dust plumes from Asia reached North America. Wu *et al.* [32], more recently, used NAAPS and remotely-sensed data from CALIPSO to study a trans-Pacific Asian dust event and its impact on the east coast of the United States. McKendry *et al.* [30] relied on the internal consistency between NAAPS and variety of other tools such as another global chemical model (GEOS-Chem) and surface and satellite observations to trace large dust plumes to their sources in areas, such as North Africa. The consistency between NAAPS and the other aforementioned resources provides confidence in the former for the purposes of source attribution of dust to Asia.

A category termed "High EC" is defined as when both $PM_{2.5}$ and EC exhibit extreme levels. These events likely stem from anthropogenic sources and biomass burning events owing to the high levels of EC (as compared to other emission sources) and predominantly accumulation mode particles in wildfires [37]. Events that do not qualify as being extreme fine soil or High EC events are considered as "Other".

It is cautioned that the number of extreme events reported between 2001 and 2014 represents an underestimate since data is used only up through August 2014 and only starts in January 2001 for three sites with the most delayed start time being for Organ Pipe in December 2002.

3. Results and Discussion

3.1. Frequency and Categorization of Events

Of the total number of days when data were available in the time ranges in Table 1 (*i.e.*, 1431–1664 depending on site), between 76 and 120 total days were characterized by some type of extreme event (*i.e.*, PM_{10}, $PM_{2.5}$, and/or fine soil) depending on the site (Table 2). This number of days of extreme events corresponds to between 5% and 7% of the total days examined. Grand Canyon exhibited the most extreme event days (120, *i.e.*, 7% of its total days), which is coincident with it being one of the most recognized tourism spots in the Southwest.

Relative to the total days with extreme $PM_{2.5}$ levels, Grand Canyon exhibited the highest percentage in the High EC category (47% *versus* 13%–25% for other sites). Of the total number of days with extreme fine soil (54–69 days depending on the site), the number of these events being linked to Asian dust ranged from 19% to 29%

(*i.e.*, 10–20 days). The total number of days with extreme events classified as Other (*i.e.*, not High EC or fine soil events) ranged from 17 to 30 days, which represents between 21% and 33% of the total extreme days depending on the site. The fact that the highest percentage of Other days were at Phoenix (31%) and Saguaro West (33%), the most urban-impacted sites among those studied, suggests that anthropogenic pollution, including anthropogenic dust, contributes to these events. Queen Valley also reached 31%, reflective of possible impact from transported pollution from the major nearby urban center Phoenix. Between 7 and 22 days in the Other category also registered extreme values of Coarse Mass (CM = PM_{10} – $PM_{2.5}$), supporting the possibility of influence from locally generated dust.

To gain a sense of the spatial extent of pollution registering as extreme events, Table 3 shows how many sites experienced an extreme event for a specific pollutant category on the same day. Locally produced aerosol would not be expected to impact multiple sites at an extreme level as compared to a transported plume such as from Asia. Grand Canyon is farther removed from the other seven sites that are clustered closer in Southern Arizona, and, thus, Grand Canyon exhibits the highest number of days where an extreme event only impacted that site. Computed as a percentage of all extreme days registered for a particular pollutant type, Grand Canyon was the only site impacted out of 65%, 64%, 51%, 100%, and 84% of its extreme events for PM_{10}, $PM_{2.5}$, fine soil, High EC, and Other, respectively. The categories with the least number of extreme events impacting five or more sites were High EC (0 days for all sites) and Other (0–1 day depending on site). This result is thought to be due to locally generated pollution from either (i) some combination of biomass burning and anthropogenic activity (for High EC) or (ii) dust (for Other) that was not regional in nature. Fine soil events conversely impacted five or more sites on between 13 and 23 days, accounting for between 16% and 38% of all fine soil extreme events, depending on the site. Therefore, for the study region, fine soil is the pollutant type that most successfully impacts multiple sites at once at an extreme level.

The region-wide average for the $PM_{2.5}:PM_{10}$ ratio was 0.37, 0.35, and 0.23 for non-Asian dust, Asian dust, and Other-CM, respectively. Unexpectedly, the ratios for non-Asian dust events exceeded those for Asian dust events for half the sites (*i.e.*, Chiricahua, Phoenix, Saguaro National Monument, Saguaro West). Also of interest is that non-Asian dust event averages for $PM_{2.5}:PM_{10}$ were well above 0.35 at the two sites in Tucson, Arizona (Saguaro National Monument = 0.57 ± 0.36; Saguaro West = 0.46 ± 0.26). A plausible explanation for these unexpected results is interference of background anthropogenic emissions at these urban-impacted sites in Tucson (Saguaro National Monument and Saguaro West), in addition to Phoenix which exhibited an average ratio of 0.35. The same explanation can be applied to the Fe:Ca results, which do not show a clear reduction in value for Asian dust events as compared to the more locally-relevant pollution categories for Chiricahua, Saguaro National Monument, and Saguaro West. However, the region-wide average for Fe:Ca was lowest for Asian dust (0.88), followed by non-Asian dust (0.96), and Other-CM (1.03). The values were generally low and close to the threshold value applied by past work to classify dust as purely Asian dust. These results suggest that caution should be exercised with the use of such ratios to distinguish between dust sources owing to mixing between distant and local sources.

Due to the nature of Asian dust pollution being more geographically widespread than other forms of pollution, this category registered the highest frequency of its events impacting $\geqslant 5$ sites. Depending on the site, 30%–57% of extreme Asian dust events (i.e., 3–8 days) impacted $\geqslant 5$ sites.

It is of interest to compare the Asian dust extreme event data to criteria used previously to distinguish Asian dust events in the study region, including mass concentration ratios of both Fe:Ca and $PM_{2.5}:PM_{10}$. Previous work showed that Fe:Ca ratios below 1 are considered to be 100% Asian dust and values above 2 are 100% local dust [38]. A threshold ratio value of 0.35 for $PM_{2.5}:PM_{10}$ has been applied in other work to remove contamination of non-local dust sources in the study region [39]. This ratio generally increases with dust plume age and, thus, values higher than 0.35 are assumed to be contaminated with sources such as transported Asian dust. Values between 0.15 and 0.26 are associated with soil dust emissions from human activities according to the EPA [39]. Table 4 examines statistics associated with the two aforementioned ratios for non-Asian dust, Asian dust, and also the subset of Other extreme events that also had extreme values of CM ($PM_{10}-PM_{2.5}$). The latter are presumed to be due to locally generated dust.

Table 2. Statistics associated with the number of days with extreme events observed in the date range shown in Table 1 for each site. (NAAPS global data were unavailable for the following dates in 2001 that are omitted from categorization into Asian and non-Asian dust: *16 October 2001, 26 August 2001, 9 November 2001; **16 October 2001, 31 October 2001, 21 November 2001; ***16 October 2001, 9 November 2001; ****9 November 2001). Values in the Total category represent days with any type of extreme event (i.e., PM_{10}, $PM_{2.5}$, fine soil).

Site Name	Total Days Data Available	Extreme Event Types							
		Total	PM_{10}	$PM_{2.5}$	Fine Soil	High EC	Non-Asian Dust	Asian Dust	Other
Chiricahua *	1664	89	61	56	54	14	41	10	23
Grand Canyon	1664	120	82	64	69	30	49	20	25
Organ Pipe	1431	76	45	55	54	7	41	13	17
Phoenix **	1628	98	54	58	63	9	50	10	30
Queen Valley ***	1628	98	65	60	60	9	43	15	30
Saguaro NM ***	1628	85	49	56	56	8	40	14	23
Saguaro West ****	1563	91	56	55	55	10	44	10	30
Tonto ***	1664	103	67	65	67	12	47	18	25

Table 3. Percentage breakdown (represented as fractions; *i.e.*, 0.1 = 10%) of the extreme events for different pollutant categories in terms of how many sites registered an extreme event for a particular pollutant on the same day. Each pollution category is separated into three columns representing extreme events occurring only at that site (1), 2–4 total sites, or 5–8 total sites.

Site Name	PM$_{10}$			PM$_{2.5}$			Fine Soil			High EC			Non-Asian Dust			Asian Dust			Other		
	1	2–4	5–8	1	2–4	5–8	1	2–4	5–8	1	2–4	5–8	1	2–4	5–8	1	2–4	5–8	1	2–4	5–8
Chiricahua	0.38	0.34	0.28	0.39	0.38	0.23	0.43	0.33	0.24	0.79	0.21	0.00	0.30	0.40	0.30	0.49	0.34	0.17	0.74	0.26	0.00
Grand Canyon	0.65	0.24	0.11	0.64	0.33	0.03	0.51	0.33	0.16	1.00	0.00	0.00	0.50	0.20	0.30	0.51	0.39	0.10	0.84	0.12	0.04
Organ Pipe	0.31	0.40	0.29	0.44	0.36	0.20	0.37	0.30	0.33	0.71	0.29	0.00	0.08	0.38	0.54	0.46	0.27	0.27	0.71	0.29	0.00
Phoenix	0.39	0.37	0.24	0.47	0.34	0.19	0.41	0.40	0.19	0.89	0.11	0.00	0.00	0.60	0.40	0.48	0.38	0.14	0.83	0.17	0.00
Queen Valley	0.25	0.45	0.31	0.27	0.47	0.27	0.22	0.40	0.38	0.56	0.44	0.00	0.00	0.47	0.53	0.30	0.40	0.30	0.57	0.40	0.03
Saguaro NM	0.20	0.45	0.35	0.27	0.50	0.23	0.27	0.36	0.38	0.75	0.25	0.00	0.21	0.21	0.57	0.30	0.43	0.28	0.74	0.22	0.04
Saguaro West	0.25	0.43	0.32	0.35	0.36	0.29	0.38	0.36	0.25	0.50	0.50	0.00	0.10	0.40	0.50	0.45	0.36	0.18	0.60	0.37	0.03
Tonto	0.18	0.51	0.31	0.20	0.57	0.23	0.13	0.55	0.31	0.75	0.25	0.00	0.11	0.44	0.44	0.15	0.62	0.23	0.60	0.36	0.04

Table 4. Average (±standard deviation) of two mass concentration ratios often applied to distinguish local dust from non-local dust (*i.e.*, Asian dust). Statistics are calculated for three extreme event categories: non-Asian dust, Asian dust, and both Other and CM simultaneously.

Site Name	PM$_{10}$:PM$_{2.5}$			Fe:Ca		
	Non-Asian Dust	Asian Dust	Other and CM	Non-Asian Dust	Asian Dust	Other and CM
Chiricahua	0.32 ± 0.17	0.30 ± 0.10	0.19 ± 0.08	0.82 ± 0.35	0.88 ± 0.24	0.95 ± 0.30
Grand Canyon	0.38 ± 0.11	0.43 ± 0.11	0.34 ± 0.06	0.83 ± 0.27	0.79 ± 0.21	0.92 ± 0.19
Organ Pipe	0.27 ± 0.10	0.39 ± 0.08	0.20 ± 0.09	0.98 ± 0.35	0.81 ± 0.16	0.96 ± 0.43
Phoenix	0.35 ± 0.16	0.30 ± 0.08	0.20 ± 0.07	1.27 ± 0.28	1.02 ± 0.25	1.22 ± 0.26
Queen Valley	0.27 ± 0.17	0.30 ± 0.10	0.18 ± 0.05	1.04 ± 0.35	0.84 ± 0.13	1.09 ± 0.40
Saguaro NM	0.57 ± 0.36	0.39 ± 0.13	0.27 ± 0.04	0.79 ± 0.28	0.80 ± 0.15	0.89 ± 0.24
Saguaro West	0.46 ± 0.26	0.31 ± 0.12	0.21 ± 0.06	0.95 ± 0.48	1.04 ± 0.38	1.02 ± 0.44
Tonto	0.35 ± 0.19	0.38 ± 0.14	0.23 ± 0.08	1.00 ± 0.31	0.86 ± 0.15	1.16 ± 0.38

3.2. "Other" Events

The Other category was investigated in more detail to gain insight about the source of these extreme events (Table 5). Between 47% and 84% of the Other events exhibited extreme PM_{10} levels, which is suggestive of the presence of locally-generated coarse matter (*i.e.*, dust) since fine soil levels did not reach extreme levels. To gain confidence in this reasoning, the percent frequency of extreme CM days was calculated and is similar to the percent frequency of extreme PM_{10} days (*i.e.*, within two days) with the exception of Grand Canyon and Phoenix, which had nine and seven fewer extreme CM days as compared to PM_{10}, respectively. Since the ratio of extreme CM:Other days ranges from 41% (Organ Pipe) to as high as 74% (Chiricahua), with the average among all sites being 56%, locally generated CM (*i.e.*, dust) accounted for a significant amount of the Other events.

$PM_{2.5}$ levels reached extreme levels in 30%–76% of the Other extreme events, with Organ Pipe being the only site with a higher percentage for $PM_{2.5}$ being extreme *versus* PM_{10}. The $PM_{2.5}$ constituents only reached extreme levels in an average of 10% (OC), 11% (K), 17% (nitrate), and 22% (sulfate) of the Other events. Among these four $PM_{2.5}$ constituents, sulfate reached extreme levels in 48% and 35% of the Other events in Tonto and Organ Pipe, respectively, which were the highest values among all species and sites. This is likely due to anthropogenic emissions near those two sites such as from smelting [40–42]. Between 0% and 28% of Other events exhibited extreme levels of nitrate, OC, and potassium, which are all associated with wintertime pollution and fine soil emissions. These relatively low percentages for $PM_{2.5}$ constituents are consistent with the majority of the Other events being due to CM.

Table 5. Percentage frequency summary (represented as fractions; *i.e.*, 0.1 = 10%) of how many of the Other events at each site exhibited extreme levels of PM_{10}, $PM_{2.5}$, coarse mass (CM = PM_{10} - $PM_{2.5}$) and individual $PM_{2.5}$ constituents (potassium, organic carbon, nitrate, sulfate).

Species	Chiricahua (N = 23)	Grand Canyon (N = 25)	Organ Pipe (N = 17)	Phoenix (N = 30)	Queen Valley (N = 30)	Saguaro NM (N = 23)	Tonto (N = 25)
PM_{10}	0.78	0.84	0.47	0.70	0.63	0.65	0.73
$PM_{2.5}$	0.30	0.52	0.76	0.50	0.43	0.43	0.30
CM	0.74	0.48	0.41	0.47	0.63	0.57	0.73
K	0.09	0.24	0.12	0.17	0.10	0.09	0.03
OC	0.04	0.04	0.12	0.23	0.13	0.00	0.07
NO_3^-	0.13	0.20	0.18	0.17	0.10	0.17	0.10
SO_4^{2-}	0.17	0.08	0.35	0.07	0.20	0.26	0.13

3.3. Temporal Nature of Events

Figure 2 displays the monthly distribution of cumulative (*i.e.*, summed for all years and sites) extreme events broken into the various pollutant categories. The month of August experienced the highest number of extreme events in the study region for PM_{10}, $PM_{2.5}$, fine soil, non-Asian dust, High EC, and Other (also had an equal peak in June). In contrast to all other pollutant categories, Asian dust events mainly occurred in the spring months of March–June (41 out of 42 days, *i.e.*, 98%) with only one event in February. The Other category exhibited a relatively constant amount in each month (12–16 days). Unlike PM_{10}, $PM_{2.5}$ exhibited a secondary mode in the winter month of January, driven mostly by High EC and Other events, suggestive of the importance of anthropogenic emissions and biomass burning, and secondary production of aerosol species that are favorably produced in wintertime conditions such as ammonium nitrate.

Figure 3 represents the interannual distribution of extreme events for different pollutant categories. It is cautioned that the time range with full years of data at all eight sites is from 2003 to 2013 (refer to Table 1 for data time ranges for each site). All categories exhibited the most events in either 2002 or 2003 with the exception of Asian dust which reached 10 events in 2007 and followed a distinctly different temporal pattern than all other categories due to its distant source. An interesting feature of Figure 3 is the cyclical pattern of there being a peak every few years for PM_{10}, $PM_{2.5}$, fine soil, non-Asian dust, and Other, specifically in the years 2002–2003, 2006–2007, 2009, and 2011–2012. It is unclear with the dataset as to what explains these recurring peaks, and future work is warranted, with a longer term record, to identify what an explanation could be for these features in the data.

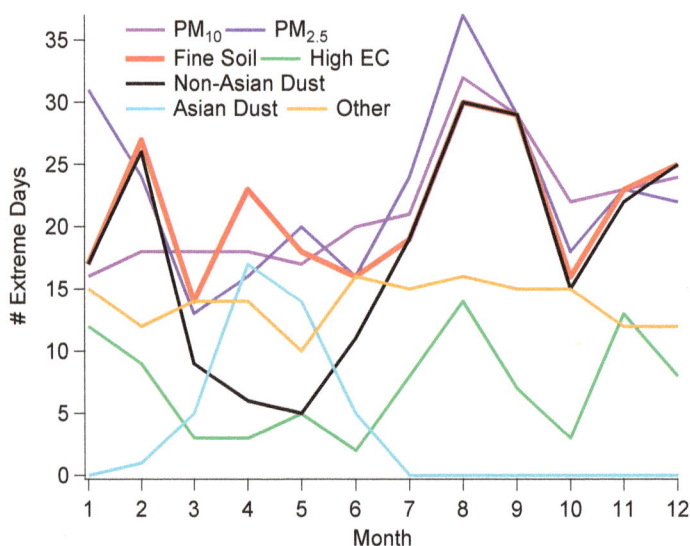

Figure 2. Monthly distribution of extreme events (cumulative for all years and sites) for different pollution categories.

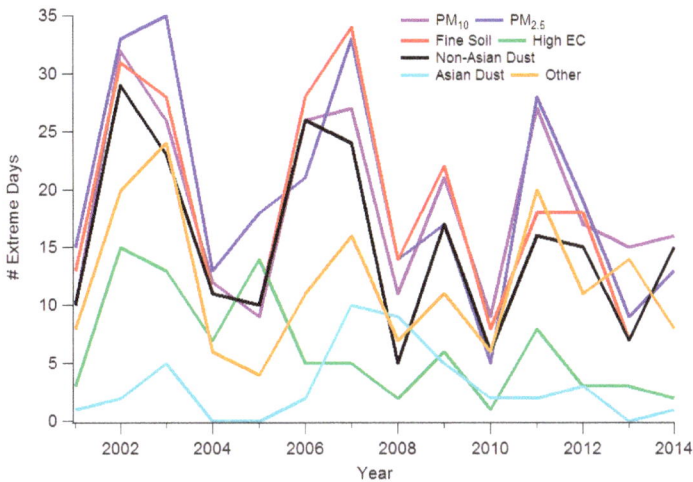

Figure 3. Time series of total extreme events as a function of year for different pollution categories. It is cautioned that the time range with full years of data at all eight sites is from 2003 to 2013 (refer to Table 1 for data time ranges for each site).

A simple linear regression was used to obtain the best-fit line for each pollutant type in Figure 3 using data between 2003 and 2013 when data were available for all sites for full years. Most all slopes were negative except for the Other category, which was only barely positive. The slopes, reported in units of number of events per year, and p values (in parenthesis) are as follows: PM_{10} = −0.19 (0.80), $PM_{2.5}$ = −1.13 (0.23), fine soil = −0.95 (0.29), High EC = −0.83 (0.03), non-Asian dust = −0.85 (0.25), Asian dust = −0.11 (0.76), Other = 0.03 (0.97). The only statistically significant trend at 95% confidence was for the High EC category. This is thought to be due to reduced anthropogenic emissions since other work for the study region examining 2005–2009 has shown that the fastest rate of decline in EC levels was in Phoenix [43], which is the most populated area. Another study analyzing IMPROVE data between 1990 and 2004 across the United States, including the Southwest, showed that there has been a ~25% reduction in EC attributed mostly to emissions controls, with the reduction being most dramatic in the winter as compared to summer [44].

The distribution of extreme events across days of the week is of interest for a few reasons. For example, EC rooted in anthropogenic emissions is thought to lead to higher concentrations around Thursday with minimum values on the weekend [44], and thus examining the frequency of High EC events as a function of the day of the week could help determine if the source of these events is anthropogenic in nature *versus* biomass burning. When normalized by total number of days on either the weekend (Saturday-Sunday) or weekday (Monday-Friday), High EC events occurred more frequently during weekdays (13.2 *versus* 10.5). All other categories exhibited more events during weekdays too. No air pollutant category exhibited a statistically significant difference in the number of events (normalized by number of either weekend or weekday days) on either the weekend or weekdays (or on any specific day) using a chi-square statistical test at the 95% confidence level. The day of the week with the most extreme events for the various pollutant categories was either Wednesday or Thursday.

Table 6. Day of week distribution of extreme events combining data from all eight sites over the entire time duration of the study. The number of extreme events on weekends and weekdays are shown with the values in parenthesis being the normalized values relative to the total number of weekend (2) and weekday (5) days. The day of the week with the most and least events are also shown with values in parenthesis being the actual number of occurrences on that particular day.

–	Total	PM_{10}	$PM_{2.5}$	Fine Soil	High EC	Non-Asian Dust	Asian Dust	Other
Weekend (Saturday–Sunday)	57 (28.5)	65 (32.5)	60 (30)	21 (10.5)	52 (26)	9 (4.5)	44 (22)	57 (28.5)
Weekday (Monday–Friday)	201 (40.2)	208 (41.6)	197 (39.4)	66 (13.2)	162 (32.4)	33 (6.6)	122 (24.4)	201 (40.2)
Day With Most Events	W (53)	Th (55)	W/Th (50/50)	Th (19)	Th (42)	W (11)	W (35)	W (53)
Day with Least Events	Tu (23)	Tu (25)	Tu (21)	M/F (11)	Tu (19)	Tu (1)	Tu (11)	Tu (23)

4. Conclusions

The study examined long-term aerosol data for the Arizona region to describe the frequency and character of extreme aerosol events. The results are as follows in order of the questions raised at the end of Section 1:

(i) Between 5% and 7% of the total days (*i.e.*, 1431–1664 depending on site) examined at the various sites exhibited an extreme aerosol event due to either extreme levels of PM_{10}, $PM_{2.5}$, and/or fine soil. Grand Canyon exhibited the most extreme event days (120, *i.e.*, 7% of its total days), which is coincident with it being one of the most recognized tourism spots in the Southwest. Relative to the total number of extreme days, Grand Canyon exhibited the highest percentage in the High EC category (47% *versus* 13%–25% for other sites). "Other" events accounted for between 2% and 33% of the total extreme days, with most of these being associated with extreme PM_{10} levels (*i.e.*, locally-generated dust). Of the total number of days with extreme fine soil (54–69 days depending on the site), the number of these events being linked to Asian dust, based on NAAPS analysis, ranged from 19% to 29% (i.e., 10–20 days). The analysis highlighted the complexity of using NAAPS and various mass concentration ratios to distinguish between transported and local dust owing to likely mixing effects, especially in urban-impacted areas, such as Tucson and Phoenix.

(ii)   Fine soil is the pollutant type that most frequently impacted multiple sites simultaneously on the same day at an extreme level. Five or more sites reached extreme fine soil levels on the same day for 16%–38% of all possible fine soil extreme events depending on the site. Within the fine soil category, Asian dust events impacted five or more sites between 30% and 57% of the time when they occurred. The pollutant categories with the least number of extreme events impacting five or more sites on the same day were High EC (0 days for all sites) and Other (0–1 day depending on site) due to locally generated emissions that were not regional in nature. Grand Canyon exhibited the highest number of days where an extreme event only impacted that site since it is farther removed from the other seven sites that are clustered closer in Southern Arizona.

(iii)   Most pollutant categories (PM_{10}, $PM_{2.5}$, fine soil, non-Asian dust, High EC, Other) exhibited the highest number of extreme events in August. The Asian dust category was unique in its monthly pattern with its events occurring in the spring months of March–June (41 out of 42 days, *i.e.*, 98%) with only one event in February. Unlike the other pollutant categories, High EC was the only one to show a statistically significant change in frequency of occurrence between 2003 and 2013. While extreme events were most frequent for the various pollutant categories on either Wednesday or Thursday, there was no statistically significant difference in the number of events on any particular day or on weekend days *versus* weekdays.

Supplementary Materials: The following are available online at http://www.mdpi.com/2073-4433/7/1/1/s1

Table S1:   Summary of criteria concentrations for PM_{10}, $PM_{2.5}$, fine soil, and elemental carbon (EC) as a function of month for eight EPA IMPROVE sites in Arizona.

Figure S1:   Case examples demonstrating how NAAPS was used as a tool to identify which extreme fine soil events qualified as Asian dust events. The examples below are for extreme fine soil events occurring on 15 May 2003, 12 April 2007, and 5 June 2008, which qualified as extreme events at four, six, and five sites, respectively.

Acknowledgments: This work was funded by Grant 2 P42 ES04940–11 from the National Institute of Environmental Health Sciences (NIEHS) Superfund Research Program, NIH and the Center for Environmentally Sustainable Mining through the TRIF Water Sustainability Program at the University of Arizona. The authors acknowledge Andrew Huerta and the UROC-PREP program in the Graduate College at the University of Arizona. The authors gratefully acknowledge data provided by EPA IMPROVE and the NRL NAAPS model. Some of the analyses and visualizations used in this study were produced with the Giovanni online data system, developed and maintained by the NASA GES DISC.

Author Contributions: David H. Lopez and Michael R. Rabbani collected and analyzed data. David H. Lopez and Armin Sorooshian wrote the manuscript. Ewan Crosbie, Avelino F. Arellano Jr. and Aishwarya Raman provided useful comments and contributed to the manuscript.

Conflicts of Interest: The authors declare no conflict of interest.

References

1.   Dockery, D.W.; Pope, C.A.; Xu, X.P.; Spengler, J.D.; Ware, J.H.; Fay, M.E.; Ferris, B.G.; Speizer, F.E. An Association between Air-Pollution and Mortality in 6 United-States Cities. *New Engl. J. Med.* **1993**, *329*, 1753–1759.

2.   Pope, C.A.; Ezzati, M.; Dockery, D.W. Fine-particulate air pollution and life expectancy in the United States. *New Engl. J. Med.* **2009**, *360*, 376–386.

3.   Porter, W.C.; Heald, C.L.; Cooley, D.; Russell, B. Investigating the observed sensitivities of air-quality extremes to meteorological drivers via quantile regression. *Atmos. Chem. Phys.* **2015**, *15*, 10349–10366.

4.   Crosbie, E.; Sorooshian, A.; Monfared, N.A.; Shingler, T.; Esmaili, O. A multi-year aerosol characterization for the greater Tehran area using satellite, surface, and modeling data. *Atmosphere* **2014**, *5*, 178–197.

5.   Sutton, L.J. Haboobs. *Q. J. Roy. Meteor. Soc.* **1925**, *51*, 25–30.

6.   Miller, S.D.; Kuciauskas, A.P.; Liu, M.; Ji, Q.; Reid, J.S.; Breed, D.W.; Walker, A.L.; Al Mandoos, A. Haboob dust storms of the southern Arabian Peninsula. *J. Geophys. Res.* **2008**, *113*.

7.   Raman, A.; Arellano, A.F.; Brost, J.J. Revisiting haboobs in the southwestern United States: An observational case study of the 5 July 2011 Phoenix dust storm. *Atmos. Environ.* **2014**, *89*, 179–188.

8.   Seager, R.; Ting, M.F.; Held, I.; Kushnir, Y.; Lu, J.; Vecchi, G.; Huang, H.P.; Harnik, N.; Leetmaa, A.; Lau, N.C.; *et al.* Model projections of an imminent transition to a more arid climate in southwestern North America. *Science* **2007**, *316*, 1181–1184.

9.   Neff, J.C.; Ballantyne, A.P.; Farmer, G.L.; Mahowald, N.M.; Conroy, J.L.; Landry, C.C.; Overpeck, J.T.; Painter, T.H.; Lawrence, C.R.; Reynolds, R.L. Increasing eolian dust deposition in the western United States linked to human activity. *Nat. Geosci.* **2008**, *1*, 189–195.

10. Field, J.P.; Belnap, J.; Breshears, D.D.; Neff, J.C.; Okin, G.S.; Whicker, J.J.; Painter, T.H.; Ravi, S.; Reheis, M.C.; Reynolds, R.L. The ecology of dust. *Front. Ecol. Environ.* **2010**, *8*, 423–430.

11. Dennison, P.E.; Brewer, S.C.; Arnold, J.D.; Moritz, M.A. Large wildfire trends in the western United States, 1984–2011. *Geophys. Res. Lett.* **2014**, *41*, 2928–2933.

12. Stavros, E.N.; Abatzoglou, J.T.; McKenzie, D.; Larkin, N.K. Regional projections of the likelihood of very large wildland fires under a changing climate in the contiguous Western United States. *Clim. Chang.* **2014**, *126*, 455–468.

13. Painter, T.H.; Barrett, A.P.; Landry, C.C.; Neff, J.C.; Cassidy, M.P.; Lawrence, C.R.; McBride, K.E.; Farmer, G.L. Impact of disturbed desert soils on duration of mountain snow cover. *Geophys. Res. Lett.* **2007**, *34*.

14. Husar, R.B.; Tratt, D.M.; Schichtel, B.A.; Falke, S.R.; Li, F.; Jaffe, D.; Gasso, S.; Gill, T.; Laulainen, N.S.; Lu, F.; *et al.* Asian dust events of April 1998. *J. Geophys. Res.* **2001**, *106*, 18317–18330.

15. Tratt, D.M.; Frouin, R.J.; Westphal, D.L. April 1998 Asian dust event: A southern California perspective. *J. Geophys. Res.* **2001**, *106*, 18371–18379.

16. VanCuren, R.A.; Cahill, T.A. Asian aerosols in North America: Frequency and concentration of fine dust. *J. Geophys. Res.* **2002**, *107*.

17. Jaffe, D.; Snow, J.; Cooper, O. The 2001 Asian Dust Events: Transport and impact on surface aerosol concentrations in the U.S. *EOS* **2003**, *84*, 501–516.

18. Wells, K.C.; Witek, M.; Flatau, P.; Kreidenweis, S.M.; Westphal, D.L. An analysis of seasonal surface dust aerosol concentrations in the western US (2001–2004): Observations and model predictions. *Atmos. Environ.* **2007**, *41*, 6585–6597.

19. Kavouras, I.G.; Etyemezian, V.; DuBois, D.W.; Xu, J.; Pitchford, M. Source reconciliation of atmospheric dust causing visibility impairment in Class I areas of the western United States. *J. Geophys. Res.* **2009**, *114*.

20. Creamean, J.M.; Spackman, J.R.; Davis, S.M.; White, A.B. Climatology of long-range transported Asian dust along the West Coast of the United States. *J. Geophys. Res.* **2014**, *119*, 12171–12185.

21. Westerling, A.L.; Hidalgo, H.G.; Cayan, D.R.; Swetnam, T.W. Warming and earlier spring increase western US forest wildfire activity. *Science* **2006**, *313*, 940–943.

22. Kavouras, I.G.; Nikolich, G.; Etyemezian, V.; DuBois, D.W.; King, J.; Shafer, D. *In situ* observations of soil minerals and organic matter in the early phases of prescribed fires. *J. Geophys. Res.* **2012**, *117*.

23. Popovicheva, O.; Kistler, M.; Kireeva, E.; Persiantseva, N.; Timofeev, M.; Kopeikin, V.; Kasper-Giebl, A. Physicochemical characterization of smoke aerosol during large-scale wildfires: Extreme event of August 2010 in Moscow. *Atmos. Environ.* **2014**, *96*, 405–414.

24. Maudlin, L.C.; Wang, Z.; Jonsson, H.H.; Sorooshian, A. Impact of wildfires on size-resolved aerosol composition at a coastal California site. *Atmos. Environ.* **2015**, *119*, 59–68.

25. Malm, W.C.; Sisler, J.F.; Huffman, D.; Eldred, R.A.; Cahill, T.A. Spatial and Seasonal Trends in Particle Concentration and Optical Extinction in the United-States. *J. Geophys. Res.* **1994**, *99*, 1347–1370.

26. IMPROVE Particulate Monitoring Network Standard Operating Procedures. Available online: http://vista.cira.colostate.edu/improve/Publications/SOPs/UCDavis_SOPs/IMPROVE_SOPs.htm (accessed on 19 December 2015).

27. Hogan, T.F.; Rosmond, T.E. The description of the navy operational global atmospheric prediction systems spectral forecast model. *Mon. Weather Rev.* **1991**, *119*, 1786–1815.

28. Hogan, T.F.; Brody, L.R. Sensitivity studies of the Navy's global forecast model parameterizations and evaluation of improvements to NOGAPS. *Mon. Wea. Rev.* **1993**, *121*, 2373–2395.

29. Witek, M.L.; Flatau, P.J.; Quinn, P.K.; Westphal, D.L. Global sea-salt modeling: Results and validation against multicampaign shipboard measurements. *J. Geophys. Res.* **2007**, *112*.

30. McKendry, I.G.; Strawbridge, K.B.; O'Neill, N.T.; Macdonald, A.M.; Liu, P.S.K.; Leaitch, W.R.; Anlauf, K.G.; Jaegle, L.; Fairlie, T.D.; Westphal, D.L. Trans-Pacific transport of Saharan dust to western North America: A case study. *J. Geophys. Res.* **2007**, *112*.

31. Cottle, P.; Strawbridge, K.; McKendry, I.; O'Neill, N.; Saha, A. A pervasive and persistent Asian dust event over North America during spring 2010: Lidar and sunphotometer observations. *Atmos. Chem. Phys.* **2013**, *13*, 4515–4527.

32. Wu, Y.H.; Han, Z.; Nazmi, C.; Gross, B.; Moshary, F. A trans-Pacific Asian dust episode and its impacts to air quality in the east coast of US. *Atmos. Environ.* **2015**, *106*, 358–368.

33. Zhang, J.L.; Reid, J.S.; Westphal, D.L.; Baker, N.L.; Hyer, E.J. A system for operational aerosol optical depth data assimilation over global oceans. *J. Geophys. Res.* **2008**, *113*.

34. Hsu, N.C.; Herman, J.R.; Torres, O.; Holben, B.N.; Tanre, D.; Eck, T.F.; Smirnov, A.; Chatenet, B.; Lavenu, F. Comparisons of the TOMS aerosol index with Sun-photometer aerosol optical thickness: Results and applications. *J. Geophys. Res.* **1999**, *104*, 6269–6279.

35. Torres, O.; Bhartia, P.K.; Herman, J.R.; Ahmad, Z.; Gleason, J. Derivation of aerosol properties from satellite measurements of backscattered ultraviolet radiation: Theoretical basis (vol 103, pg 17099, 1998). *J. Geophys. Res.* **1998**, *103*, 23321–23321.

36. Gkikas, A.; Hatzianastassiou, N.; Mihalopoulos, N. Aerosol events in the broader Mediterranean basin based on 7-year (2000–2007) MODIS C005 data. *Ann. Geophys.-Germany* **2009**, *27*, 3509–3522.

37. Reid, J.S.; Koppmann, R.; Eck, T.F.; Eleuterio, D.P. A review of biomass burning emissions part II: Intensive physical properties of biomass burning particles. *Atmos. Chem. Phys.* **2005**, *5*, 799–825.

38. VanCuren, R.A.; Cliff, S.S.; Perry, K.D.; Jimenez-Cruz, M. Asian continental aerosol persistence above the marine boundary layer over the eastern North Pacific: Continuous aerosol measurements from Intercontinental Transport and Chemical Transformation 2002 (ITCT 2K2). *J. Geophys. Res.* **2005**, *110*.

39. Tong, D.Q.; Dan, M.; Wang, T.; Lee, P. Long-term dust climatology in the western United States reconstructed from routine aerosol ground monitoring. *Atmos. Chem. Phys.* **2012**, *12*, 5189–5205.

40. Prabhakar, G.; Sorooshian, A.; Toffol, E.; Arellano, A.F.; Betterton, E.A. Spatiotemporal distribution of airborne particulate metals and metalloids in a populated arid region. *Atmos. Environ.* **2014**, *92*, 339–347.

41. Sorooshian, A.; Csavina, J.; Shingler, T.; Dey, S.; Brechtel, F.J.; Saez, A.E.; Betterton, E.A. Hygroscopic and Chemical Properties of Aerosols Collected near a Copper Smelter: Implications for Public and Environmental Health. *Environ. Sci. Technol.* **2012**, *46*, 9473–9480.

42. Sorooshian, A.; Shingler, T.; Harpold, A.; Feagles, C.W.; Meixner, T.; Brooks, P.D. Aerosol and precipitation chemistry in the southwestern United States: Spatiotemporal trends and interrelationships. *Atmos. Chem. Phys.* **2013**, *13*, 7361–7379.

43. Sorooshian, A.; Wonaschutz, A.; Jarjour, E.G.; Hashimoto, B.I.; Schichtel, B.A.; Betterton, E.A. An aerosol climatology for a rapidly growing arid region (southern Arizona): Major aerosol species and remotely sensed aerosol properties. *J. Geophys. Res.* **2011**, *116*.

44. Murphy, D.M.; Chow, J.C.; Leibensperger, E.M.; Malm, W.C.; Pitchford, M.; Schichtel, B.A.; Watson, J.G.; White, W.H. Decreases in elemental carbon and fine particle mass in the United States. *Atmos. Chem. Phys.* **2011**, *11*, 4679–4686.

Windblown Dust Deposition Forecasting and Spread of Contamination around Mine Tailings

Michael Stovern, Héctor Guzmán, Kyle P. Rine, Omar Felix, Matthew King, Wendell P. Ela, Eric A. Betterton and Avelino Eduardo Sáez

Abstract: Wind erosion, transport and deposition of windblown dust from anthropogenic sources, such as mine tailings impoundments, can have significant effects on the surrounding environment. The lack of vegetation and the vertical protrusion of the mine tailings above the neighboring terrain make the tailings susceptible to wind erosion. Modeling the erosion, transport and deposition of particulate matter from mine tailings is a challenge for many reasons, including heterogeneity of the soil surface, vegetative canopy coverage, dynamic meteorological conditions and topographic influences. In this work, a previously developed Deposition Forecasting Model (DFM) that is specifically designed to model the transport of particulate matter from mine tailings impoundments is verified using dust collection and topsoil measurements. The DFM is initialized using data from an operational Weather Research and Forecasting (WRF) model. The forecast deposition patterns are compared to dust collected by inverted-disc samplers and determined through gravimetric, chemical composition and lead isotopic analysis. The DFM is capable of predicting dust deposition patterns from the tailings impoundment to the surrounding area. The methodology and approach employed in this work can be generalized to other contaminated sites from which dust transport to the local environment can be assessed as a potential route for human exposure.

Reprinted from *Atmosphere*. Cite as: Stovern, M.; Guzmán, H.; Rine, K.P.; Felix, O.; King, M.; Ela, W.P.; Betterton, E.A.; Sáez, A.E. Windblown Dust Deposition Forecasting and Spread of Contamination around Mine Tailings. *Atmosphere* **2016**, *7*, 16.

1. Introduction

Wind erosion, transport and deposition of particulate matter from contaminated sites may have significant effects on the surrounding environment, especially in arid and semi-arid regions, which are especially susceptible to erosion because of the dry climate and lack of vegetation. Wind erosion occurs on a variety of spatial scales from very large dust storms that can travel thousands of kilometers in the atmosphere [1] to small local sources whose impact is regionally confined. Some of the human health concerns associated with elevated concentrations of particulate matter include fungi and bacteria transport [2] and respiratory stress and cardiovascular disease [3]. Local sources of windblown dust such as dry lake beds, plowed fields and mine tailings can regularly produce windblown particulate matter [4,5]. In semi-arid mining regions, such as the US Southwest, mine tailings impoundments can be a significant anthropogenic local source of windblown particulate matter [4].

Modeling wind erosion, transport and deposition of particulate matter from mine tailings impoundments on scales of a few kilometers is challenging. Typical regulatory models, such as CALPUFF and AERMOD, have difficulty simulating aerosol transport in topographically complex regions on such small scales. Recent advancements in computational capabilities have made Computational Fluid Dynamic (CFD) modeling a viable approach for simulating the erosion and transport of aerosols but minimal research has been done on the modeling of wind erosion and aerosol transport from mine tailings impoundments. Previous studies that have investigated wind erosion of tailings impoundments through the use of CFD modeling [6–9] focused on the erosion process and not the transport and deposition of the windblown particulate matter. In a previous work [10], we utilized a CFD model to simulate wind erosion of a tailings impoundment in a topographically complex region, in order to understand the mechanisms that influence deposition.

The prediction of transport and deposition of windblown dust from mine tailings impoundments is vital in determining the exposure risks in neighboring communities. In a previous work [11] we developed a windblown dust Deposition Forecasting Model (DFM) that was designed to be used in conjunction with operational weather models to forecast deposition of windblown particulate matter for the Iron King Mine (IK) and Humboldt smelter tailings impoundments in Dewey-Humboldt, Arizona. The DFM is a hybrid model that uses both empirical relations derived from direct observations and physical model simulations of aerosol trajectories. The IK and Humboldt smelter tailings impoundments are part of a US Environmental Protection Agency Superfund site that has elevated concentrations of toxic species such as lead and arsenic [12].

In this work, we assess predictions of the DFM by validating the deposition forecasts against a variety of field measurements at the mine site and surrounding areas, including: (i) spatially distributed measurements of dust deposition; (ii) metals composition; and (iii) lead isotope analysis. The spatially distributed measurements of dust deposition were collected using inverted-disc samplers during two month-long field sampling campaigns in April and June 2014. The sampling periods were selected to coincide with spring and early summer, which represent the windy season in southern Arizona. Dust collected during the sampling campaigns was analyzed for total weight, chemical composition and lead isotopes and directly compared to forecasted deposition patterns generated by the DFM. We hypothesize

257

that forecast spatial deposition patterns generated by the DFM will agree with arsenic and lead tracers captured by the inverted-disc samplers, which will help in establishing routes of contaminant exposure in local communities. In addition, we investigate how topsoil measurements of arsenic and lead performed in this work based on techniques developed in a previous work [13] correlate with the dust sampler contaminant concentrations and isotopic signatures, which serves to quantify the spread of contamination from the mine tailings site.

2. Methodology

2.1. Site Description

The Iron King Mine Tailings and Humboldt Smelter Superfund area is located in central Arizona in the vicinity of Dewey-Humboldt (Figure 1). The smelter produced lead, gold, silver, zinc and copper in the period 1906–1969. The area was classified as a Superfund site by the US EPA in 2008 [12] after it was discovered that it is contaminated with lead and arsenic.

The region is classified as semiarid with an annual rainfall of about 480 mm. The vegetation is Pinyon Juniper woodlands with limited desert grasses and other bushes [14]. Most of the land adjacent to the northern, western and southern edges of the tailings and mine operations property is publicly owned state trust and grazing land. Arizona State Highway 69 separates the mine tailings area from the town of Dewey-Humboldt, which has an elevation of 1396 m.

The tailings impoundment consists of two areas: the main tailings impoundment with a total aerial extent of 96,000 m^2 and the lower tailings region located directly adjacent to the main tailings impoundment with a total aerial extent of 84,000 m^2. The tailings impoundment is devoid of vegetation except where a phytostabilization project is attempting to reestablish native vegetation. The revegetation project has been in progress since May 2010 and is confined to an area of 7200 m^2 on top of the tailings [15]. The surface of the impoundment is made up of 34.7% sand, 44.8% silt, and 20.4% clay and has patches of reddish coloration attributed to iron minerals, mostly ferrihydrite [15]. This type of tailings material composition is highly skewed towards the fine particle size range with 65.2% being silt size (<50 μm). The tailings are typically covered by crusted soil with patchy efflorescence that usually forms following rain events. The crust can be broken up and results in very fine powdery material that is easily eroded by the wind. The average arsenic and lead concentrations measured in bulk samples of the mine tailing material are about 0.12% and 0.10% by mass, respectively.

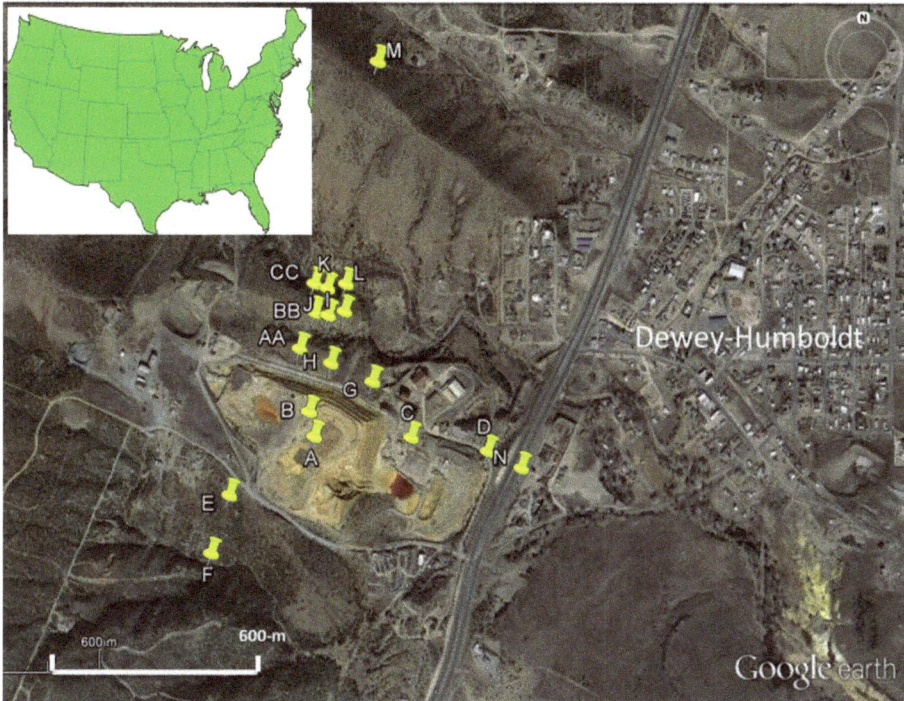

Figure 1. Google Earth visible satellite image of the Iron King Mine tailings impoundment and the town of Dewey-Humboldt (March 2014). The locations of the inverted-disc samplers are denoted by yellow pins (image from Google Earth).

A suite of meteorological and dust monitoring instruments was installed on the tailings impoundment in 2009. The setup consists of two eddy flux towers, equipped with six DUSTTRAK dust monitors for PM_{27} (particulate matter with an aerodynamic diameter less than 27 μm,) anemometers, wind vanes, thermometers and hygrometers. In addition, a Micro-Orifice Uniform-Deposit Impactor (MOUDI) was deployed to measure size fractionated distribution of aeolian dust from April 2013 to January 2014. Four month-long MOUDI samples were collected during this period. The MOUDI (MSP Corp) has eleven stages that collect particles with aerodynamic cut point diameters of: 18-μm, 9.9-μm, 6.2-μm, 3.1-μm, 1.8-μm, 1-μm, 0.55-μm, 0.32-μm, 0.18-μm, 0.1-μm, 0.054-μm and ultrafine particles collected on an "after filter". The MOUDI was located along the northern edge of the tailings impoundment with the inlet located 1 m above ground level. Observations collected by the DUSTTRAK and the MOUDI were used to generate empirical relations between dust concentration (discriminated by particle size), wind speed at 10-m height and relative humidity. These correlations were employed in the calculation of dust fluxes in the DFM [11]. In addition, the meteorological observations are

259

used to determine systematic biases between observed conditions and operational WRF forecasts.

2.2. Deposition Forecasting Model Description

The DFM predicts the deposition of PM_{27} generated by windblown dust from the Iron King tailings impoundment. This size range has been chosen since it corresponds with the measurements made by the DUSTTRAK dust monitors. The hybrid model utilizes both empirically derived relations and particle transport simulations. Details of model development and validation are presented elsewhere [11].

The nearly two years (2012–2013) of meteorological and dust monitoring data collected from the eddy flux towers were used to derive empirical relationships between meteorological conditions and windblown dust generation. These empirical relations include the effects of wind speed and relative humidity on airborne dust, particle size fractionation and the vertical profile of dust concentration measured on the tailings and wind speed effects on wind direction. Here, we will summarize the empirical correlations presented in our previous work [11]. We correlated dust concentrations in the atmosphere over the tailings surface to obtain

$$C_{dust} = A_1 A_2 U^B \tag{1}$$

where C_{dust} is the mass concentration $(\mu g/m^3)$ of dust in the atmosphere at a specific height and for a specific particle size range, U is the 10-m height wind speed, A_1 and B are constants that are fitted to measurements for each specific height and particle size range (see [11] for values and details), and A_2 is a function of atmospheric relative humidity, empirically found to be:

$$A_2 = \begin{cases} 1 \ for \ RH \leqslant 25\% \\ (50 - (RH - 25))/50 \ for \ 25\% < RH \leqslant 75\% \\ 0 \ for \ RH > 75\% \end{cases} \tag{2}$$

where RH is relative humidity (%). Concentrations, wind speeds and relative humidities in these empirical equations correspond to 1-h averages.

The DFM forecasts a dust emission flux from the tailings as proportional to the integral of the concentration given by Equation (1) over height. The flux is assumed to be uniform over the whole surface area of tailings (95,000 m^2).

The DFM forecasts deposition in three particle size ranges. The size ranges are defined by particles that have aerodynamic diameters (D_p) that fall within: 27 μm > D_p > 18 μm (coarse), 18 μm > D_p > 3.1 μm (medium), and D_p < 3.1 μm (fine). The size ranges were determined by MOUDI and DUSTTRAK observation capabilities: The coarse range represents the difference between DUSTTRAK and total MOUDI mass measured in a given event (that is, for the same amount of air treated). The medium and fine ranges were obtained by adding the mass of collected sample in corresponding stages in the MOUDI. The cut point (3.1 μm) was selected based on the fact that, on average, total mass collected by the MOUDI above and below that particle size was approximately the same. In addition, the same size ranges were used in deriving the empirical relationships between meteorological conditions and wind erosion flux within the DFM model [11].

The DFM also includes the effect that topography has on aerosol deposition. Stovern *et al.* [10] simulated windblown transport of fugitive aerosols from the Iron King Mine tailings using computational fluid dynamics. They found that windblown dust preferentially deposits in regions of topographic upslope relative to the mean flow. Due to the complex topography of the site, these effects are included in the model.

The DFM results for each forecast period (see Section 3) are produced for three particle size ranges and are modelled on a spatial grid that is approximately 25 km^2 in area with 10.3 m spatial resolution. These simulations include the effects that a convectively turbulent boundary layer has on particle trajectories. Simulation area includes all of the Iron King tailings and most of the adjacent town of Dewey-Humboldt. The model is initialized using the 48-h forecasts from an operational version of the WRF model produced by the department of Atmospheric Sciences at the University of Arizona.

2.3. WRF Model Forecasts

The WRF model is configured with two nested grids. The inner grid covers the entire state of Arizona, portions of Southeastern California, Western New Mexico and Northern Mexico. The outer grid covers from Northern Nevada to the tip of the Baja California peninsula, and from Western Texas to portions of the Eastern Pacific Ocean. The inner and outer domains have horizontal resolutions of 1.8-km and 5.4-km, respectively. Model forecasts are produced daily at 12Z and 6Z, using both Global Forecast System (GFS) and North American Mesoscale Forecast System (NAM) initializations. Each forecast run is 48 h long at 1-h intervals. The DFM was initialized using the WRF forecast conditions predicted 28–39 h in advance.

We used the daily 12Z GFS WRF runs during the periods 21 April to 22 May and 11 June to 9 July 2014 to tabulate the horizontal components of the 10-m height winds, 2-m specific humidity, the 2-m height temperature and the surface pressure for each hour between 9 A.M. and 9 P.M. We then calculated the wind speed and wind direction and relative humidity for each hourly interval by averaging the five nearest-neighbor WRF grid points at the tailings location, which was used to initialize the deposition forecasting model.

WRF model forecasts and observed meteorological conditions on the tailings were compared to test for systematic biases over a period of 163 days (June 2012

261

to August 2013). This period coincides with the same observing period used to determine the empirical relations used in the creation of the deposition forecasting model [11]. As described later, forecast and observed wind speed and relative humidity were corrected for systematic bias. Results will be summarized in Section 3.1. To calculate directional bias we first calculated the residual for each model and observation pair which fall within the range of -180 to $+180$ degrees. A positive residual indicates counter-clockwise rotation between the WRF model and observed wind direction and *vice versa*. A perfectly nonbiased data set would produce an average residual of zero.

2.4. Dust Deposition Sampling

Inverted-disc samplers have been used in dust deposition experiments, including aeolian deposition near an eroding source field [16] and dry deposition of polychlorinated organics [17]. The collection efficiencies of inverted-disc samplers have been studied extensively [18–22] and are dependent on wind speed and particle diameter. However, there appears to be general consensus that the collection efficiency falls within the range of 5%–40% for all grain sizes [21]. For this study, collection efficiency is unimportant because we base our comparisons on relative deposition amounts (Section 3.3.2). Hall and Waters [19] showed that an inverted-disc sampler has a significantly higher blowout wind speed than both the flat disc and the British Standard deposit gauge. This reduces resuspension and loss of particulate matter.

The plastic discs (Frisbees) were purchased from discountmugs.com. They have a diameter of 233 mm and a depth of 25 mm. The discs were glued to the lids of high density polyethylene 500-mL sample bottles (Thermo Scientific) using Loctite plastic bonder. An 8.5 mm hole was drilled through the disc into the bottle. The samplers were mounted on iron stakes and placed 1 m above the ground.

A set of 20 inverted-disc samplers was placed around the Iron King tailings impoundment and surrounding area. The disc samplers were placed along three transects to measure deposition in the northward, eastward and southward directions, starting from the main tailings and extending up to 1 km away. A majority of the samplers were located north of the main tailings, along the dominant wind direction (Figure 1). There is significant topographic variation north of the tailings, which is used for testing the effect that topography has on dust deposition.

2.5. Soil Sampling

Soil samples were taken during a single sampling campaign in 2014 at different distances from the mine tailings to assess the extent of contamination. A total of nine soil samples were collected with the first sampling point located on the mine tailings and eight more sampling points in a straight line NE transect, which corresponds to the prevailing wind direction. Samples at different depths were taken in order to obtain a vertical profile of the contamination at the following depth intervals: 0–3 mm, 3–6 mm, 6–9 mm and finally 100 mm. Details of the sampling technique are provided elsewhere [13]. Final samples at each site and depth were a composite of three different samples. The composite samples were dried for 10 h at a temperature of 110 °C and then sieved through a 0.84 mm sieve to discard coarser fractions. Distances from the tailings to the sampling points along the NE transect were 13 m, 70 m, 130 m, 355 m, 1100 m, 1128 m, 1150 m and 5032 m.

2.6. Sample Analysis

In the field, deionized water was used to flush all the dust captured on the inverted-disc samplers into the attached 500 mL bottle. The bottles were unscrewed and capped for transportation to the laboratory where each sample was partially dried in the 500 mL bottle in an oven (60 °C) and then transferred into a pre-weighed 50 mL glass vial. The vial samples were then completely dried and sieved to remove particles >500 μm. The samples were then weighed using a Mettler AE100 balance ($\pm$0.1 mg). The sample masses were normalized to the area of the disc, which yields the mass deposition per unit area for each sample location. The handling of MOUDI samples has been described in detail elsewhere [4].

Both inverted-disc and soil samples were prepared for metal and lead isotopes analysis by extraction with 15 mL of *aqua regia* (1.03 M HNO_3/2.23 M HCl, trace-metal grade) with sonication at 80 °C for 60 min. Aliquots of 1.2 mL of solution were extracted and diluted to 4 mL with deionized water before the analysis. Due to the relatively low concentrations of lead in some of the samples, the lead isotope analysis samples were concentrated on a hot plate [13]. An ICP-MS (Agilent 7700X with an Octopole Reaction System) was used to analyze for metal concentrations and lead isotopic composition. MiliQ water, 0.669 HCl (Fisher, trace-metal grade) and 0.309 M HNO_3 (EMD, Omnitrace) were used to create the certified calibration standards from Accustandard. In addition to each sample, the National Institute of Standards and Technology (NIST) standard reference material (SRM 1643e trace elements in water) was also analyzed. We used the same operating condition for the analysis of both elemental concentrations and the lead isotopic ratios. NIST SRM 981 (Lead isotopic standard) was used for validation and calibration and the analytical precision of lead isotopic ratios was under 0.5% [13].

3. Results

3.1. WRF Model Verification

Biases in the 24-h WRF forecast of wind speed, wind direction and relative humidity are determined by direct comparison to *in situ* observations. The forecast of relative humidity showed good agreement with the observations (Figure 2). The results show that the WRF model slightly under predicts the observed relative humidity, which may be caused by precipitation events that the WRF model failed to forecast since potential dust mitigation due to remnant soil moisture from past rain events are not included in the model. To adjust for this slight bias in our calculations, the model-predicted relative humidity is multiplied by a correction factor of 1/0.97, product of a least-squared error fit of the parity plot in Figure 2.

The forecast and observed hourly averaged wind speeds are also correlated but with a high degree of scatter (Figure 3A). The results show that the WRF forecast winds are systematically lower than those observed (slope < 1). The low R^2 value is indicative of the difficulties when comparing model forecasts using 1.8-km grid spacing to point observations where boundary layer mechanics and surface roughness play an important role. The frequency distribution of the model forecast and the observed wind speeds, Figure 3B, shows that over multiple hourly periods the distributions are very similar. The model-predicted wind speed is corrected in our calculations by multiplying by 1/0.84, which is the fitted slope of the parity plot (Figure 3A).

The histogram of the hourly wind direction residuals is shown in Figure 4. The histogram was generated using 10 degree bins from $-180°$ to $180°$. The histogram is shifted to the right of zero, which means the model-predicted wind direction is systematically biased counter clockwise from the observed. By adding $14.7°$ of clockwise rotation to the forecast wind direction (first moment of the histogram) we account for the apparent bias between model forecast and observed conditions.

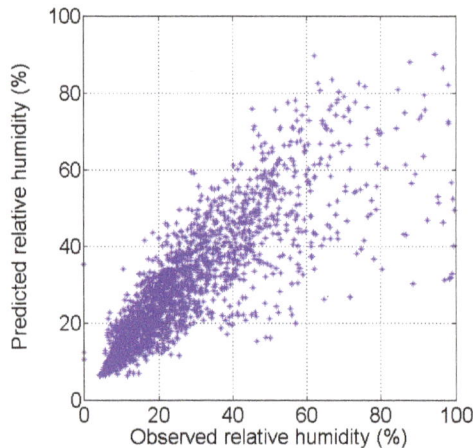

Figure 2. Comparison of hourly averaged observed relative humidity and WRF model forecast predictions for the period of 29 May 2012 to 4 August 2013. A linear fit yields a slope of 0.97 with $R^2 = 0.59$.

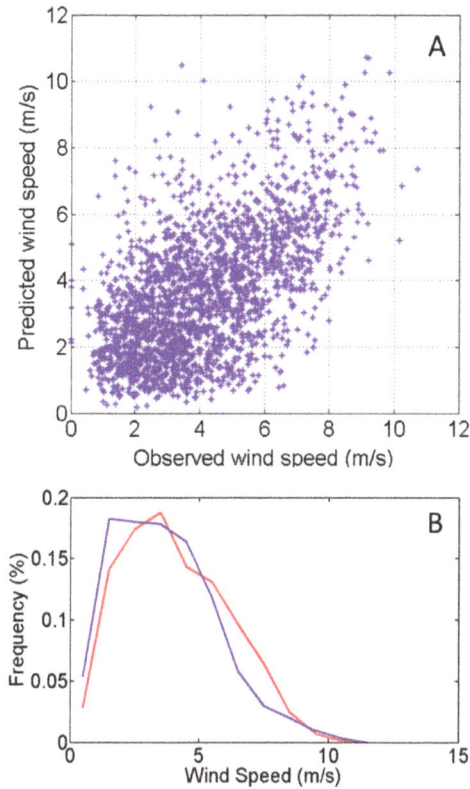

Figure 3. (A) Direct comparison of hourly averaged 10-m height observed wind speed and WRF model predictions for the period of 29 May 2012 to 4 August 2013. A linear fit yields a slope of 0.84 with $R^2 = 0.23$; **(B)** Frequency distribution of hourly averaged 10-m height observed wind speed and WRF model predictions for the same period.

3.2. Deposition Model Predictions

The DFM simulations were initialized using the corrected WRF model forecasts. Figure 5 shows deposition patterns predicted by the DFM for the fine ($PM_{3.1}$), medium (PM_{18}–$PM_{3.1}$), coarse (PM_{27}–PM_{18}) and total suspended fine particulate (PM_{27}) for the period 21 April to 22 May 2014. A majority of the deposition occurs in the northward direction, which is the prevailing wind direction for the period. However, there is significant predicted deposition in the southeastward and southwestward directions, which was a result of several synoptic scale troughs that shifted the daytime wind direction. One of these troughs was accompanied by precipitation on 27 April 2014. The WRF model predicted the strong wind speeds associated with the trough but failed to accurately forecast the precipitation that was

265

observed at the site. The increased soil moisture caused by the precipitation greatly reduces windblown erosion. Hence, the deposition in the southwesterly direction is thought to be significantly overestimated.

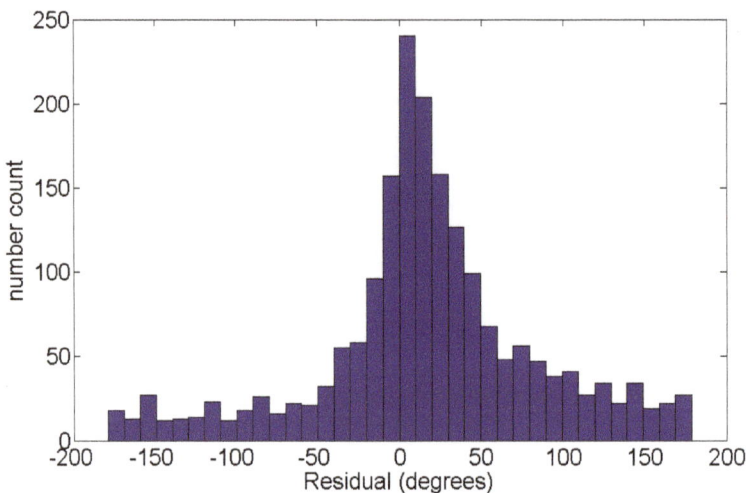

Figure 4. Histogram of residuals of measured wind direction (1-h averages) and WRF model predictions for the period of 29 May 2012 to 4 August 2013. The displaced maximum is at +14.7°, which indicates a counter clockwise bias in the model results.

The maximum coarse particle deposition was mainly constrained to the immediate vicinity of the tailings due to their large size and fast settling times, which do not allow for long distance transport. The fine particles have a much smaller terminal velocity and are transported much further downwind. Particles of 3-μm size have a maximum deposition location that is about three hundred meters from the tailings in the northward direction. Smaller particles may travel longer distances. There are slight variations in the deposition patterns caused by the impact of topography of the surrounding region.

The resulting DFM simulations for the sampling period 11 June to 9 July 2014 can be seen in Figure 6. Compared to the May sampling period, the forecast weather conditions at the site were consistent with predominantly southerly winds that caused almost all the deposition to be in the northward direction.

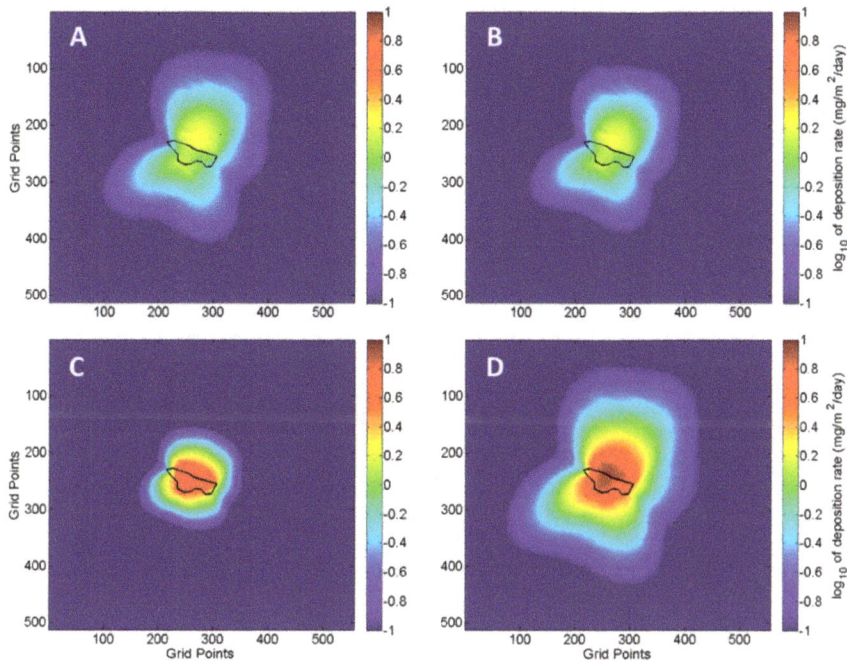

Figure 5. Maps of predicted dust deposition around the IK mine tailings for three size fractions and PM_{27} predicted by the DFM for the forecast period 21 April to 22 May 2014. Particle size ranges: (**A**) $PM_{3.1}$; (**B**) $PM_{3.1}$–PM_{18}; (**C**) PM_{18}–PM_{27}; (**D**) PM_{27}. The tailings impoundment is outlined in black. The grid points are spaced by 10.3 m and the domain has a total horizontal extent of 34.47639° to 34.52380° latitude and −112.27639° to −112.22491° longitude. Calculated dust deposition corresponds to dust emitted from the mine tailings.

3.3. Inverted Disc and Soil Sampling Results

3.3.1. Mass Analysis

Inverted-disc sampling occurred during the sampling periods 21 April to 22 May 2014; and 11 June to 9 July 2014. Table 1 shows the deposition fluxes (mass deposited per unit time and sampler surface area) for each disc sampler during both periods and the corresponding model predicted PM_{27} deposition. It is important to point out that the model developed here predicts deposition of dust generated on the mine tailings and not total deposition, since the region has multiple potential dust sources. It is clear from the data in Table 1 that the model under predicts total deposition, as expected.

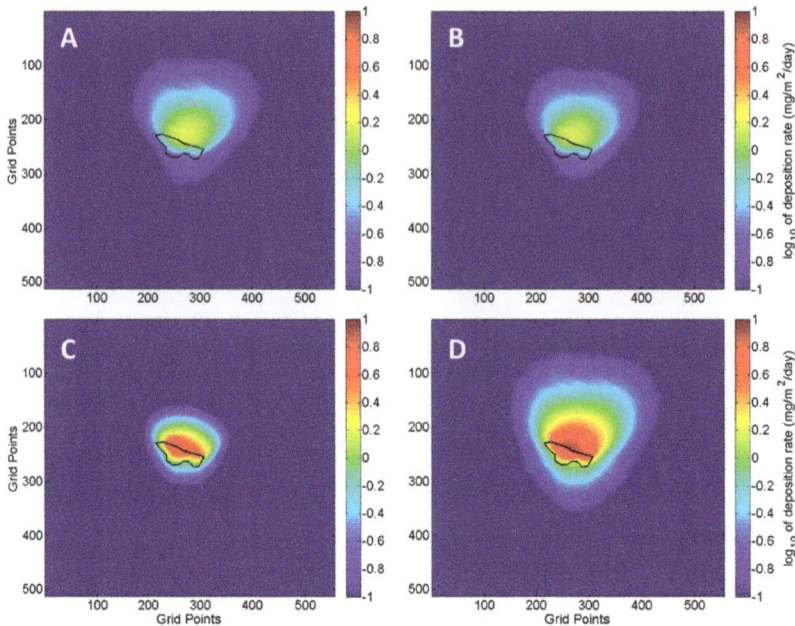

Figure 6. Maps of predicted dust deposition around the IK mine tailings for three size fractions and PM$_{27}$ predicted by the DFM for the forecast period 11 June to 9 July 2014. Particle size ranges: (**A**) PM$_{3.1}$; (**B**) PM$_{3.1}$–PM$_{18}$; (**C**) PM$_{18}$–PM$_{27}$; (**D**) PM$_{27}$. The tailings impoundment is outline in black. The grid points are spaced by 10.3 m and the domain has a total horizontal extent of 34.47639° to 34.52380° latitude and −112.27639° to −112.22491° longitude. Calculated dust deposition corresponds to dust emitted from the mine tailings.

The highest deposition fluxes were measured at location N (Figure 1) for the May sampling period and location B for the June sampling period. Sample N is located adjacent to the highway while sample B is located on the main tailings impoundment. Because roadways are well documented as production source of atmospheric aerosols, it follows that we would expect larger amounts of deposition to be captured with the N sampler.

The observed deposition flux for the two sampling periods is within an order of magnitude of the forecast deposition fluxes. The peak forecast deposition flux is approximately half of that measured using the samplers. However, the model forecast deposition fluxes drops dramatically for samplers located far from the tailings source (*i.e.*, samplers F, M and N) while the total mass of dust measured by the samplers remains similar no matter their location. It is important to consider the fact that wind erosion occurs from a variety of sources within the region and each deposition sampler is collecting dust from all of them not just aerosols resulting from the tailings, although the exposed, elevated tailings are thought to be an important contributor.

Table 1. Measured dust (PM$_{27}$) mass deposition fluxes in inverted-disc samplers.

Location	4/21 to 5/22 Observed Deposition (mg/m^2/day)	4/21 to 5/22 Predicted Deposition (mg/m^2/day)	6/11 to 7/09 Observed Deposition (mg/m^2/day)	6/11 to 7/09 Predicted Deposition (mg/m^2/day)
A	15.6	7.69	17.0	4.48
B	24.2	7.96	37.1	6.11
C	6.9	6.04	26.1	4.66
D	26.0	1.82	29.7	1.36
E	9.4	4.48	12.1	0.53
F	14.2	2.45	10.6	0.27
G	12.3	8.11	8.5	8.21
H	18.2	8.75	8.1	8.67
I	10.1	7.39	12.6	7.02
J	16.2	7.76	9.0	7.54
K	30.0	5.47	17.3	5.16
L	16.3	5.45	15.5	5.15
M	16.0	1.06	N/A	N/A
N	32.5	1.16	6.4	0.68
AA	N/A	N/A	12.0	7.77
BB	N/A	N/A	8.7	7.43
CC	N/A	N/A	11.0	4.81

As mentioned, the deposition forecasting model only simulates the transport of windblown particulate matter from the tailing impoundment and for particles with an aerodynamic diameter ⩽27 µm. Comparing forecasts of PM$_{27}$ to bulk deposition samples has inherent errors caused by the potential of the samplers to collect larger particles and skewing the mass fluxes. In order to minimize these issues certain steps were taken to reduce the impact of large particles. The inverted-disc samplers were shown in wind tunnel tests to have the highest collection efficiency, up to 60%, for particles with diameter 10 to 31 µm. Their collection efficiency significantly drops for particles up to 89-µm [21]. The samplers were placed at 1-m height to minimize the capture of large saltating particles. Additionally, mass fractions of dust measured by the MOUDI at 1-m height [11] were 30.1%, 30.0%, and 39.9% for the >18-µm, 18 to 3.1-µm and <3.1-µm size fractions. The large percentage of mass observed in the smaller size fractions gives us confidence that the inverted-disc samplers are not significantly affected by larger particles and their results are reasonable approximations of PM$_{27}$ aerosol deposition.

Additional errors in the forecast depositions can arise from the DFM dependency on the WRF wind speed forecast. The WRF model provides us with the highest resolution weather forecasts of the region but it is still susceptible to errors including those experienced during short term high wind events. The compounding factors of multiple dust sources, DFM forecast of PM$_{27}$ and errors in the WRF forecast makes the under estimated DFM deposition fluxes measured by the inverted-disc samplers reasonable, Table 1. Through the use of elemental analysis we can partition the captured dust and determine the relative influence the tailings have on each sampler and the relative spatial distributions.

3.3.2. Lead and Arsenic Analysis

The tailings and surrounding soil have significantly elevated arsenic and lead concentrations when compared to the natural surroundings [14]. This is illustrated by the results in Figure 7, where topsoil concentrations of lead and arsenic are shown as a function of distance from the tailings along a NE transect. Significant arsenic and lead contamination extend at least to 1 km from the tailings, while the 5 km site exhibits relatively low concentrations, which could correspond to background levels in the region (around 10 ppm for both arsenic and lead).

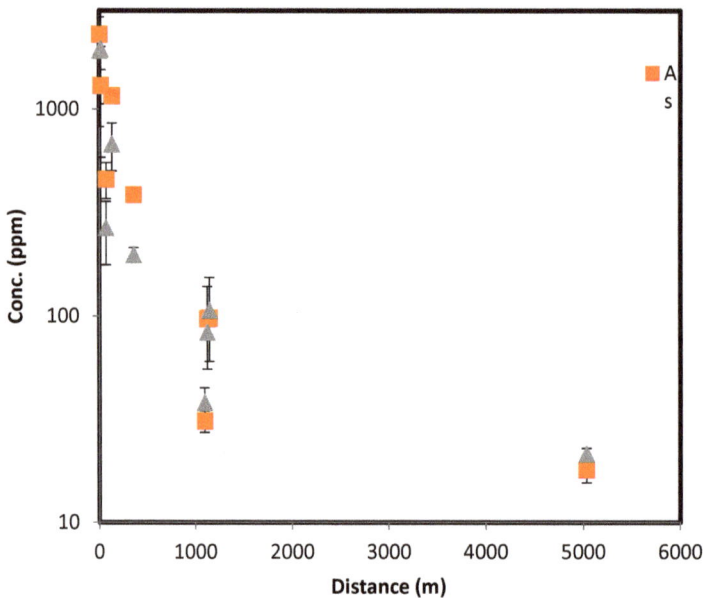

Figure 7. Arsenic and lead concentrations in topsoil (0–3 mm depth) samples at different distances from the mine tailings following a NE transect. Error bars show standard deviation of triplicate samples at the same location.

The arsenic and lead concentrations measured in inverted-disc samplers are presented in Table 2. Figures 8 and 9 show the DFM predictions of average deposition fluxes for the May and June sample periods. A way to assess the results is to compare transects of relative observed concentrations of As and Pb with forecast deposition of PM_{27}. Three transects are evaluated in the southwestward, eastward and northward directions. The relative Pb and As concentration transects are calculated by normalizing each sampler to the average concentration measured by inverted-disc samplers A and B located on the tailings pile. Equation (1) shows the normalization process for a sample concentration (C_{sample}) where C_A and C_B are the concentrations measured at samplers A and B, respectively. Transects of the DFM PM_{27} are normalized by the forecast deposition at the location of sampler B ($34.50087°$ latitude and $-112.25305°$ longitude). All distances of the sample locations are calculated from sample point B.

$$C_{sample\ norm} = \frac{C_{sample}}{(C_A + C_b)/2} \tag{3}$$

The southward transect is generated using the normalized samples A and B, E (366-m downwind) and F (538-m downwind). The eastward transect was generated using the normalized samples A and B, C (377-m downwind), D (657-m downwind) and N (786-m downwind). For the June sample period, the northward transect was generated by first averaging the sample concentrations collected at different downwind ranges, which included the AA samples (~205-m downwind), I, J, and BB samples (~300-m downwind) and the K, L and CC samples (~379-m downwind). The sample averaged concentrations at tailings, 205-m, 300-m and 379-m downwind were then normalized and used to generate the northward transect.

Table 2. Measured total As and Pb concentrations measured in the dust collected by the inverted-disc samplers. Locations are shown in Figure 1.

Location	4/21 to 5/22 Period Pb (ppm)	6/11 to 7/09 Period Pb (ppm)	6/11 to 7/09 Period As (ppm)
A	332	903	1826
B	981	1743	3856
C	317	1037	1406
D	454	775	1389
E	130	317	132
F	39.3	71.2	46.4
G	506	1614	1311
H	482	838	588
I	217	435	586
J	218	383	504
K	234	342	465
L	292	439	649
M	41.9	N/A	N/A
N	41.2	138	156
AA	N/A	384	449
BB	N/A	453	595
CC	N/A	302	435

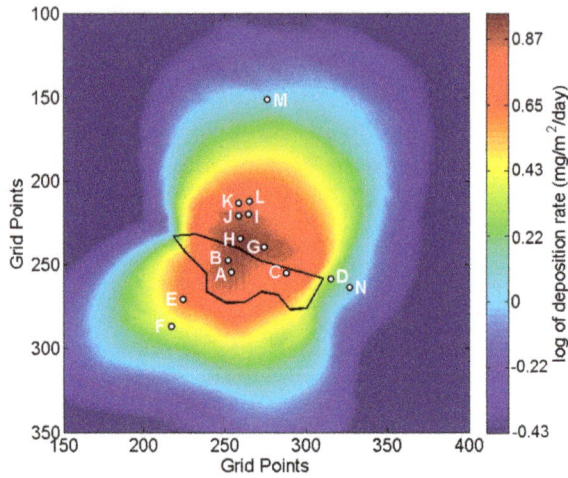

Figure 8. Map of PM$_{27}$ deposition predicted by the DFM for the forecast period 21 April to 22 May 2014. The color scale represents the natural log of the deposition flux. The tailings impoundment is outlined in black and the locations of the inverted-disc samplers are indicated. The grid points are spaced by 10.3-m and the domain has a total horizontal extent of 34.49142° to 34.51452° latitude and −112.26247° to −112.23939° longitude.

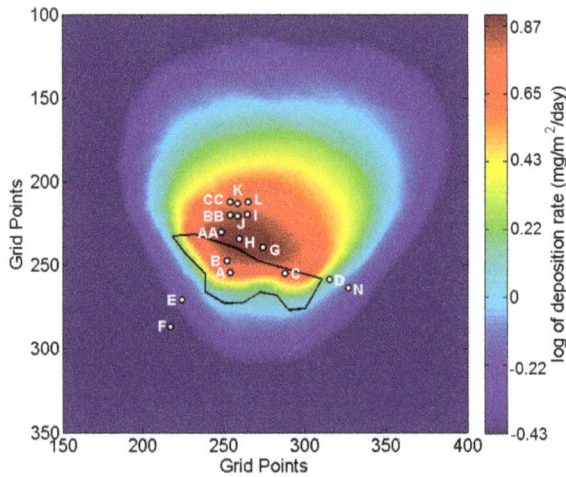

Figure 9. Map of PM$_{27}$ deposition predicted by the DFM for the forecast period 11 June to 9 July 2014. The color scale represents the natural log of deposition. The tailings impoundment is outlined in black and the location and sample label of the inverted-disc samplers are indicated. The grid points are spaced by 10.3-m and the domain has a total horizontal extent of 34.49142° to 34.51452° latitude and −112.26247° to −112.23939° longitude.

For the May sampling period we first averaged the sample concentrations of I and J (~300-m downwind) and K and L (~379-m downwind). The northward transect was then generated using the normalized samples A and B, H (205-m downwind), 300-m downwind average, 379-m downwind average, and M (1027-m downwind).

Results for the southward, eastward and northward cross sections for the May and June sample periods can be seen in Figure 10. The southward cross sections of the model-predicted fractional reduction in deposition follow similar trends to the observed fractional reduction of As and Pb concentrations measured in the inverted-disc samplers. For the May sampling period, the forecast overestimated the relative amount of deposition located at the sample locations E and F. This was most likely caused by strong northeasterly winds associated with a synoptic scale weather system. This weather system produced precipitation in the region, which the WRF model failed to predict. The erroneous WRF weather forecast yielded an overestimation of windblown dust transport in the southwestward direction. For June, the DFM model accurately forecasts the relative reduction in As and Pb at the E and F sample locations.

The eastward transect shows that the model-predicted fractional reduction in deposition was similar to the observed fractional reduction of As and Pb concentrations for both the May and June sampling periods for samplers C and N. However, the DFM underestimated the fractional reduction sampler D for both monthly sampling periods (Table 1). Sampler D is located just 10-m from the eastern edge of the lower tailings area and approximately 4-m lower in elevation. This proximity and lower elevation of sampler D to the lower tailings area increased the likelihood of eroded particles to gravitationally settle into this sampler. Table 1 shows that sampler D had systematically large deposition fluxes measured during both sample periods. Size distributions of the collected dust could not be determined due to the lack of total material captured by the sampler. The increased likelihood of capturing tailing material most likely skewed the fractional As and Pb deposition measured by the sampler.

The northward transect shows a similar downwind pattern of reduction in the fractional As and Pb concentration (Figure 10). The DFM overestimates the fractional reductions of Pb for the May sampling period and both As and Pb for the June sampling period. However, the DFM was significantly better estimating the relative reduction of As and Pb for the May sampling period when compared to June. For the May sampling period the DFM accurately predicted the fractional reduction in Pb concentrations for sampler M located approximately 1-km north of the tailings.

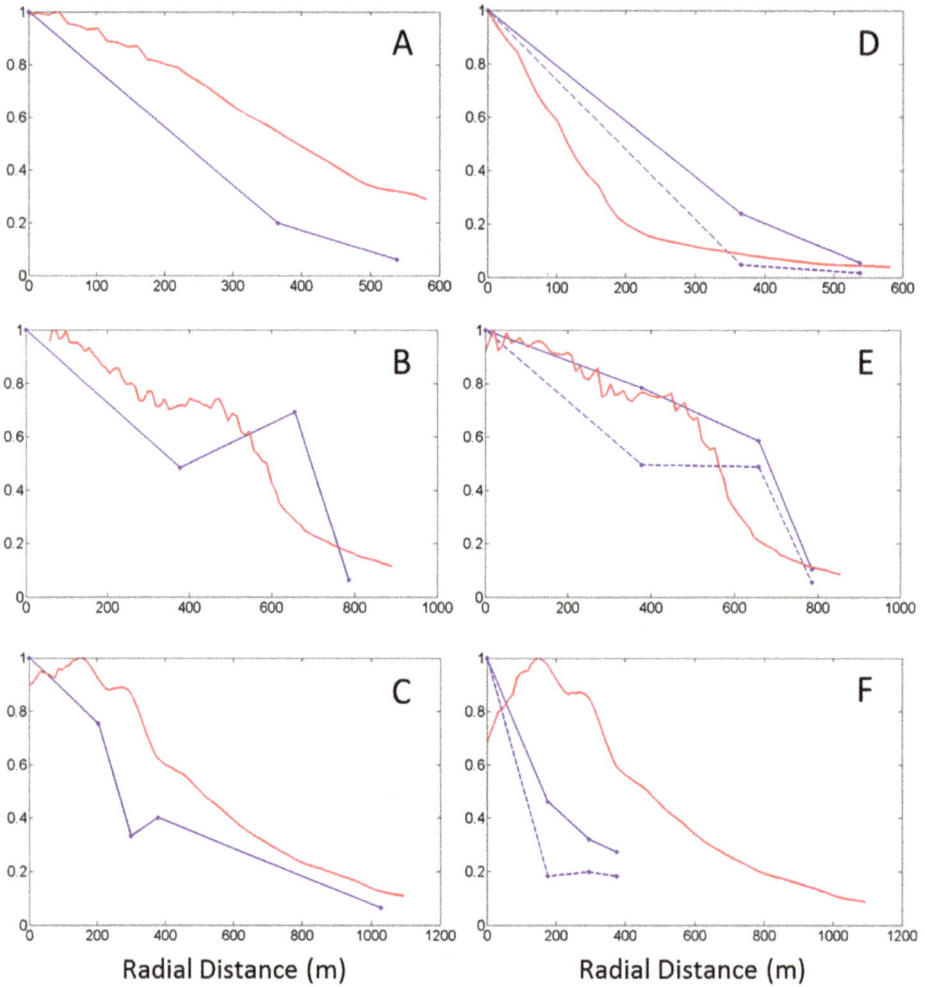

Figure 10. Comparison of relative decreases in arsenic (blue dashed lines) and lead (blue solid lines) mass concentrations measured by the samplers *versus* the relative decreases of deposition flux forecast by the DFM (red lines) for the southward (**A**) eastward (**B**) and northward (**C**) cross sections for the 21 April to 22 May 2014 sample period and the southward (**D**) eastward (**E**) and northward (**F**) cross sections for the 11 June to 9 July 2014 sample periods.

The highest concentrations of both As and Pb were measured by the inverted-disc samplers located on the tailings (A and B). In comparison, the DFM model predicts that the highest amount of tailing dust deposition should occur approximately 150 m north of sample point B. However the inverted-disc samplers show significant reduction in relative As and Pb between the tailings (A and B) and the points located about 205-m north (H and AA).

The DFM includes topographic slopes when calculating deposition rates to the surface. Using computational fluid dynamics modeling, Stovern *et al.* [10] showed that the slope of the ground significantly impacts deposition in this topographically complex region. The inverted-disc samplers were strategically placed in locations that were sloped for the northward transect. The samples located at 205-m, 300-m and 379-m were in down-sloping, up-sloping and down-sloping regions, respectively. The slope of the ground at 205-m, 300-m and 379-m is $-5.7°$, $+15.1°$, and $-22.3°$ respectively with a slope length of 50 m. Thus, we would expect that the samples collected at 300 m should have systematically more deposition than the samplers at 379-m. However, we appear to see the opposite occurring in the May sampling period and relatively equal amounts of deposition in the June sampling period. One reason why the effects of topographic slope are not evident in the deposition patterns may be due to changes in surface roughness between the sampler locations at 300-m and 379-m. The samplers located on the up-sloping terrain at 300 m are surrounded by very sparse vegetation, usually less than a meter in height with large barren patches of soil. On the other hand, the samplers located in the down sloping region at 379-m are surrounded by significantly more vegetation including shrubs, bushes and trees that are typically 2–3 m in height. The model, however, uses a constant surface roughness of 0.1-m. Large surface roughness and obstructing objects capture airborne dust in two ways, it removes momentum from the mean flow slowing transport of airborne particulates allowing them to gravitationally settle, as well as directly capturing airborne dust through direct contact and impaction of the airborne dust. This severe change in surface roughness may explain the counterintuitive results from the samplers. Another possible reason might be the high natural variability in deposition may overwhelm the topographic effect.

3.3.3. Lead Isotope Analysis

Figures 11 and 12 show the lead isotopic ratios for the dust collected on the inverted-disc samplers for the May and June sampling periods, respectively. Dust samples A and B located on the tailings themselves have the same isotopic composition as the bulk tailings sample, implying the airborne lead captured in the samplers originated exclusively from the tailings, as expected.

Samples with lower isotopic ratios are the consequence of mixing between the tailings source and regional background. For the May sampling period, the samples that have the lowest isotope ratio are F and N. Sample F is the most southern sample located approximately 300-m from the southern edge of the tailings while N the eastern most sample is located approximately 150-m from the eastern-most edge of the lower tailings area. The small contribution of tailings lead measured in these samplers matches the monthly wind patterns that predominantly transported dust northward, away from the samplers. Samples M and E also have a significantly lower isotopic ratios than the tailing samples. It is interesting to note that sample E which is located only about 200 m from the southern edge of the tailings has the same fractional contribution of tailings lead as the sampler located 1 km north, as expected from the prevailing winds. Also, the source of lead captured by the inverted-disc sampler N, located along AZ highway 69 which separates the Iron King tailings and the town of Dewey-Humboldt, had a smaller tailings contribution than the sampler located 1 km north of the tailings.

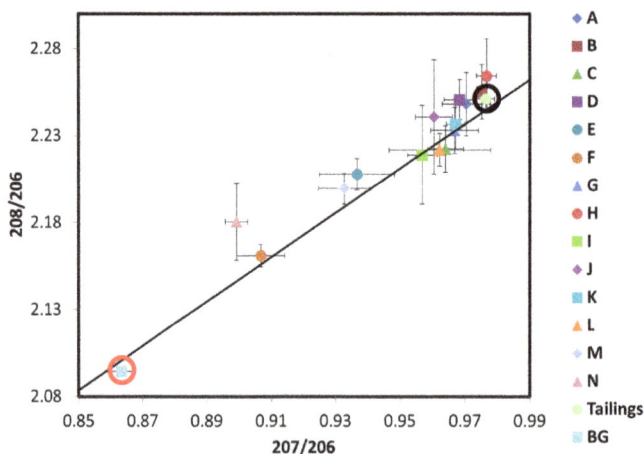

Figure 11. Lead isotopic ratios ^{208}Pb/^{206}Pb and ^{207}Pb/^{206}Pb for each inverted-disc sampler and a bulk sample of tailings material for the 21 April to 22 May 2014 sampling period. The letters represent the inverted-disc sample locations from Figure 1. Tailings (surrounded by a black circle), represent "fingerprint" ratios of the source. The background sample (BG, surrounded by a red circle) corresponds to topsoil (0–3 mm) collected 5 km from the source and represents the natural Pb isotopic "fingerprint" of the region. Error bars represent standard deviations from triplicate samples. Additionally included is the growth curve (solid line) adapted from Chen *et al.* [23], that represents changes in lead isotopic composition with time due to radiogenic production from isotopes of uranium and thorium, which encompasses all possible Earth samples.

For the June sampling period, the lead isotopic signatures were significantly closer to the tailings bulk sample when compared to the May sampling period. The samples with the lowest isotopic ratios included samples F, N and E. Sample F had the lowest lead contribution from the tailings, this matches the results from the May sampling period. However, more of the lead measured in sample F was sourced from the tailings compared to May. For the eastern most sampler, N and southern sampler E, the isotopic ratios were closer to the tailings signature as well, while still maintaining the lowest isotopic ratios of all the June samplers.

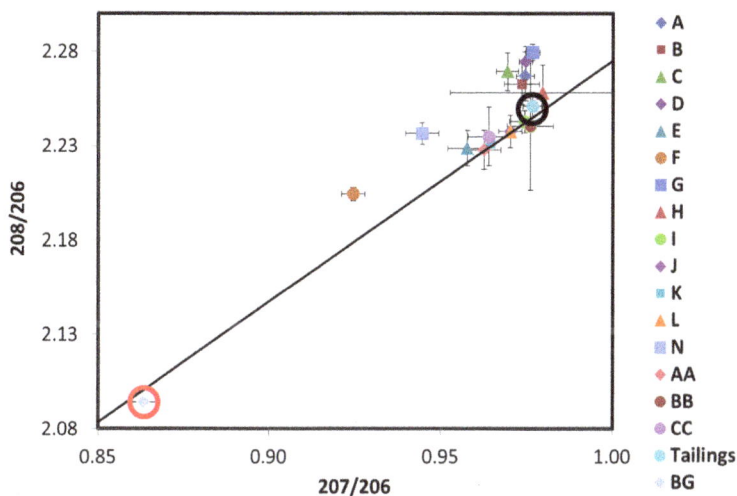

Figure 12. Lead isotopic ratios $^{208}Pb/^{206}Pb$ and $^{207}Pb/^{206}Pb$ for each inverted-disc sampler and a sample of tailings topsoil for the 11 June to 9 July 2014 sampling period. The letters represent the inverted-disc sample locations from Figure 1. Tailings (surrounded by a black circle), represent "fingerprint" ratios of the source. The background sample (BG, surrounded by a red circle), corresponds to topsoil (0–3 mm) collected 5 km from the source and represents the natural Pb isotopic "fingerprint" of the region. Error bars represent standard deviations from triplicate samples. Additionally included is the growth curve (solid line) adapted from Chen et al. [23].

It is interesting to note that the June samplers had significantly higher lead concentrations and isotopic ratios when compared to May. This was caused by a precipitation event that occurred on April 27. This precipitation event significantly increased the tailings moisture content, minimizing wind erosion and reducing windblown lead deposition for the May sample period. In the June sample period the tailings had not received precipitation in over a month which significantly increasing the erosion potential, which resulted in more tailings sourced lead deposition causing the increase in lead concentrations and also shifting isotopic fingerprints closer to the tailings isotopic signature. This shows that local weather patterns including predominant wind directions and precipitation have a significant effect on the deposition of windblown dust from the Iron King tailings impoundment.

The lead isotope ratios can be used to calculate the contribution of deposited dust that originates in the tailings. Assuming only two different lead sources (tailings and background), the calculated fractional Pb tailings contribution was defined here as the ratio between the difference in the 207/206 ratio between the samples and the background, divided by the difference between the tailings and the background. Results are plotted as a function of predicted dust deposition fluxes in Figure 13, for the two different sampling periods. Despite the scatter, a clear increasing relation is obtained between the proportion of dust originating in the tailings and the total deposition flux of tailings materials predicted by the model.

Lead isotopic ratios for the soil samples at different levels collected along the NE transect (Figure 7) are shown in Figure 14. Samples located 130 m from the tailings were collected in up-sloped terrain and their lead signatures are close to the tailings, indicating that the relatively high concentrations of lead (Figure 7) are a consequence of dust transport from the tailings. At this distance, results from the DFM model also point to a high deposition of tailings dust (Figure 10). It is interesting to point out that even at a depth of 100 mm, lead isotope analysis points to the tailings as main contributor of the metal in the soil. At longer distances from the tailing (1150 m), topsoil and 100-mm depth soil have significantly different signatures, with deep soil reflecting a higher contribution from the 5-km background. The sampling point located 13 m from the tailing has the same lead isotope ratios as the tailing topsoil, as expected, but isotopic ratios decrease monotonically with distance towards the background site located 5 km from the tailings.

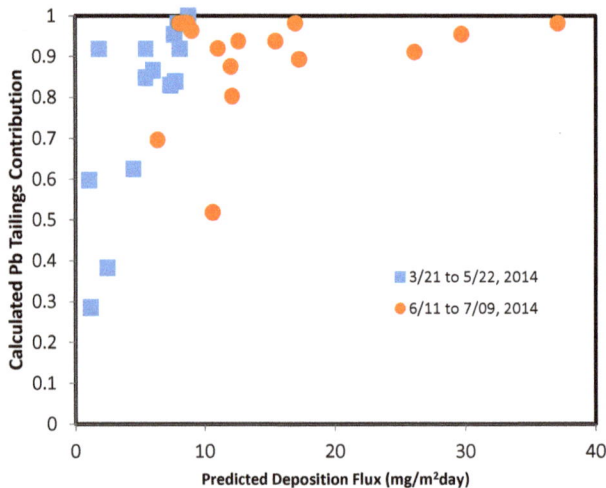

Figure 13. Calculated tailings contribution of deposited lead in inverted-disc samplers as a function of the model-predicted dust deposition flux for the two different sampling periods in 2014. Tailings contribution (1 for tailings, 0 for background) are calculated from the 207/206 isotopic ratios reported in Figures 11 and 12.

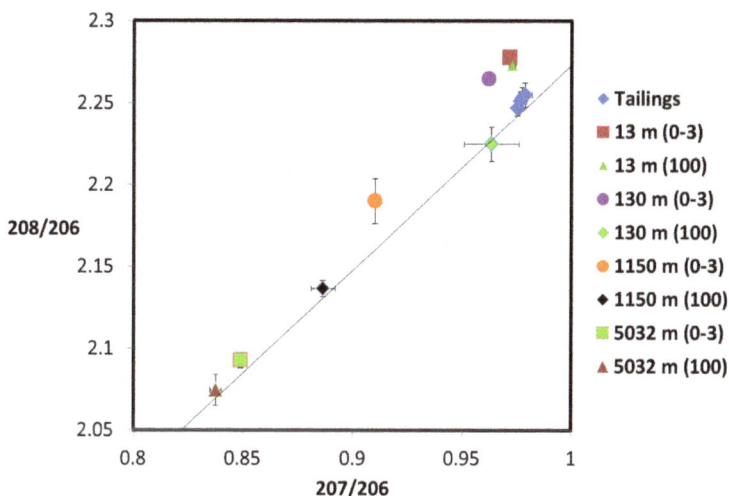

Figure 14. Lead isotopic ratios for soil samples at different distances from the mine tailings (indicated). Numbers in parentheses represent the depth of the sample in mm. Error bars represent standard deviations from triplicate samples. Additionally included is the growth curve (solid line) adapted from Chen *et al.* [23].

4. Conclusions

The DFM is designed to utilize weather forecasts to predict the deposition of fugitive PM_{27} dust originating from the Iron King tailings impoundment. By comparing the DFM predicted PM_{27} deposition to arsenic and lead tracers collected by the inverted-disc samplers, it has been shown that the DFM captures trends on spatial variations of the deposition patterns in the surrounding region up to 1 km distance from the tailings. The effects of topography on deposition still need adjustment due to the complex variations of surface roughness within the region. In addition, the difficulties associated in directly quantifying PM_{27} deposition using inverted-disc samplers leaves room for additional investigation into the absolute deposition quantities provided by the DFM. However, combining the deposition patterns generated by the DFM and the known concentrations of arsenic and lead in tailing dust we can provide relative estimates of arsenic and lead deposition rates near the tailings impoundment. These estimates of deposition should improve the characterization of potential health impacts caused by windblown transport from the tailings. The methodology employed in this work can be generalized to other contaminated sites from which dust transport to surrounding communities can be assessed in terms of a potential route for human exposure. Dust emission quantification relied on weather forecasting and empirical relations for emitted dust fluxes while transport and deposition were predicted by following particle trajectories in the forecast wind field. This approach led to a model that can be used to forecast the rate of transport and deposition of contaminants from the tailings to the local environment.

Acknowledgments: This work was supported by grant number P42 ES04940 from the National Institute of Environmental Health Sciences (NIEHS), National Institutes of Health (NIH). The views of authors do not necessarily represent those of the NIEHS, NIH.

Author Contributions: Michael Stovern developed the model, performed model calculations, designed and collected inverted-disc sample data and contributed to the writing and analysis. Hector Guzman and Omar Felix collected and analyzed soil samples. Kyle Rine and Matthew King contributed to field data collection and experimental design. Wendell Ela, Eric Betterton and Eduardo Sáez contributed to experiment and model design, data analysis and writing.

Conflicts of Interest: The authors declare no conflict of interest.

References

1. Wilkening, K.E.; Barrie, L.A.; Engle, M. Atmospheric science: Trans-Pacific air pollution. *Science* **2000**, *290*, 65–67.
2. Sprigg, W.A.; Nickovica, S.; Galgianic, J.N.; Pejanovic, G.; Petkovic, S.; Vujadinovic, M.; Vukovic, A.; Dacic, M.; DiBiase, S.; Prasad, A.; *et al.* Regional dust storm modeling for health services: The case of valley fever. *Aeolian Res.* **2014**, *14*, 53–73.
3. Brook, R.D.; Rajagopalan, S.; Pope, C.A.; Brook, J.R.; Bhatnagar, A.; Diez-Roux, A.V.; Holguin, F.; Hong, Y.; Luepker, R.V.; Mittleman, M.A.; *et al.* Particulate matter air pollution and cardiovascular disease an update to the scientific statement from the American Heart Association. *Circulation* **2010**, *121*, 2331–2378.
4. Csavina, J.; Field, J.D.; Taylor, M.P.; Gao, S.; Landázuri, A.; Betterton, E.A.; Sáez, A.E. A review on the importance of metals and metalloids in atmospheric dust and aerosol from mining operations. *Sci. Total Environ.* **2012**, *433*, 58–73.
5. Csavina, J.; Field, J.D.; Félix, O.; Corral-Avitia, A.Y.; Sáez, A.E.; Betterton, E.A. Effect of wind speed and relative humidity on atmospheric dust concentrations in semi-arid climates. *Sci. Total Environ.* **2014**, *487*, 82–90.
6. Badr, T.J.; Harion, L. Numerical modeling of flow over stockpiles: Implications on dust emissions. *J. Atmos. Environ.* **2005**, *39*, 5576–5584.
7. Turpin, C.; Harion, J.L. Effect of the topography of an industrial site on dust emissions from open storage yards. *Environ. Fluid Mech.* **2010**, *10*, 677–690.
8. Turpin, C.; Harion, J.L. Numerical modelling of flow structures over an industrial site: Effect of the surrounding buildings on dust emissions. *Glob. NEST J.* **2010**, *12*, 40–45.
9. Diego, I.; Pelegry, A.; Torno, S.; Torana, J.; Menendez, M. Simultaneous CFD evaluation of wind flow and dust emission in open storage piles. *J. Appl. Math. Model.* **2009**, *33*, 3197–3207.
10. Stovern, M.; Felix, O.; Csavina, J.; Rine, K.P.; Russell, M.R.; Jones, R.M.; King, M.; Betterton, E.A.; Sáez, A.E. Simulation of windblown dust transport from a mine tailings impoundment using a computational fluid dynamics model. *Aeolian Res.* **2014**, *14*, 75–83.

11. Stovern, M.K.; Rine, K.P.; Russell, M.R.; Felix, O.; King, M.; Sáez, A.E.; Betterton, E.A. Development of a dust deposition forecasting model for a mine tailings impoundment using *in situ* observations and idealized particle simulations. *Aeolian Res.* **2015**, *18*, 155–167.

12. EPA. Iron King Mine and Humboldt Smelter. Available online: Http://yosemite.epa. gov/r9/sfund/r9sfdocw.nsf/db29676ab46e808188825742600743734/316161edfc4699a 5882574ab0001d1c0!OpenDocument#descr (accessed on 2 July 2014).

13. Félix, O.I.; Csavina, J.; Field, J.; Rine, K.P.; Sáez, A.E.; Betterton, E.A. Use of lead isotopes to identify sources of metal and metalloid contaminants in atmospheric aerosol from mining operations. *Chemosphere* **2015**, *122*, 219–226.

14. EA. Engineering, Science and Technology. *Remedial Investigation Report Iron King Mine—Humboldt Smelter Superfund Site Dewey-Humboldt, Yavapai County, Arizona, 2010*; US Environmental Protection Agency Region 6: Dallas, TX, USA, 2010.

15. Solís-Dominguez, F.A.; White, S.A.; Hutter, T.B.; Amistadi, M.K.; Root, R.A.; Chorover, J.; Maier, R.M. Response of key soil parameters during compost-assisted phytostabilization in extremely acidic tailings: Effect of plant species. *Environ. Sci. Technol.* **2012**, *46*, 1019–1027.

16. Hagen, L.J.; Pelt, S.V.; Zobeck, T.M.; Retta, A. Dust deposition near an eroding source field. *Earth Surf. Process. Landf.* **2007**, *32*, 281–289.

17. Koester, C.J.; Hites, R.A. Wet and dry deposition of chlorinated dioxins and furans. *Environ. Sci. Technol.* **1992**, *26*, 1375–1382.

18. Hall, D.J.; Upton, S.L. A wind tunnel study of the particle collection efficiency of an inverted Frisbee used as a dust deposition gauge. *Atmos. Environ.* **1988**, *22*, 1383–1394.

19. Hall, D.J.; Waters, R.A. An improved readily available dustfall gauge. *Atmos. Environ.* **1986**, *20*, 219–222.

20. Vallack, H.W. A field evaluation of Frisbee-type dust deposit gauges. *Atmos. Environ.* **1995**, *29*, 1465–1469.

21. Sow, M.; Goossens, D.; Rajot, J.L. Calibration of the MDCO dust collector and of four versions of inverted Frisbee dust deposition sampler. *Geomorphology* **2006**, *82*, 360–375.

22. Goossen, D.; Rajot, J.L. Techniques to measure the dry aeolian deposition of dust in arid and semi-arid landscapes: A comparative study in West Niger. *Earth Surf. Process. Landf.* **2008**, *33*, 178–195.

23. Chen, J.; Tan, M.; Li, Y.; Zheng, J.; Zhang, Y.; Shan, Z.; Zhang, G.; Li, Y. Characteristics of trace elements and lead isotope ratios in PM(2.5) from four sites in Shanghai. *J. Hazard. Mater.* **2008**, *156*, 1–3.

Decreasing Aerosol Loading in the North American Monsoon Region

Aishwarya Raman, Avelino F. Arellano Jr. and Armin Sorooshian

Abstract: We examine the spatio-temporal variability of aerosol loading in the recent decade (2005–2014) over the North American Monsoon (NAM) region. Emerging patterns are characterized using aerosol optical depth (AOD) retrievals from the NASA Terra/Moderate Resolution Imaging Spectroradiometer (MODIS) instrument along with a suite of satellite retrievals of atmospheric and land-surface properties. We selected 20 aerosol hotspots and classified them into fire, anthropogenic, dust, and NAM alley clusters based on the dominant driver influencing aerosol variability. We then analyzed multivariate statistics of associated anomalies during pre-, monsoon, and post-monsoon periods. Our results show a decrease in aerosol loading for the entire NAM region, confirming previous reports of a declining AOD trend over the continental United States. This is evident during pre-monsoon and monsoon for fire and anthropogenic clusters, which are associated with a decrease in the lower and upper quartile of fire counts and carbon monoxide, respectively. The overall pattern is obfuscated in the NAM alley, especially during monsoon and post-monsoon seasons. While the NAM alley is mostly affected by monsoon precipitation, the frequent occurrence of dust storms in the area modulates this trend. We find that aerosol loading in the dust cluster is associated with observed vegetation index and has only slightly decreased in the recent decade.

Reprinted from *Atmosphere*. Cite as: Raman, A.; Arellano, A.F., Jr.; Sorooshian, A. Decreasing Aerosol Loading in the North American Monsoon Region. *Atmosphere* **2016**, *7*, 24.

1. Introduction

Aerosols play a critical role in global and regional climate, monsoonal circulation, hydrological cycle, air quality, and public health (e.g., [1–4]). Such is the case for aerosols in the semi-arid regions of North America. There is growing concern about aerosols in the North American monsoon (NAM) region as studies have shown (and projected) a warmer and drier southwest United States [5], leading to increased wildfire risks and occurrence of dust storms ([6–8], and references therein). Studies have also reported the ability of aerosols in modifying NAM precipitation through direct or indirect effects [9–11] and epidemiological outbreaks from pollution [12]. While there has been increased attention in recent decades directed to aerosols in the Asian monsoon region [2,3,13,14], limited studies have examined local-to-regional characteristics and trends of aerosols in the NAM region, precluding our ability to accurately predict its response to projections of changes in environmental conditions.

NAM is a notable feature in the atmospheric circulation over North America. It is characterized by a shift in the circulation pattern due to warmer land surfaces in the southwestern United States and northwestern Mexico during May and June, resulting in an upper-level anti-cyclone over western Mexico and a pronounced increase in precipitation from convective storms over these regions (including western Texas) during July and August. Although the NAM location is centered at the Sierra Madre Occidental in Mexico, its influence on monsoon precipitation extends widely into the areas of Arizona and New Mexico. In fact, NAM provides 70% of the annual precipitation in the region [15]. We refer the reader to several studies [16–18] on further details of its spatial extent and the underlying meteorological processes. For this study, we define the NAM region to include: northern Mexico, Arizona, southern California (SoCal), New Mexico, and western Texas. In terms of aerosols, the NAM region exhibits distinct spatio-temporal patterns due to diversity in aerosol sources and sinks during its pre-monsoon (PRM: May–June), monsoon (MON: July–August), and post-monsoon (POM: September–October) phases. During May and June, the warmer and drier conditions result in fires in vegetated areas of NAM. Large dust events typically occur in the desert areas (and abandoned agricultural fields) during MON as a result of mesoscale convective storms. NAM is not only characterized by natural aerosols (dust and fire) but also by anthropogenic aerosols [19,20]. This region in fact has the top five most polluted cities for particulate matter [21]. In terms of aerosol removal, NAM experiences significant rainfall events between July and August that are quite distinct across the year for this semi-arid environment.

Past studies of aerosol trends in the United States (US) have mainly focused on the entire North American continent or the eastern US (e.g., [22,23]). For example, a recent study [24] examined the decadal trend (2000–2009) of aerosol optical depth (AOD) in the US based on NASA Terra Moderate Resolution Imaging Spectroradiometer (MODIS) and from Advanced Very high Resolution Radiometer (AVHRR) aerosol retrievals along with simulations from the Goddard Chemistry Aerosol Radiation and Transport (GOCART) model as part of a global model/data trend analyses. They found a significant decrease in AOD [24], which they attributed to the reductions in anthropogenic (combustion-related) emissions. This finding is consistent with an earlier model study [25] using the Model of Atmospheric Transport and Chemistry and the Dust Entrainment and Deposition models that reported a significant decline of the AOD trend from anthropogenic aerosol sources over the period 1980–2006. This study also revealed that the long-term trends in natural aerosol sources over the US were not significant for the given time period. Thus, more local-to-regional studies are needed as competing sources can obscure the reported global/regional trend [24].

Over the southwest US, a climatology of aerosol loading in several areas across the state of Arizona using a suite of aerosol ground-based measurements (e.g., Interagency Monitoring of PROtected Visual Environments or IMPROVE), remotely-sensed aerosol products (e.g., MODIS), as well as model output from GOCART, showed a significant range in spatio-temporal variations of aerosols within the state, depending on the proximity of the measurement site to the dominant drivers (e.g., dust, rural activities, urban pollution, *etc.*) [19]. Recently, another study in this region noted that AOD has increased by 19% over Tucson, Arizona, in the last 35 years based on two sets of AOD (400–900 nm) measurements over Tucson (*i.e.*, 1975–1977 *versus* 2010–2012) [26]. They postulated that the increase might have been contributed by urbanization and the near tripling of population in the city. These studies highlight the importance of examining the aerosol variations across the southwestern US in detail, given mixtures of drivers on aerosol loading across the region and its unique environmental conditions (including complex topography) relative to the southeast US. In light of decadal records of remotely-sensed retrievals of atmospheric composition, precipitation, and land-surface properties, there is also a unique opportunity to conduct a multivariate analysis for a more comprehensive and consistent picture of aerosol trends in the region.

In this study, we focus on assessing the spatio-temporal trends in aerosol loading across the NAM region by examining several aerosol hotpots and their associated trends during pre-monsoon (PRM), monsoon (MON), and post-monsoon (POM) seasons over the recent decade. In particular, we aim to: (1) analyze the trends in MODIS AOD from 2005–2014 over areas characterized by a dominant source or sink (*i.e.*, fire, anthropogenic, dust, and a region affected by monsoon rainfall); (2) elucidate controlling factors of these trends using correlative information from Tropical Rainfall Measuring Mission (TRMM) precipitation products (on aerosol removal), MODIS vegetation index (on dust aerosols), fire products (on fire aerosols), and CO from Measurement of Pollution in The Troposphere (MOPITT) instrument (on anthropogenic aerosols).

2. Methodology

2.1. Satellite Products

All satellite datasets were downloaded through NASA GIOVANNI web portal developed by Goddard Earth Sciences Data and Information Science Center (GES DISC) [27]. These datasets are monthly averages and were regridded to $1° \times 1°$ to match the MODIS spatial resolution. A brief description of each data is given below, together with Table 1 which summarizes relevant information about the datasets.

Table 1. Analysis datasets.

Instrument and Dataset	Resolution	Relevance to Study (Main Product Reference)
NASA Terra and Aqua L3 MODIS Aerosol Optical Depth (AOD) 550 nm	$1° \times 1°$	aerosol loading (Levy *et al.*, 2007 [29])
NASA Terra L3 MODIS Fire Radiative Power (FRP) and fire counts	$1° \times 1°$	fire sources (Wooster *et al.*, 2005 [36])
NASA Terra L3 MODIS Normalized Vegetation Index (NDVI)	$0.05° \times 0.05°$	biogenic and dust sources (Lunetta *et al.*, 2006 [38])
NASA OMI L2G UV Aerosol Index (AI) 354 nm	$0.25° \times 0.25°$	aerosol cluster identification (Torres *et al.*, 2007 [39])
NASA MOPITT L3 TIR/NIR Total Column CO	$1° \times 1°$	combustion sources (Deeter *et al.*, 2012 [41])
NASA TRMM Best Estimate Precipitation Rate (BEPR)	$0.25° \times 0.25°$	aerosol removal (Huffman *et al.*, 2007 [40])
UMBC Anthropogenic Biomes V2 (2000) (ecotope.org)	$0.083° \times 0.083°$	aerosol cluster identification (Ellis *et al.*, 2010 [45])
NASA SEDAC Global Rural Urban Mapping Project version 1 (GRUMPv1) Population Density (sedac.ciesin.columbia.edu)	1 km $\times$ 1 km	aerosol cluster identification (Balk *et al.*, 2009 [43])

2.1.1. MODIS

The MODIS instrument is an imaging spectroradiometer on board two polar orbiting satellites, NASA EOS/Terra (Febraury 2000—present) and NASA EOS/Aqua (June 2002—present). It provides global coverage every one to two days with an equatorial overpass time around ~10:30 am (descending) for Terra and ~1:30 pm (ascending) for Aqua. MODIS acquires data in 36 spectral bands from 0.41 to 14 µm. It has a swath of 2330 km at cross-track and 10 km at nadir [28].

Aerosol Optical Depth (AOD). We use the Level 3 (gridded) MODIS $1° \times 1°$ Collection 5.1 (M3 mean_mean) AOD retrievals at 550 nm for both Terra and Aqua (MOD08_M3_V051) ([28] and references therein; [29,30]) to investigate trends of aerosol loading across the period 2005–2014. The Level 2 (swath) version of these retrievals are produced using three spectral wavelengths (470, 650, 2100 nm) and two algorithms ("dark target" for dark/vegetated surfaces and "deep blue" for bright surfaces such as the desert). The main aerosol product at 550 nm is derived by matching the reflectance values at different channels to the atmospheric properties using a look-up table. These retrievals have been widely used in the studies pertinent to the spatio-temporal variability of aerosols ([31,32] and references therein). We note, however, that evaluation against AERONET *in-situ* measurements of Collection 5.1 AOD products as derived using the "dark target" algorithm show overestimation of AOD [29,30]. We use the dark target for this study since the Collection 5.1 AOD retrievals using the "deep blue" algorithm were not available after December 2007 due to polarization issues in the sensor [33]. Our initial comparison of AOD between "deep blue" (from Collection 6) and "dark target" over the NAM region shows similarity in spatio-temporal patterns with "deep blue" AOD values (on average) being less by 0.1. We further note that previous studies report a shift from high to low bias of AOD from the Terra satellite after 2004 due to the degradation of Terra/MODIS's optical response [29]. Further information about these products is discussed in detail in [29,34].

Fire. We use Terra/MODIS monthly mean and cloud-corrected Fire Radiative Power (FRP in units of million watts) along with MODIS fire counts (FC, counts m^{-2} day^{-1}) at $1° \times 1°$ resolution (MODIS Active Fire Product version 005 MOD14CM1) to indicate fires in the region. MODIS identifies a candidate pixel as being affected by fires if the 4 μm brightness temperature and the difference between 4 μm and 11 μm brightness temperature depart substantially from non-fire pixels. The estimated accuracy of FRP from MODIS is 15% [35,36].

Normalized Difference Vegetation Index (NDVI). Aerosol loading in regions dominated by dust aerosols highly depend on the landcover characteristics (e.g., large-scale vegetation reduces dust lofting) (e.g., [37]). We use the information from NDVI to indicate potential dust mobilization in the region. MODIS/Terra provides cloud-free composites of NDVI obtained using corrected surface reflectances in visible and near infrared channels. Here, we use MODIS/Terra Level 3 monthly mean NDVI at 0.05 degree (MOD13C2) [38]. This product is based on the spatial and temporal averages of 16 day 1 km NDVI retrievals.

2.1.2. Ozone Monitoring Instrument (OMI) Ultraviolet Aerosol Index (UV AI)

OMI, which is on board the NASA EOS-Aura satellite, uses backscattered UV radiation measured at two wavelengths (354 and 388 nm) to derive UV AI. OMI has a local equator crossing time at ~1:45 pm. OMI has an advantage of being more sensitive to atmospheric aerosol loading since the reflectance of most terrestrial surfaces are low at UV wavelengths [39]. We use the Level 2G OMI UV AI retrievals at 354 nm mainly to identify aerosol hotspots in the region and partly to supplement MODIS AOD for aerosol loading trend analysis. The parent resolution of the product is $0.25° \times 0.25°$. Due to its higher spatial resolution compared to MODIS, OMI AI is more useful in identifying localized high aerosol loading across our domain. Also, OMI UV AI is more sensitive to smoke and dust compared to MODIS AOD.

2.1.3. Tropical Rainfall Measuring Mission (TRMM) Precipitation Rate

We use the monthly mean gridded "best estimate precipitation rate" (BEPR) product obtained from TRMM multi-satellite precpitation analyses (3B43 Version 7) to understand the potential impact of rainfall on the aerosol loading in the NAM region. TRMM best estimate precipitation rate dataset is a combination of data from rain gauge stations and satellite sensors providing precipitation estimates [40]. The spatial resolution of this product is $0.25° \times 0.25°$.

2.1.4. Measurements of Pollution in the Troposphere (MOPITT) Carbon Monoxide (CO)

The MOPITT instrument is a gas correlation radiometer also onboard the NASA EOS/Terra satellite. We use MOPITT Level 3 gridded ($1° \times 1°$) monthly total column multispectral (2.3 and 4.7 µm) CO retrievals [41,42] to indicate fire and/or anthropogenic aerosols in the region. As a product of incomplete combustion, CO is a useful tracer of fire and anthropogenic pollution. We can therefore distinguish between fire and anthropogenic aerosol signatures using the combination of MOPITT and MODIS fire products. This is further facilitated by enhanced sensitivity of MOPITT CO multispectral retrievals (thermal+near-infrared) to CO in the lowermost troposphere [41].

2.2. Ancillary Datasets

We use data on population density from Global Rural-Urban Mapping project version 1 (GRUMP V1 2000), which is provided by Socio Economic Data and Application Center (SEDAC) [43,44], to identify populated areas in the region. The data is available at 30 arc-second resolution. Along with GRUMP, we also use the anthropogenic biomes version 2 for year 2000 (also downloaded from SEDAC) to identify urban and rural areas in the region. Anthropogenic biomes or anthromes provide useful information on the alterations to the global ecosystem and biotic communities by human population [45,46]. The anthrome product has 21 anthropogenic biomes with the following six major classes: dense settlements, villages, croplands, rangelands, forested and wildlands (see Figure 1C for the different anthromes in NAM).

2.3. Data Analysis

Aerosol hotspots and clusters. We have selected a region of study bounded between the coordinates (120.5°W, 95.5°W) and (22.5°N, 37.5°N). We use the monthly decadal mean from OMI AI, along with population density and anthromes, to identify 20 aerosol hotspots in the NAM region. First, we find the climatological maximum in OMI UV AI across all six months (May to October) for each grid cell in the NAM domain. The corresponding month at which the maximum occurs for a grid cell is also identified (see Figure 1a,b). Second, we determine if a grid cell is an aerosol hotspot based on the following criteria: (1) grid value of OMI UV AI is greater than a high aerosol loading threshold value set here as AI of 1.0; (2) population density is higher than the neighboring grid cells; (3) grid cell is unique in terms of anthropogenic biome classification, topography, and/or meteorology (see Table 2). For example, although a lot of grid cells in the northwestern Mexico area satisfy the first criteria, we chose only those high aerosol cells which (a) may be affected by monsoon; (b) represent a relatively populated area that can be impacted by high aerosol loading; and (c) have distinct topographic features or anthrome classification. Our goal is to find aerosol hotspots representative of different aerosol environments. OMI UV AI is particularly useful for this purpose as it can locate smaller areas with high aerosol loading that cannot be easily detected using MODIS AOD. We note also that OMI UV AI has higher sensitivity to absorbing aerosols (e.g., black carbon and dust). Once a hotspot is identified, we re-grid OMI AI (using simple averaging) to match the $1° \times 1°$ resolution of the MODIS AOD that we are using for this study. We define the hotspot region to include the nearest grid cells surrounding the hotspot grid cell that has climatological maximum within ±2 standard deviations of the value of the hotspot grid cell. Finally, we group the hotspot regions into four main aerosol clusters, where each cluster exhibits a distinct characteristic of aerosol variability within the NAM region. Our four clusters include: dust, NAM alley, fire, and anthropogenic (see Figure 1 and Table 2 for locations of these hotspot regions).

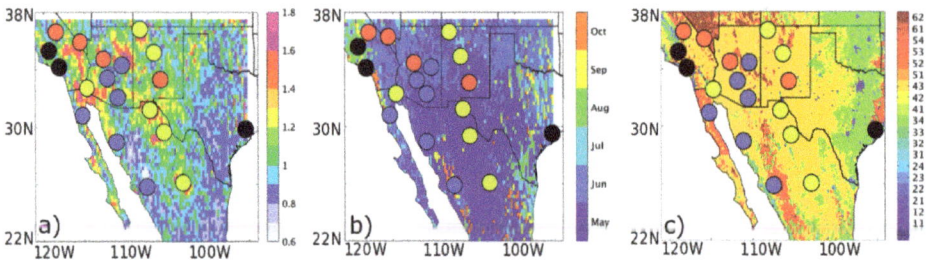

Figure 1. Maps of hotspot regions and aerosol clusters in NAM region. The three panels represent (**a**) the OMI UV AI climatological maximum across 2005–2014; (**b**) the month of climatological maximum; and (**c**) the anthropogenic biomes where different numbers on the colorbar indicate biome categories (11–12 are urban, 21–26 are villages, 31–35 are croplands, 41–43 are rangelands, 51–52 are forests, and 61–63 are considered wildlands). Superimposed on all three plots are the locations of 20 hotspot regions (colored circles) with the color fill indicating the aerosol cluster they belong to (red: fire, yellow: dust, blue: NAM alley, and black: anthro).

Anomalies. We start our analysis by first calculating the decadal monthly mean across 2005–2014 and the corresponding monthly standardized anomalies for each year of OMI UV AI, MODIS AOD, MODIS fire counts and FRP, MODIS NDVI, MOPITT CO, and TRMM precipitation rate. These are calculated as follows:

$$\bar{x}_{m,n} = \frac{1}{N} \sum_{y=1}^{N} \left(x_{m,n,y} \right) \tag{1}$$

$$\sigma_{m,n} = \sqrt{\frac{\sum_{y=1}^{N} \left(x_{m,n,y} - \bar{x}_{m,n} \right)^2}{N-1}} \tag{2}$$

$$x'_{m,n,y} = \frac{\left(x_{m,n,y} - \bar{x}_{m,n} \right)}{\sigma_{m,n}} \tag{3}$$

where $\bar{x}_{m,n}$ is the 10-year average ($y = 1$ to $N = 10$) of a certain quantity $\left(x_{m,n,y} \right)$ in each $1° \times 1°$ grid (n) in our domain for a particular month (m) (May to October) of a given year (y). The standardized anomaly $\left(x'_{m,n,y} \right)$ is calculated as the difference between $x_{m,n,y}$ and $\bar{x}_{m,n}$ normalized by the 10-year standard deviation ($\sigma_{m,n}$). Positive (negative) anomalies for each grid indicate values higher (lower) than its mean. We standardized the anomalies to facilitate comparison with quantities having different units (similar to [25]).

Table 2. Hotspot sites and associated aerosol cluster information.

Site No.	Site Name	Latitude (deg N)	Longitude (deg W)	Aerosol Cluster
1	Mt. Whitney, CA	36.557	118.50	Fire
2	Charleston Peak, CA	36.29	115.69	Fire
3	Tucson, AZ	32.22	110.93	NAM alley
4	Baja, CA	30.98	115.38	NAM alley
5	Phoenix, AZ	33.45	112.08	NAM alley
6	Yuma, AZ	32.69	114.63	Dust
7	LA County, CA	34.35	118.37	Anthro
8	Bakersfield, CA	35.37	119.02	Anthro
9	Prescott, AZ	34.54	112.46	Fire
10	Petrified Forest, AZ	34.41	110.65	NAM alley
11	White Mountain, NM	33.41	105.74	Fire
12	Farmington, NM	36.73	108.22	Dust
13	Albuquerque, NM	35.01	106.61	Dust
14	Ejido El Vergel, Mex	31.20	106.59	Dust
15	Chihuahuan Desert, Mex	29.52	105.48	Dust
16	Hermosillo, Mex	29.07	110.97	NAM alley
17	Sierra Madre Occidental, Mex	25.96	107.53	NAM alley
18	Sierra Madre Oriental, Mex	26.12	103.10	Dust
19	Houston, TX	29.74	95.36	Anthro
20	Waco, TX	31.55	97.15	Dust

3. Results and Discussion

In this section, we present our results on the spatial variability in aerosol loading and decadal trends of the changes in aerosol loading over different aerosol clusters in the NAM region. The temporal trend is examined for the period 2005–2014. We express these changes as standardized anomalies to account for the variability in aerosol loading. We can interpret these anomalies to represent local rather than regional to global changes. This attempts to minimize the influence of the changes in aerosol loading due to background or transported aerosols in the region. Our analysis focuses on intra-seasonal periods corresponding to pre-monsoon, monsoon, and post-monsoon phases since the aerosol patterns vary within the monsoon season. The overall trend is further elucidated by specific trends exhibited by each cluster (fire, anthropogenic, NAM alley, and dust). We note, however, that the only aerosol dataset that shows statistically significant decadal trends in aerosol anomalies is Terra/MODIS AOD. Other aerosol datasets, such as Aqua/MODIS and Aura/OMI which both have local overpass times in the early afternoon, show considerable interannual variations of aerosol anomalies with trends (not shown) that are not statistically significant. The inability of Aqua/MODIS and Aura/OMI to exhibit significant trends can be associated with its large variability possibly due to sampling issues, especially on observing aerosols at times when their abundance is most sensitive to cloud formation, boundary layer mixing, and convection in the region (e.g., [39]). The use of a longer period should provide sufficient samples for decadal trend analysis. For this study, however, we focus on the trends as can be inferred from Terra/MODIS.

3.1. Spatial Variability

The decadal mean AODs $(\bar{x}_{Aod})$ for PRM, MON, and POM are shown in Figure 2. Three regions with the highest AOD patterns are apparent in these plots. The first region lies in northern California and Nevada bounded between 36°N and 38°N, and 118°W and 115°W. This region is mostly dominated by fires, especially during PRM and POM. Also, the fire cluster in this region shows the maximum mean AOD and maximum variability during PRM (Table 3). The second maxima in aerosol loading is seen in the southwestern Arizona and northern Mexico regions (30°N, 33°N, and 108°W, 41°N, 109°W). This region has been identified as an important source of dust by other studies (e.g., [19]). This region is characterized by haboob-type dust storms during MON and frontal dust storms during POM. The third source region is found between 25°N and 31°N, and 105°W and 107°W. This includes the Chihuahuan desert and Sierra Madre Oriental mountains. In all the three source regions, aerosol loading peaks in PRM and MON and decreases in POM (Table 3). In addition to these source regions, a minor region of activity, centered on LA county (34.35°N and 118.37°W), is also observed. While this region is noted for its anthropogenic pollution, recent investigations have suggested the mixing of anthropogenic and biomass burning pollution in Southern California regions (e.g., [47]).

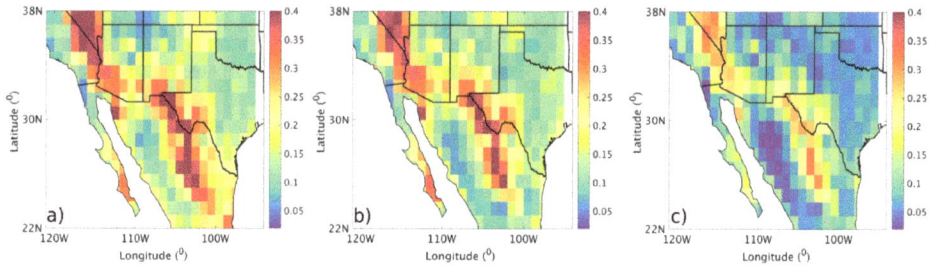

Figure 2. Maps of decadal mean (2005–2014) AOD from Terra-MODIS. The three panels represent (**a**) mean AOD during PRM; (**b**) MON; and (**c**) POM.

Table 3. Decadal mean and standard deviation of AOD for cluster regions.

Cluster	Season	Mean ($\bar{x}_{Aod}$)	Standard Deviation (σ_{AOD})
Fire	PRM	0.23	0.12
	MON	0.21	0.10
	POM	0.15	0.09
NAM alley	PRM	0.19	0.08
	MON	0.18	0.09
	POM	0.11	0.07
Dust	PRM	0.23	0.11
	MON	0.23	0.11
	POM	0.16	0.08
anthro	PRM	0.13	0.02
	MON	0.14	0.02
	POM	0.10	0.01

We also estimated the spatial correlation for these source regions by correlating AOD anomalies over a few selected places with AOD over the rest of the domain. The spatial patterns in correlation coefficients vary with season and cluster (Figure 3). The places selected in these source regions are: (1) Charleston peak (fire); (2) Tucson (NAM alley); (3) Hermosillo (dust); (4) Bakersfield (anthropogenic). Plots of correlation coefficients for these places are shown in Figure 3. For example, the peak of the regions of high correlation in AOD anomalies around Charleston is highest during PRM and decreases in MON and POM. In the case of Tucson and Hermosillo, regions of high correlation extend further into Mexico during MON and the higher correlations follow the monsoon track. We notice that, in the case of Bakersfield, regions of higher correlation extend into northern California in all the seasons. This suggests the potential similarity in frequency or magnitude of sources in between these regions. We also show spatial correlation matrices between selected aerosol hotspots (identified in Figure 1) in the supplementary material (Figure S1).

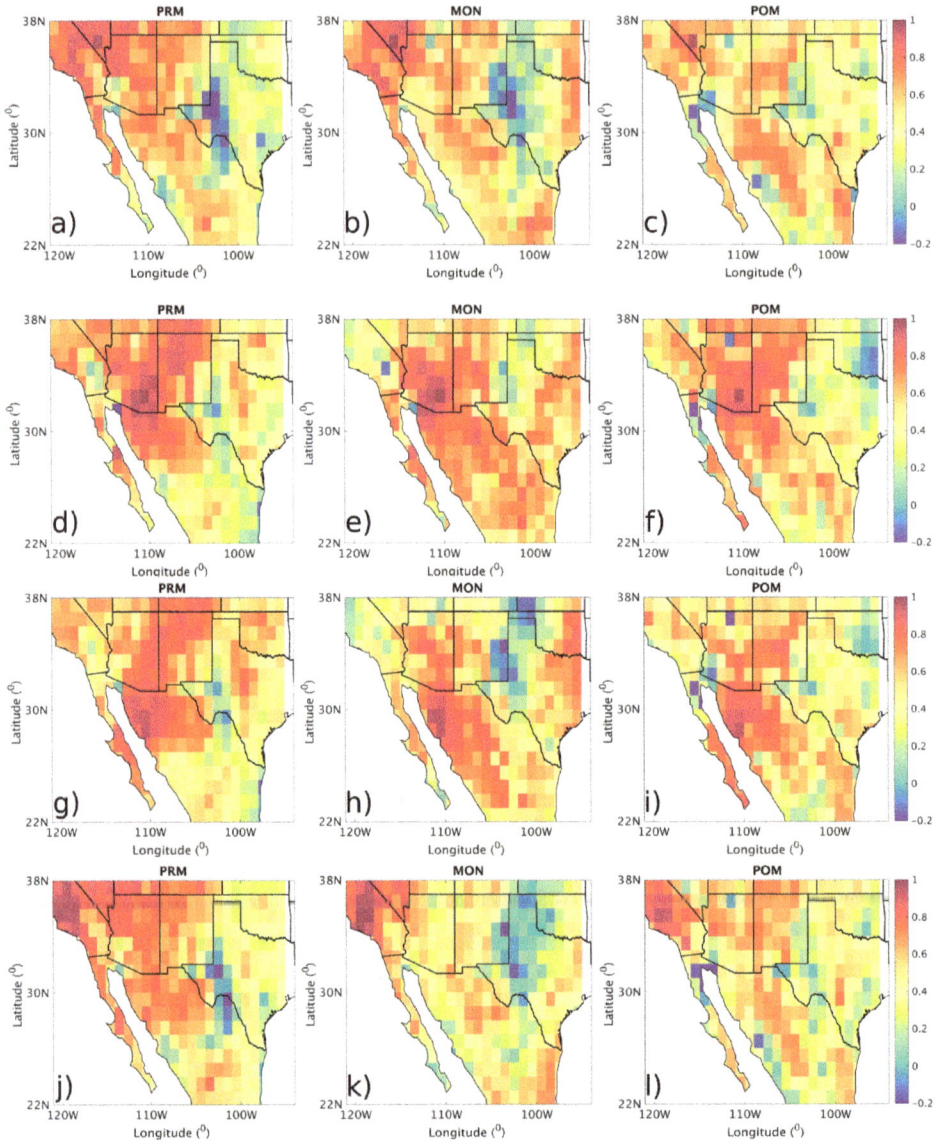

Figure 3. Correlation coefficient of AOD anomalies during PRM, MON, and POM. The panels (**a–c**) correspond to spatial correlation of all the grid points with respect to Charleston peak; (**d–f**) Tucson; (**g–i**) Hermosillo; (**j–l**) Bakersfield.

3.2. Overall Aerosol Trend

We show in Figure 4 the yearly statistics of anomalies (x'_{aod}) across the entire land region (region as shown in Figure 1) during PRM, MON, and POM. The statistics are summarized here using box plots where the top, central, and bottom edges of the box correspond to the 75th (q_3), 50th (median), and 25th (q_1) percentiles, respectively, while the whiskers correspond to ± 1.5 of the interquartile range ($IQR = q_3 - q_1$). The interannual variability in these AOD anomalies during all seasons mostly arises from natural aerosol sources such as dust and fires. We explain this further in the following sections for different aerosol clusters. The median of the anomalies during the latter part of the decade (2010–2014) is generally negative (and lower) compared to the earlier part of the decade (2005–2009), which is generally positive (and higher) (Figure 4). This is most evident for the monsoon season. This is also the case for the interquartile range, where a shrinking of the spread in x'_{aod} is observed in the latter part of the decade. Although there appears to be an oscillating pattern across the decade, there is a clear decreasing trend in the median and spread pointing to relatively lower aerosol loading and variability in recent years. Here, we estimate the linear trends of x'_{aod} to be -0.14 for PRM, -0.12 for MON, and -0.15 for POM. In other words, the yearly average decrease in x'_{aod} is greater than 10% for PRM, MON and POM, with larger decreases during PRM and POM.

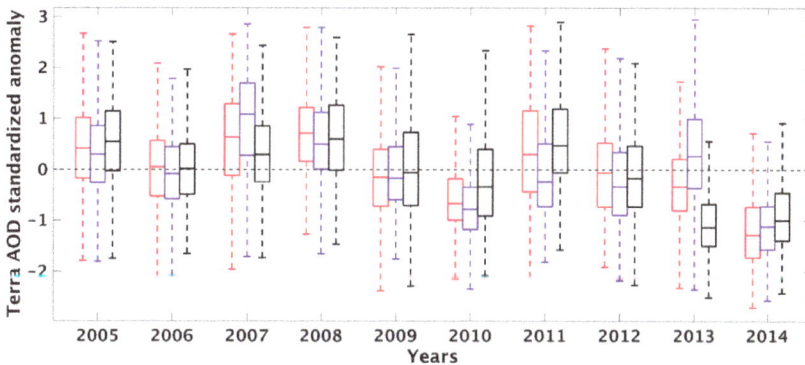

Figure 4. Box plots of yearly Terra/MODIS AOD standardized anomaly over the entire NAM region for the years 2005–2014. PRM (May–June), MON (July–August), and POM (September–October) periods are indicated in red, blue, and black. Whiskers represent 1.5 (IQR) above/below the upper/lower quartiles.

3.3. Aerosol Trends Across Clusters

We can further examine this decadal change by looking at the trends in the aerosol hotspot clusters. We expect more enhanced signatures of aerosol changes at these hotspot sites. In Figure 5 we show the decadal variations in x'_{aod} within fire, dust, NAM alley, and anthro clusters. We also summarize our estimates of linear trends in Table 4. The interannual variability of x'_{aod} is more pronounced in the fire, dust, and NAM alley, which are mostly driven by natural sources/sinks, than in the anthro cluster. This is expected as the "natural" aerosol clusters are mainly influenced by periodic atmospheric variations such as droughts and El Nino Southern Oscillation (ENSO) [48–50]. The fire and dust clusters, in particular, exhibit this periodic pattern, which can be linked to larger sources of aerosols (organic and black carbon in fires) under more arid and warmer conditions (PRM and POM). We also find a more pronounced decreasing trend (>20%) in all clusters during PRM relative to the overall trend (>10%). However, during MON, the difference is less pronounced, except for the anthro cluster, which is consistently >20% and appears to not be influenced by the monsoon. Except for the NAM alley cluster during the monsoon season, all the linear trends are significant at the 95% confidence interval (see Table 4).

The maximum decrease in x'_{aod} can be seen in the fire cluster (27%) followed by anthro (25%). We also see a larger difference in x'_{aod} statistics between the early and latter part of the decade in these two clusters. We see a more positive x'_{aod} in 2005–2009 and a more negative x'_{aod} in 2010–2014 for all three periods (PRM, MON, and POM). This decrease is most evident in the anthro cluster, which shows an increase until 2007 and then a continuous drop (both median and spread) after 2007, regardless of the monsoon season. This is consistent with the findings of [24] and [25] (albeit from a different study period) of a decreasing trend in aerosol loading which they attributed to a decreasing trend in anthropogenic aerosols. Further, a recent study on extreme events in the southwestern US also pointed out a decrease in extreme elemental carbon events between 2003 and 2013 [51], which compliments the decreasing trend in AOD anomalies in the NAM region. Our results show that the x'_{aod} trend for the entire NAM region is mainly tied to aerosol hotspots driven mostly by anthropogenic sources that have become weaker in recent years. This is also supported (not shown here) by the decreasing trend in combustion-related aerosol emissions based on recent emission inventories (e.g., [52]) and surface measurements of particulate matter ([53]).

Table 4. Linear trends in Terra/MODIS AOD anomalies for different aerosol clusters.

Clusters	PRM ($n = 10$)	MON ($n = 10$)	POM ($n = 10$)
Fire	−0.27 (0.0002)	−0.19 (0.02)	−0.21 (0.01)
NAM alley	−0.21(0.01)	−0.15 (0.11)	−0.13 (0.06)
Dust	−0.21 (0.007)	−0.17 (0.03)	−0.17 (0.02)
Anthro	−0.25 (0.0007)	−0.21 (0.009)	−0.22 (0.0001)
entire domain	−0.14 (0.02)	−0.12 (0.07)	−0.15 (0.01)

Note: values in the brackets refer to the p-value for the trends.

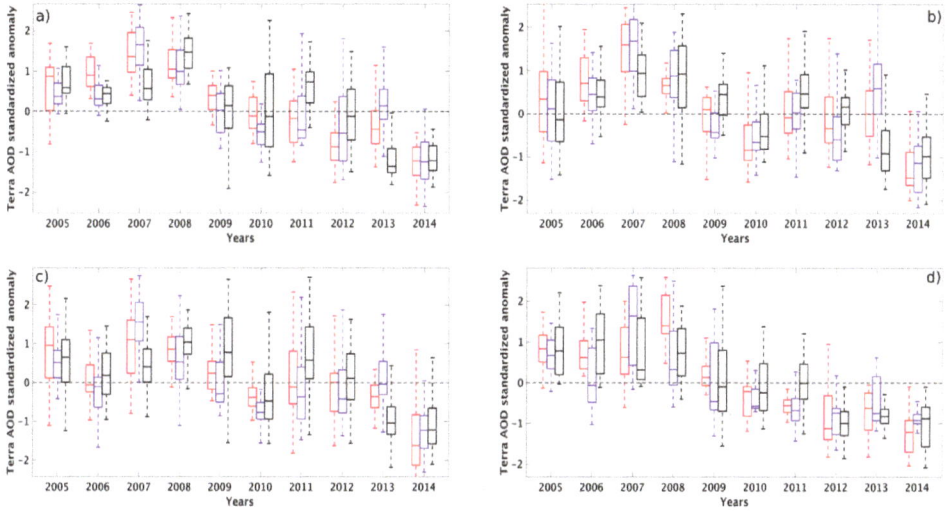

Figure 5. Similar to Figure 4 but for (**a**) fire; (**b**) NAM alley; (**c**) dust; and (**d**) anthro clusters.

In the NAM alley and dust cluster, we see a very similar trend, especially during PRM. This trend, however, is lower than in the fire and anthro clusters. Differences in the trends between NAM alley and dust clusters can be seen during MON and POM, where the trend in x'_{aod} in the dust cluster is relatively higher than in the NAM alley (see Table 4). The oscillating pattern (as seen in [54]) is also less pronounced in the NAM alley than in the dust cluster. This difference can be attributed to an apparent modulation of x'_{aod} in the NAM alley cluster by more frequent occurrences of convective dust storms or haboobs in this region than in the dust cluster. These storms are typically produced from isolated thunderstorms that merge to form cold pools during monsoon season. These cold pools result in severe downburst winds that lift massive quantities of dust off the surface ([8] and references therein). Based on the storm event database [55] over Arizona, we find that, although the trends in haboob-type dust storms across this study period are not statistically significant, there appears to be an increase in the frequency of dust storms over Arizona in most recent years, especially during the monsoon period. Despite the increase in rainfall (aerosol removal) during this period (which is common to both NAM alley and dust, albeit with a slight shift in timing), we infer that the increase in dust sources in the NAM alley obscures the decrease in the trend as seen in the dust cluster.

3.4. Multivariate Correlations

As noted earlier, we use a suite of satellite products (see Table 1) to corroborate the trends that we found in NAM and aerosol clusters. We present here the correlations between x'_{aod} and the following anomalies across different aerosol clusters and monsoon periods: Aqua/MODIS AOD (x'_{aod_aqua}), Aura/OMI AI (x'_{ai}), Terra/MODIS FRP (x'_{frp}), FC (x'_{fc}), and NDVI (x'_{ndvi}), Terra/MOPITT CO (x'_{co}), and TRMM BEPR (x'_{rain}). These quantities provide unique first-order information on aerosol source/sink types. We note that the trends in x'_{aod} can be influenced by several confounding factors other than those mentioned here, such as trans-Pacific transport of aerosols [56,57], mixing of aerosol emissions from within North America, removal efficiency, injection height, and frequency of emissions. Although we expect that the actual relationship between x'_{aod} and these quantities may be nonlinear, we only examine the linear component of this relationship as a first-order approximation. More robust and quantitiative assessments of aerosol trends including source attribution require modeling of aerosol sources, transport, and sinks, which is beyond the scope of this study. We note, however, that our analysis can be made useful to show observational constraints of these trends.

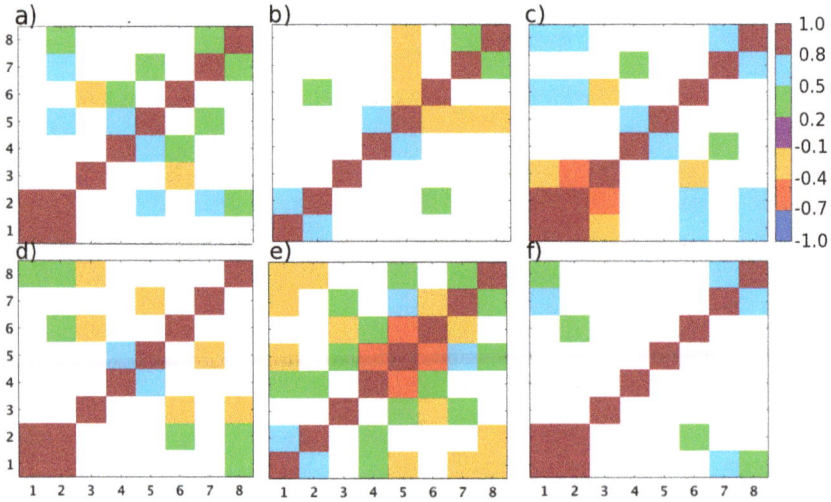

Figure 6. Anomaly correlation matrices between different aerosol-related quantities. The plots show correlation only during months that show significant correlation between the variables. The standardized anomalies considered here are the following: (1) Terra/MODIS AOD; (2) Aqua/MODIS AOD; (3) Aura/OMI UV AI; (4) Terra/MODIS FRP (in MW); (5) Terra/MODIS fire counts; (6) Terra/MOPITT CO (in ppbv); (7) Terra/MODIS NDVI; and (8) TRMM precipitation rate (in mm/day). Positive correlations are shown only if they are greater than 0.3 and significant at 90% confidence. Negative correlations are shown if they are significant at 90% confidence. The plots correspond to (a) fire cluster in May; (b) dust cluster in July; (c) anthro cluster in July; (d) NAM alley cluster in July; (e) dust cluster in September; and (f) anthro cluster in October.

Our results are presented in Figure 6. The correlation coefficient indicates how closely the changes in aerosol loading mirror the changes in these quantities. Here, we focus our analysis on large and significant correlations between anomalies. While high correlation between x'_{aod} and x'_{aod_aqua} is expected, we find a lesser correlation in the dust cluster (Figure 6b,e). This may be due to differences in retrievals of aerosols in Terra and Aqua. In the fire cluster, however, x'_{aod_aqua} is correlated with x'_{fc} and x'_{ndvi} (>0.5) in May (see Figure 6a). This is also seen from the decrease in the fire counts during June for the period 2005–2014 (Figure S2). The high correlation implies that most of the anomalies in aerosol loading in this cluster are directly tied to fire aerosols. As the fuel available from biomass increases, the probability of fires also increases, especially when the region tends to get drier in the period. This is usually the case for the summer fires in California [58]. There is also high correlation between x'_{fc} and x'_{frp} and to a lesser extent between x'_{ai} and x'_{co} which may indicate higher organic (and black) carbon aerosol emissions during mixtures of intense/less intense (flaming/smoldering) fires. This is fairly consistent with the previous studies that suggest that the interannual variability in natural AOD across the US is due to the variability of organic carbon emissions from droughts and biomass burning events [24,50]. However, this result needs to be supported by surface measurements and other datasets.

During MON, correlations in the NAM alley cluster show a strong positive relationship between x'_{aod}, x'_{rain}, and x'_{aod_aqua} (July: Figure 6b,d) which is not apparent in the dust cluster. This implies that the anomalies in aerosol loading and precipitation rates are moving in the same direction in the NAM alley cluster during the monsoon season. Although, in general, we expect negative correlation between x'_{rain} and x'_{aod}, this result suggests that the dust sources, especially from dust storms during the monsoon period, offset the aerosol removal due to rain. This can be contrasted with the dust cluster in September (Figure 6e) where there is a strong negative correlation between x'_{rain}, x'_{ndvi} (see Figure S3 for time series of NDVI) and x'_{aod}, implying the stronger influence of precipitation in the absence of dust storms (and lesser vegetation) (e.g., [59,60]) and despite potentially more intense fires in this region (i.e., negative correlation between x'_{co}, x'_{fc} and x'_{frp}, positive correlation between x'_{aod} and x'_{frp}, and between x'_{frp} and x'_{ai} implies more intense flaming fires—less CO, high FRP at lower fire counts, and high black carbon (e.g., [61]).

In the anthro cluster during July, we find strong positive correlations between x'_{aod}, x'_{aod_aqua} and x'_{co}, x'_{rain} and a negative correlation between x'_{aod} and x'_{ai} (Figure 6c). Although UV AI has been noted for its sensitivity to both dust and black carbon (BC) aerosols, previous studies have also indicated that UV AI is more sensitive to dust aerosols than smoke or BC [19]. In the absence of a positive correlation with fire indicators (FRP and FC), it is evident that anomalies in aerosol loading in the anthro cluster are directly related to anthropogenic combustion. A positive correlation between x'_{rain} and x'_{aod} in MON (Figure 6c) and POM (Figure 6f) explains, to some extent, the lesser impact of precipitation to x'_{aod} in the anthro cluster. This indicates that this region is mostly driven by anthropogenic (x'_{co}) and biogenic (x'_{ndvi}) aerosol sources than precipitation (aerosol sinks) in October.

3.5. AOD Sensitivity

Here, we examine the difference in standardized anomalies between the maximum within the early (2005–2009) and latter parts of the decade (2010–2014). This is shown in Figure 7 for selected clusters and years with distinct sensitivities. We focus on contrasting the local maximum between these two segments of the decadal period to support the trend and correlation analysis previously discussed. Together, they provide information on the main factors of change in aerosol loading for different clusters. In Figure 7a we show the sensitivity of x'_{aod} to x'_{fc} in the fire cluster, suggesting a similar shift in x'_{aod} and x'_{fc} from positive to negative anomalies in the latter part of the decade (2010–2014) for all monsoon periods (also see Figure S2). This is consistent with the strong correlation between x'_{aod_aqua} and x'_{fc} shown in Figure 7. Although the largest decrease in x'_{fc} occurs in PRM, the maximum local sensitivity (steeper slope) in x'_{fc} with x'_{aod} is observed during POM (*i.e.*, large change in fire counts is tied with smaller change in AOD). This reveals that AOD is more associated with smoldering fire aerosols (more emissions) during POM than in PRM. This is also indicated in Figure 7d where x'_{co} shifts from a negative to positive anomaly during POM and shows high sensitivity with x'_{aod}. However, the apparent decrease in fire counts shown in Figure 7a cannot be directly linked to a decrease in intensity or frequency of fires since this scatterplot is only a snapshot of this change.

During MON and POM in the NAM alley cluster, the x'_{aod} decrease is associated with an increase in x'_{rain} (Figure 7b and Figure S4). This sensitivity is mostly influenced by data in August (for MON) rather than July where dust storms may modulate this relationship (as we have seen in Figure 6d). Similarly, we find the decrease in x'_{aod} is associated with the decrease in x'_{ndvi}, with the highest sensitivity during POM in the dust cluster. The small change in AOD despite a large change in NDVI points to confounding factors such as fires, aerosol transport, and atmospheric moisture during POM in this region.

Finally, in the anthro cluster, we see a consistent shift in x'_{aod} and x'_{co} from positive to negative anomalies across the monsoon season (Figure 7e and Figure S5). This is very consistent with the trend and correlation results for this cluster.

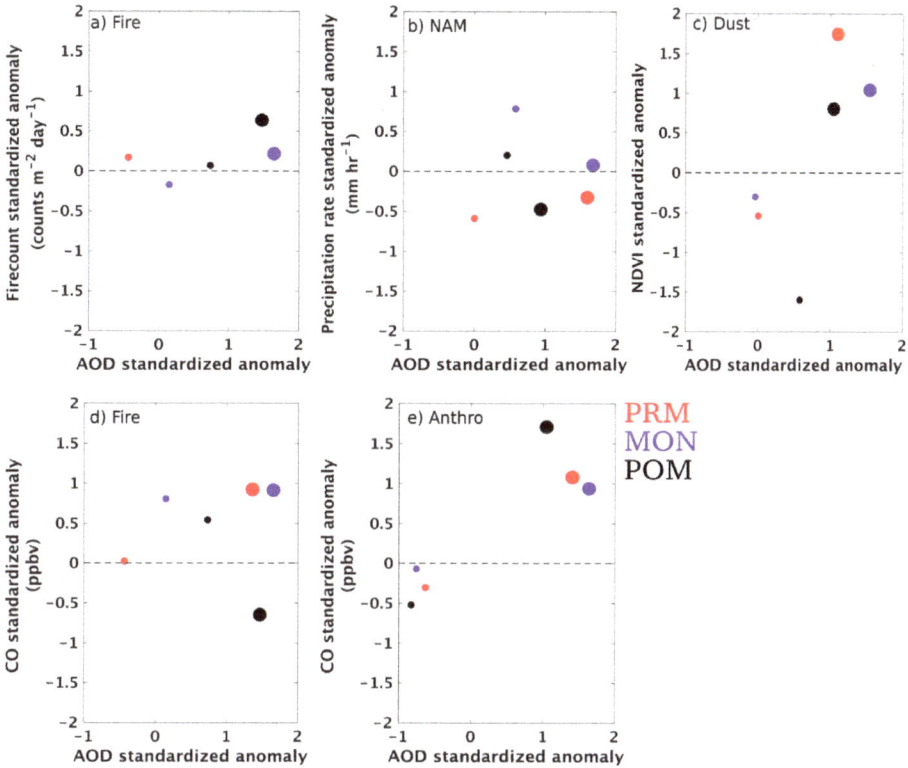

Figure 7. Scatterplots of standardized anomalies for selected clusters and monsoon periods. The filled circles represent local maxima values, with the larger and smaller circles corresponding to local maxima within 2005–2009 and 2010–2014, respectively. The colors represent different periods (red for PRM, blue for MON and black for POM). The different plots correspond to AOD anomalies *versus* (a) fire counts for fire cluster; (b) precipitation for NAM alley cluster; (c) NDVI for dust cluster; (d) CO for fire cluster; and (e) CO for anthro cluster.

3.6. Comparison between OMI UV AI and MODIS AOD during MON

As noted already, the UV AI product is useful in aerosol hotspot identification because of its finer spatial resolution. Although UV AI did not show significant trends, previous studies have demonstrated the capability of UV AI in capturing smoke and dust aerosols. We present in Figure 8 the scatterplot between x'_{ai} and x'_{aod} during MON to demonstrate the utility of UV AI for periods exhibiting large AOD variability. The different symbols represent different clusters with the smaller sizes indicating the last local maxima value in the time segment 2010–2014, and larger sizes indicating the first local maxima values in the time segment 2005–2010. Although these two quantities cannot be directly compared in terms of magnitude (hence the standardized anomalies), we expect these variables to move in the same direction. In contrast, x'_{ai} and x'_{aod} show an inverse relationship during MON in all the source regions. One possible reason for the apparent increase in x'_{ai} during MON (and POM) could again be the difference in equatorial crossing times of the satellites. Convection builds up in the morning and winds increase in intensity in the afternoon over these regions during the monsoon season. Therefore, windblown dust captured by Aura/OMI UV AI is not seen in Terra/MODIS AOD. Other potential reasons for the discrepancies between Terra/AOD and OMI UV AI are differences in spatial resolution, sampling differences due to overpass times, retrieval characteristics and sensitivities (OMI UV AI retrieval is sensitive to black carbon and dust) (e.g., [62] and references therein). However, the results in Figure 8 show that OMI UV AI can provide additional constraints on aerosols in this region where mixed (confounding) processes are involved, making it challenging to infer aerosol trends. Future studies, based on the emission database and particulate matter concentrations, are required in the NAM region to provide a deeper understanding of the discrepancies in between these products.

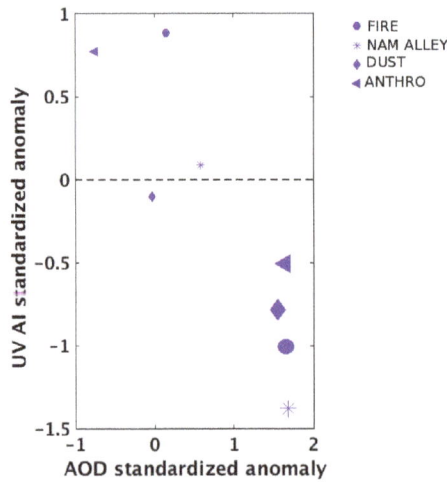

Figure 8. AOD *versus* UV AI standardized anomalies during the monsoon period (July–August). As in Figure 7, the larger (smaller) size symbols correspond to local maximum within 2005–2009 (2010–2014).

4. Conclusions

We investigate the spatial and temporal variability of aerosol loading in the NAM region using retrievals of AOD from Terra-MODIS for the years 2005–2014 during the monsoon season (May–October). We interpret the trends in AOD using other correlative information such as NDVI, CO, rainfall rate and active fire products.

The decadal average of AOD over the study area shows that the maximum aerosol loading occurs during May and June (PRM) in active source regions. Major aerosol source regions are northern California (fires), southwestern Arizona and northern Mexico (dust), and southern California (anthropogenic). Monsoonal rainfall in Arizona and Mexico acts as a major sink during this period.

We identified four aerosol clusters in the NAM region and selected 21 aerosol hotspots in the entire study area. We conducted a series of statistical analyses for these four cluster regions (fires, dust, NAM alley, and anthropogenic) and for three periods of the monsoon season (pre-, monsoon, and post-monsoon). We correlated the anomalies of AOD to fire counts and FRP (for fires), CO (for combustion), NDVI (for dust) and precipitation rates (for aerosol removal) to explain the trends in aerosol loading. Our results show a significant interannual variability in AOD anomalies in fire, NAM alley and dust clusters which could be linked to climatic events. However, these speculations warrant further analyses of climatic products. The temporal trends in AOD anomaly exhibit a statistically significant decreasing trend in fire (19%–27%) and anthropogenic clusters (17%–21%). The trends in the fire cluster can be attributed to the decreasing trend in the fire counts. The decreasing trends in the NAM alley (13%–21%) and dust clusters although statistically significant, are influenced by compensating or nullifying processes such as haboobs, increased moisture during the monsoon period, rainfall, *etc*. While rainfall removes aerosol in this region, processes such as convective dust storm cause massive enhancements of dust loading.

In summary, this study highlights the need for augmenting and integrating observing systems of aerosols (and correlative measurements) in this region with high spatio-temporal resolution datasets. While anthropogenic aerosols show a clear statistically significant decreasing trend, trends in natural aerosol clusters such as dust and the NAM alley are still unclear due to the complex interplay between sources and sinks in this region. A next step would be to corroborate these results with surface observations of PM_{10} and $PM_{2.5}$ and the emission database. We would also like to note that we have not fully considered the associated biases of the retrieval products, which is an important limitation of this study. We plan to investigate these limitations in our future work using chemical transport models and ground observations.

Supplementary Materials: The following are available online at http://www.mdpi.com/2073-4433/7/2/24/s1.

Figure S1: Correlation matrices for AOD anomalies all the hotspots between the months May–October. The hotspots considered here are: (1) Mt.Whitney, CA; (2) Charleston Peak, CA; (3) Tucson, AZ; (4) Baja, CA; (5) Phoenix, AZ; (6) Phoenix, AZ; (7) Yuma, AZ; (8) LA county, CA; (9) Bakersfield, CA; (10) Prescott, AZ; (11) Petrified forest, AZ; (12) White mountain, NM; (13) Farmington, NM; (14) Albuquerque, NM; (15) Ejido El Vergel, Mex; (16) Chihuahuan Desert, Mex; (17) Hermosillo, Mex; (18) Sierra Madre Occidental,

Mex; (19) Sierra Madre Oriental, Mex; (20) Houston, TX; (21) Waco, TX. The panels correspond to (A) May; (B) June; (C) July; (D) August; (E) September; (F) October.

Figure S2: Box plots of Terra/MODIS firecount standardized anomaly over the fire cluster for the years 2005–2014 between the months May–October. Whiskers represent 1.5 (IQR) above/below the upper/lower quartiles. The panels correspond to (A) May; (B) June; (C) July; (D) August; (E) September; (F) October.

Figure S3: Box plots of Terra/MODIS NDVI standardized anomaly over the dust cluster for the years 2005–2014 between the months May–October. The panels correspond to (A) May; (B) June; (C) July; (D) August; (E) September; (F) October.

Figure S4: Box plots of TRMM rainfall rate standardized anomaly over the NAM alley for the years 2005–2014 between the months May–October. The panels correspond to (A) May; (B) June; (C) July; (D) August; (E) September; (F) October.

Figure S5: Box plots of MOPITT CO standardized anomaly over the anthropogenic cluster for the years 2005–2014 between the months May–October. The panels correspond to (A) May; (B) June; (C) July; (D) August; (E) September; (F) October.

Acknowledgments: We acknowledge all teams (MODIS, OMI, MOPITT, TRMM, Anthromes, GRUMP) and GIOVANNI for the data. The work performed at the Department of Atmospheric Sciences, University of Arizona and was funded by NASA Grant NNX13AK24G. We also acknowledge support from Grant 2 P42 ES04940–11 from the National Institute of Environmental Health Sciences (NIEHS) Superfund Research Program, NIH. The authors are grateful to the support of National Weather Service, Tucson, for their insightful discussions on North American Monsoon and dust storms. The authors also thank Wenfu Tang and Nick Dawson (Department of Atmospheric Sciences, University of Arizona) for discussions on the topic.

Author Contributions: A.R. downloaded and analyzed the data. A.R. wrote the manuscript and A.F.A. provided guidance and edited the manuscript. A.S. provided comments and suggestions for the manuscript.

Conflicts of Interest: The authors declare no conflicts of interest.

References

1. Hansen, J.; Sato, M.; Ruedy, R.; Lacis, A.; Oinas, V. Global warming in the twenty-first century: An alternative scenario. *Proc. Natl. Acad. Sci. USA* **2000**, *97*, 9875–9880.

2. Lau, K.M.; Tsay, S.C.; Hsu, C. The Joint Aerosol-Monsoon Experiment: A new challenge for monsoon climate research. *Bull. Am. Meteorol. Soc.* **2008**, *89*, 369–383.

3. Meehl, G.A.; Arblaster, J.M.; Collins, W.D. Effects of black carbon aerosols on the Indian monsoon. *J. Clim.* **2008**, *21*, 2869–2882.

4. Ramanathan, V.; Carmichael, G. Global and regional climate changes due to black carbon. *Nat. Geosci.* **2008**, *1*, 221–227.

5. Seager, R.; Ting, M.; Held, I.; Kushnir, Y.; Lu, J.; Vecchi, G.; Huang, H.P.; Harnik, N.; Leetmaa, A.; Lau, N.C.; *et al.* Model projections of an imminent transition to a more arid climate in southwestern North America. *Science* **2007**, *316*, 1181–1184.

6.   Westerling, A.L.; Bryant, B.P. Climate change and wildfire in California. *Clim. Chang.* **2008**, *87*, 231–249.

7.   Dennison, P.E.; Brewer, S.C.; Arnold, J.D.; Moritz, M.A. Large wildfire trends in the western United States, 1984–2011. *Geophys. Res. Lett.* **2014**, *41*, 2928–2933.

8.   Raman, A.; Arellano, A.F.; Brost, J.J. Revisiting haboobs in the southwestern United States: An observational case study of the 5 July 2011 Phoenix dust storm. *Atmos. Environ.* **2014**, *89*, 179–188.

9.   Zhao, Z.; Pritchard, M.S.; Russell, L.M. Effects on precipitation, clouds, and temperature from long-range transport of idealized aerosol plumes in WRF-Chem simulations. *J. Geophys. Res. Atmos.* **2012**, *117*.

10. Crosbie, E.; Youn, J.S.; Balch, B.; Wonaschütz, A.; Shingler, T.; Wang, Z.; Conant, W.; Betterton, E.; Sorooshian, A. On the competition among aerosol number, size and composition in predicting CCN variability: A multi-annual field study in an urbanized desert. *Atmos. Chem. Phys.* **2015**, *15*, 6943–6958.

11. Sorooshian, A.; Shingler, T.; Harpold, A.; Feagles, C.W.; Meixner, T.; Brooks, P.D. Aerosol and precipitation chemistry in the southwestern United States: Spatio-temporal trends and interrelationships. *Atmos. Chem. Phys.* **2013**, *13*, 7361–7379.

12. Sprigg, W.A.; Nickovic, S.; Galgiani, J. N.; Pejanovic, G.; Petkovic, S.; Vujadinovic, M.; Vukovic, A.; Dacic, M.; DiBiase, S.; Prasad, A.; *et al.* Regional dust storm modeling for health services: the case of valley fever. *Aeolian Res.* **2014**, *14*, 53–73.

13. Gautam, R.; Hsu, N.C.; Lau, K.M.; Kafatos, M. Aerosol and rainfall variability over the Indian monsoon region: Distributions, trends and coupling. *Ann. Geophys.* **2009**, *27*, 3691–3703.

14. Krishnamurti, T.N.; Chakraborty, A.; Martin, A.; Lau, W.K.; Kim, K.M.; Sud, Y.; Walker, G. Impact of Arabian Sea pollution on the Bay of Bengal winter monsoon rains. *J. Geophys. Res. Atmos.* **2009**, *114*, D06213.

15. Douglas, M.W.; Maddox, R.A.; Howard, K.; Reyes, S. The Mexican monsoon. *J. Clim.* **1993**, *6*, 1665–1677.

16. Adams, D.K.; Comrie, A.C. The North American Monsoon. *Bull. Am. Meteorol. Soc.* **1997**, *78*, 2197–2213.

17. Gochis, D.J.; Shuttleworth, W.J.; Yang, Z.L. Sensitivity of the modeled North American monsoon regional climate to convective parameterization. *Mon. Weather Rev.* **2002**, *130*, 1282–1298.

18. Higgins, R.W.; Yao, Y.; Wang, X.L. Influence of the North American monsoon system on the US summer precipitation regime. *J. Clim.* **1997**, *10*, 2600–2622.

19. Sorooshian, A.; Wonaschütz, A.; Jarjour, E.G.; Hashimoto, B.I.; Schichtel, B.A.; Betterton, E.A. An aerosol climatology for a rapidly growing arid region (southern Arizona): Major aerosol species and remotely sensed aerosol properties. *J. Geophys. Res. Atmos.* **2011**, *116*, D19205.

20. Wise, E.K.; Comrie, A.C. Meteorologically adjusted urban air quality trends in the Southwestern United States. *Atmos. Environ.* **2005**, *39*, 2969–2980.

21. American Lung Association Ranking. Available online: http://www.stateoftheair.org/2015/city-rankings/most-polluted-cities.html (accessed on 13 November 2015).

22. Hand, J.L.; Schichtel, B.A.; Pitchford, M.; Malm, W.C.; Frank, N.H. Seasonal composition of remote and urban fine particulate matter in the United States. *J. Geophys. Res. Atmos.* **2012**, *117*.

23. Malm, W.C.; Schichtel, B.A.; Pitchford, M.L.; Ashbaugh, L.L.; Eldred, R.A. Spatial and monthly trends in speciated fine particle concentration in the United States. *J. Geophys. Res. Atmos.* **2004**, *109*.

24. Chin, M.; Diehl, T.; Tan, Q.; Prospero, J.M.; Kahn, R.A.; Remer, L.A.; Yu, H.; Sayer, A.M.; Bian, H.; Geogdzhayev, I.V.; *et al.* Multi-decadal aerosol variations from 1980 to 2009: A perspective from observations and a global model. *Atmos. Chem. Phys.* **2014**, *14*, 3657–3690.

25. Streets, D.G.; Yan, F.; Chin, M.; Diehl, T.; Mahowald, N.; Schultz, M.; Wild, M.; Wu, Y.; Yu, C. Anthropogenic and natural contributions to regional trends in aerosol optical depth, 1980–2006. *J. Geophys. Res. Atmos.* **2009**, *114*.

26. Prabhakar, G.; Betterton, E.A.; Conant, W.; Herman, B.M. Effect of Urban Growth on Aerosol Optical Depth—Tucson, Arizona, 35 Years Later. *J. Appl. Meterol. Clim.* **2014**, *53*, 1876–1885.

27. GIOVANNI. Available online: http://giovanni.gsfc.nasa.gov/giovanni (accessed on 13 November 2015).

28. Remer, L.A.; Kaufman, Y.J.; Tanré, D.; Mattoo, S.; Chu, D.A.; Martins, J.V. The MODIS aerosol algorithm, products, and validation. *J. Atmos. Sci.* **2005**, *62*, 947–973.

29. Levy, R.C.; Remer, L.A.; Kleidman, R.G.; Mattoo, S.; Ichoku, C.; Kahn, R.; Eck, T.F. Global evaluation of the Collection 5 MODIS dark-target aerosol products over land. *Atmos. Chem. Phys.* **2010**, *10*, 10399–10420.

30. Levy, R.C.; Remer, L.A.; Mattoo, S.; Vermote, E.F.; Kaufman, Y.J. Second-generation operational algorithm: Retrieval of aerosol properties over land from inversion of Moderate Resolution Imaging Spectroradiometer spectral reflectance. *J. Geophys. Res. Atmos.* **2007**, *112*.

31. Papadimas, C.D.; Hatzianastassiou, N.; Mihalopoulos, N.; Querol, X.; Vardavas, I. Spatial and temporal variability in aerosol properties over the Mediterranean basin based on 6-year (2000–2006) MODIS data. *J. Geophys. Res. Atmos.* **2008**, *113*.

32. Remer, L.A.; Kleidman, R.G.; Levy, R.C.; Kaufman, Y.J.; Tanre, D.; Mattoo, S. Global aerosol climatology from the MODIS satellite sensors. *J. Geophys. Res. Atmos.* **2008**, *113*.

33. MODIS-Terra Collection 5.1 Atmospheric Data. Available online: http://disc.sci.gsfc.nasa.gov/gesNews/new_modis_version_5.1_data (accessed on 13 November 2015).

34. Hubanks, P.; King, M. NASA MODIS Moderate Resolution Imaging Spectroradiometer, Atmosphere Home Page [Fact Sheet]. 2007. Available online: http://modis-atmos.gsfc.nasa.gov/MOD04_L2/index.html (accessed on 13 November 2015).

35. Giglio, L.; Csiszar, I.; Justice, C.O. Global distribution and seasonality of active fires as observed with the Terra and Aqua Moderate Resolution Imaging Spectroradiometer (MODIS) sensors. *J. Geophys. Res. Biogeosci.* **2006**, *111*, G02016.

36. Wooster, M.J.; Roberts, G.; Perry, G.L. W.; Kaufman, Y.J. Retrieval of biomass combustion rates and totals from fire radiative power observations: FRP derivation and calibration relationships between biomass consumption and fire radiative energy release. *J. Geophys. Res. Atmos.* **2005**, *110*, D24311.

37. Peng, S.; Piao, S.; Ciais, P.; Fang, J.; Wang, X. Change in winter snow depth and its impacts on vegetation in China. *Glob. Chang. Biol.* **2010**, *16*, 3004–3013.

38. Lunetta, R.S.; Knight, J.F.; Ediriwickrema, J.; Lyon, J.G.; Worthy, D. Land-cover change detection using multi-temporal MODIS NDVI data. *Remote. Sen. Environ.* **2006**, *105*, 142–154.

39. Torres, O.; Tanskanen, A.; Veihelmann, B.; Ahn, C.; Braak, R.; Bhartia, P.K.; Veefkind, P.; Levelt, P. Aerosols and surface UV products from Ozone Monitoring Instrument observations: An overview. *J. Geophys. Res. Atmos.* **2007**, *112*, D2108.

40. Huffman, G.J.; Adler, R.F.; Bolvin, D.T.; Gu, G.; Nelkin, E.J.; Bowman, K.P.; Hong, Y.; Stocker, E.F.; Wolff, D.B. The TRMM multisatellite precipitation analysis (TMPA): Quasi-global, multiyear, combined-sensor precipitation estimates at fine scales. *J. Hydrometeorol.* **2007**, *8*, 38–55.

41. Deeter, M.N.; Martínez-Alonso, S.; Edwards, D.P.; Emmons, L.K.; Gille, J.C.; Worden, H.M.; Pittman, J.E.; Daube, B.C.; Wofsy, S.C. Validation of MOPITT Version 5 thermal-infrared, near-infrared, and multispectral carbon monoxide profile retrievals for 2000–2011. *J. Geophys. Res. Atmos.* **2013**, *118*, 6710–6725.

42. Worden, H.M.; Deeter, M.N.; Edwards, D.P.; Gille, J.; Drummond, J.; Emmons, L.K.; Francis, G.; Martínez-Alonso, S. 13 years of MOPITT operations: Lessons from MOPITT retrieval algorithm development. *Ann Geophys.* **2014**, *56*.

43. Balk, D.; Montgomery, M.R.; McGranahan, G.; Kim, D.; Mara, V.; Todd, M.; Buettner, T.; Dorélien, A. Mapping urban settlements and the risks of climate change in Africa, Asia and South America. In *Population Dynamics and Climate Change*; Guzmán, J.M., Martine, G., McGranahan, G., Schensul, D., Tacoli, C., Eds.; International Institute for Environment and Development: London, UK, 2009.

44. SEDAC Data Portal: Population Density. Available online: http://sedac.ciesin.columbia.edu/data/set/ grump-v1-population-density (accessed on 13 November 2015).

45. Ellis, E.C.; Klein Goldewijk, K.; Siebert, S.; Lightman, D.; Ramankutty, N. Anthropogenic transformation of the biomes, 1700 to 2000. *Global Ecol. Biogeogr.* **2010**, *19*, 589–606.

46. SEDAC Data Portal: Anthropogenic Biomes. Available online: http://sedac.ciesin.columbia.edu/data/set/anthromes-anthropogenic-biomes-world-v2–1900 (accessed on 13 November 2015).

47. Singh, H.B.; Anderson, B.E.; Brune, W.H.; Cai, C.; Cohen, R.C.; Crawfold, J.H. Pollution influences on atmospheric composition and chemistry at high northern latitudes: Boreal and California forest fire emissions. *Atmos. Environ.* **2010**, *44*, 4553–4564.

48. Van der Werf, G.R.; Randerson, J.T.; Giglio, L.; Collatz, G.J.; Kasibhatla, P.S.; Arellano, A.F. Interannual variability in global biomass burning emissions from 1997 to 2004. *Atmos. Chem. Phys.* **2006**, *6*, 3423–3441.

49. Marlier, M.E.; DeFries, R.S.; Voulgarakis, A.; Kinney, P.L.; Randerson, J.T.; Shindell, D.T.; Chen, Y.; Faluvegi, G. El Nino and health risks from landscape fire emissions in southeast Asia. *Nat. Clim. Chang.* **2013**, *3*, 131–136.

50. Wang, Y.; Xie, Y.; Cai, L.; Dong, W.; Zhang, Q.; Zhang, L. Impact of the 2011 Southern US Drought on Ground-Level Fine Aerosol Concentration in Summertime. *J. Atmos. Sci.* **2015**, *72*, 1075–1093.

51. Lopez, D.H.; Rabbani, M.R.; Crosbie, E.; Raman, A.; Arellano, A.F.; Sorooshian, A. Frequency and Character of Extreme Aerosol Events in the Southwestern United States: A Case Study Analysis in Arizona. *Atmosphere* **2016**, *7*, 1.

52. Emissions of Atmospheric Compounds and Compilation of Ancilliary Data. Available online: http://eccad.sedoo.fr (accessed on 13 November 2015).

53. Air trends. Available online: http://www3.epa.gov/airtrends/pm.html (accessed on 13 November 2015).

54. Tong, D.Q.; Dan, M.; Wang, T. Long-term dust climatology in the western United States reconstructed from routine aerosol ground monitoring. *Atmos. Chem. Phys.* **2012**, *12*, 5189–5205.

55. Storm Events Database. Available online: https://www.ncdc.noaa.gov/stormevents (accessed on 13 November 2015).

56. Yu, H.; Remer, L.A.; Chin, M.; Bian, H.; Tan, Q.; Yuan, T.; Zhang, Y. Aerosols from overseas rival domestic emissions over North America. *Science* **2012**, *337*, 566–569.

57. Lin, M.; Fiore, A.M.; Horowitz, L.W.; Cooper, O.R.; Naik, V.; Holloway, J.; Johnson, B.J.; Middlebrook, A.M.; Oltmans, S.J.; Pollack, I.B.; *et al.* Transport of Asian ozone pollution into surface air over the western United States in spring. *J. Geophys. Res. Atmos.* **2012**, *117*, D00V07.

58. Jin, Y.; Randerson, J.T.; Faivre, N. Contrasting controls on wildland fires in Southern California during periods with and without Santa Ana winds. *J. Geophys. Res Biogeosci.* **2014**, *119*, 432–450.

59. Huang, M.; Tong, D.; Lee, P.; Pan, L.; Tang, Y.; Stajner, I.; Pierce, R.B.; MacQueen, J.; Wang, J. Toward enhanced capability for detecting and predicting dust events in the western United States: The Arizona case study. *Atmos. Chem. Phys.,* **2015**, *15*, 12595–12610.

60. Moreno-Rodríguez, V.; Del Rio-Salas, R.; Adams, D.K.; Ochoa-Landin, L.; Zepeda, J.; Gómez-Alvarez, A.; Palafox-Reyes, J.; Meza-Figueroa, D. Historical trends and sources of TSP in a Sonoran desert city: Can the North America Monsoon enhance dust emissions? *Atmos. Environ.* **2015**, *110*, 111–121.

61. Kondo, Y.; Matsui, H.; Moteki, N.; Sahu, L.; Takegawa, N.; Kajino, M.; Zhao, Y.; Cubison, M.J.; Jimenez, J.L.; Vay, S.; *et al.* Emissions of black carbon, organic, and inorganic aerosols from biomass burning in North America and Asia in 2008. *J. Geophys. Res. Atmos.* **2011**, *116*.

62. OMI Version 4.2x Data Quality Document. Available online: http://disc.sci.gsfc.nasa.gov/Aura/additional/ documentation/Aura-MLS_DataQuality_v4-2x-rev1-0.pdf (accessed on 15 January 2016).

Characterization of Fine Particulate Matter Emitted from the Resuspension of Road and Pavement Dust in the Metropolitan Area of São Paulo, Brazil

Ivan Gregorio Hetem and Maria de Fatima Andrade

Abstract: Many studies have been performed in order to characterize the sources of airborne particles in the Metropolitan Area of São Paulo (MASP), in Brazil. Those studies have been based on receptor modeling and most of the uncertainties in their results are related to the emission profile of the resuspended road dust particles. In this study, we analyzed the composition of resuspended road dust particles in different environments: local streets, paved roads inside traffic tunnels, and high traffic streets. We analyzed the samples to quantify the concentrations of trace elements and black carbon. On the basis of that analysis, we developed emission profiles of the resuspended road dust that are representative of the different types of urban pavement in the MASP. This study is important given the international efforts in improving emissions factors with local characteristics, mainly in South America and other regions for which there is a lack of related information. This work presents emission profiles derived from resuspended road dust samples that are representative of the different types of urban pavement in the Metropolitan Area of São Paulo.

Reprinted from *Atmosphere*. Cite as: Hetem, I.G.; de Fatima Andrade, M. Characterization of Fine Particulate Matter Emitted from the Resuspension of Road and Pavement Dust in the Metropolitan Area of São Paulo, Brazil. *Atmosphere* **2016**, 7, 31.

1. Introduction

Airborne particles that originate specifically from vehicle emissions are known to cause a variety of deleterious health effects, such as cardiorespiratory diseases and intrauterine mortality [1,2]. The particles emitted by vehicles can originate from exhaust and non-exhaust emissions. Non-exhaust particles comprise those generated from brake and tire wear; road surface abrasion; and corrosion. The resuspended road dust identified in the source apportionment of the atmospheric aerosol is composed of particles originating from the abrasion of different pavements, including bare soil and asphalt. In urban areas, road surfaces are contaminated by the deposition of pollutants from anthropogenic sources, mainly vehicle emissions. The resuspension

307

of urban road dust affects not only the concentration of particulate matter (PM) smaller than 10 microns (PM_{10}) but also that of fine particles—those smaller than 2.5 microns ($PM_{2.5}$). It has been estimated that urban road dust resuspension is responsible for 8% of the total $PM_{2.5}$ concentration in the Metropolitan Area of São Paulo (MASP), in southeastern Brazil, [3]. Pant and Harrison [4] presented a review of road traffic emissions of particulate matter. The authors concluded that road traffic can make a significant contribution to airborne concentrations of particulate matter, and that the particles arise not only from engine exhaust but also from the abrasion of tires, of the road surface, and of brake components. The particles arising from dust emissions can be classified as urban dust, which is comprised of soluble inorganics, carbonaceous compounds, and inert species (including metals). Some compounds derive from the composition of the soil itself, whereas others are derived from the deposition on the soil of particles and gases emitted by anthropogenic sources. Abbasi [5] performed experiments simulating different tires and driving patterns, as well as testing the brake system, to determine the mass and composition of the particles produced. Although most of those particles are in the coarse fraction of particulate matter, the abrasion process can generate the fine fraction ($PM_{2.5}$). Studies performed in road tunnels in the MASP showed a large contribution of metals that are not known to be emitted during the exhaust process: Ba, Cd, and Sb, from brake wear [6]; and Zn, Sr, Co, and W from tire wear [7,8]. Studies have shown that a significant proportion of the coarse fraction of particulate matter is road dust produced by the vehicles traveling over paved or unpaved roads [9].

In the MASP, there has long been a need to quantify the contributions that vehicle emissions make to the total concentration of particulate matter, differentiating between that coming from exhaust and that generated by the mechanical process of road dust resuspension. It has been a challenge to determine the contribution of each of these sources to the $PM_{2.5}$ concentration using trace-elements because of the difficulty in finding element characteristics of one specific source. Determining the contribution that vehicle emissions make to the concentration of particles in the atmosphere of the MASP has been the theme of many studies. Many authors have evaluated the role that urban sources play in determining the concentration of pollutants, mainly particulate matter. In studies applying multivariate analysis, specifically factor analysis, Ynoue and Andrade [10] and Sanchez-Ccoyllo et al. [11], considering the elements Fe, Al, Si, and Ti as tracers of road dust, found that the proportion of resuspended road dust in the $PM_{2.5}$ varied from 15% to 25%. These trace elements are also emitted by other processes, such as the mechanical action of tires on road surfaces, evaporation, and fuel combustion. To distinguish between particles derived from exhaust and those derived from mechanical abrasion, we analyzed samples of urban road dust from inside and just outside road tunnels. These analyses were performed in the context of a more comprehensive project that had the objective of evaluating the emission factors of gaseous and particulate emissions by vehicles. In that project, the emission factors of pollutants were determined using measurements taken inside and outside road tunnels. Tunnel measurements were taken in order to characterize the true nature of the vehicle fleet in the MASP. In one tunnel study, Pérez-Martínez *et al.* [12] showed that the emission factors of regulated pollutants were higher than those presented in official reports (CETESB, 2014). The authors compared their data with those of analyses performed under the same conditions in 2004 [11,13]. The CO and nitrogen oxide (NO_x) emission factors reported by Pérez-Martínez *et al.* [12] for light-duty vehicles in 2011 were both significantly (2.2 times) lower than those reported by Martins *et al.* [13] for light-duty vehicles in 2004, whereas they were five times and 2.5 times lower for CO and NO_x, respectively, from heavy-duty vehicles. Analyzing the number and mass size distribution of particles, Pérez-Martínez *et al.* [12] calculated the $PM_{2.5}$ emission factor to be 20 mg/km for light-duty vehicles and 277 mg/km for heavy-duty vehicles, showing the great contribution of diesel to the emission of fine particles. The authors also showed that black carbon accounted for 40% of the $PM_{2.5}$ diesel emissions, compared with 50% in 2004 [13]. The decreases in the emission factors were due to the implementation of a program for controlling vehicle emissions in Brazil. Although the vehicle fleet in the MASP has increased from 4 million in 2004 to almost 8 million in 2011, total emissions have decreased [11–14]. Brito *et al.* [15] showed that the burning of diesel by heavy-duty vehicles produced particles that were more numerous and for which the mean geometric diameter was smaller than that reported for particles produced by light-duty vehicles: the average particle count found for light-duty vehicles was 73,000 cm^{-3} with an average diameter of 48 nm, compared with 366,000 cm^{-3}, with an average diameter of 39 nm, for heavy-duty vehicles.

Quantification of the contributions that road dust and vehicle emissions make to the concentration of particulate matter is a major step in developing a road dust profile for the urban area of the MASP. The objective is to distinguish, within the $PM_{2.5}$ fraction, between the metals and black carbon that originate from the exhaust process and those that originate from the resuspension of dust from paved roads. The same approach was applied by [16], who presented an analysis of the enrichment factors of atmospheric particulate pollutants at the roadside, calculating the contribution that each different source made to the total particle concentration.

In this study, we attempted to provide emission profiles for road dust contaminated by vehicle emissions in the MASP. We present the composition of particles collected on a local street near an air quality monitoring station, as well as inside and outside two road tunnels. These data are a rich resource for the characterization of urban dust profiles that are highly influenced by vehicle emissions. The road dust resuspension source profile will be used in receptor modeling studies in the MASP and can also be a reference for other urban areas that are particularly affected by vehicle emissions.

2. Material and Methods

2.1. Sample Collection

Road dust samples were collected in bags at five different sites: inside and just outside the Jânio Quadros road tunnel (JQ_i and JQ_o, respectively; 23°35'S, 46°41'W), where only light-duty and small diesel utility vehicles are allowed to travel; inside and just outside the Rodoanel road tunnel (RA_i and RA_o, respectively; 23°27'S, 46°47'W), which is part of the beltway running outside the main area of the city, with a significant contribution by heavy-duty vehicles, mainly trucks; and on a local street, traveled by a mix of light- and heavy-duty vehicles, with a significant contribution by buses, near the Institute of Astronomy, Geophysics, and Atmospheric Sciences (IAG; 23°33'S, 46°44'W), which is on the main campus of the University of São Paulo, within the MASP.

Light-duty vehicles consist of those running on gasohol, ethanol, or diesel, with a gross weight of less than 3900 kg. In Brazil, most of the light-duty vehicles run on gasohol and ethanol. Passenger cars are not allowed to use diesel as fuel. Heavy-duty vehicles in Brazil consist of diesel vehicles with a gross weight of more than 3900 kg. The collection of samples inside and outside tunnels was part of a project to determine the emission factors of vehicles [12]. The JQ_i/JQ_o and RA_i/RA_o samples were collected in May and June 2011, respectively. Table 1 provides details of the sampling sites. Samples were collected by broom sweeping paved roads at five different spots, 10 m apart, collectively representing one site. The samples were collected in plastic bags, totaling 200 g of material per site.

Table 1. Characteristics of the roadways sampled.

Characteristic	JQ	RA	IAG
Location	23°35'S; 46°41'W	23°27'S; 46°47'W	23°33'S; 46°44'W
LDVs per hour, mean	2356	1511	140
HDVs per hour, mean	18	634	32
Hierarchy	Arterial	Freeway	Local

JQ, Jânio Quadros (road tunnel, inside and outside measurements); RA, Rodoanel (road tunnel, inside and outside measurements); IAG, Institute of Astronomy, Geophysics, and Atmospheric Sciences (local street); LDVs, light-duty vehicles; HDVs, heavy-duty vehicles.

2.2. Sample Preparation

The samples were stored in bags at 4 °C until use, and the material in each bag was manually sieved. The structure of the resuspension chamber and dichotomous sampler are illustrated in Figure 1, together with the protocol for the procedure. After being sieved, each sample was transferred to a resuspension chamber belonging to the

São Paulo State Environmental Protection Agency. The chamber consists of a virtual impactor (dichotomous sampler; Sierra-Andersen, Smyrna, GA, USA), coupled to a stainless steel vat (Figure 1). The vat is cylindrical, with a diameter of 40 cm and a height of 100 cm. The dichotomous sampler uses a PM_{10} inlet operating at 16.7 L/min to provide the D50 particle size cutoff at 10 microns in diameter. The virtual impactor is located after the inlet, and there are two separate flow controllers that maintain the fine particle air flux at 15.0 L/min and the coarse particle air flux at 1.6 L/min.

We inserted 100 mg of the dust sample in the vat and then scattered it with filtered laboratory compressed air. Using laboratory procedures, we determined that, after 3 min of scattering, the volume within the vat was homogeneously filled with the dust particles. The dust was collected on 37-mm Teflon-membrane filters. Each filter was exposed to four resuspension cycles, and samples from each site were collected on four filters, corresponding to one sample for each resuspension cycle. Each filter sampled a total of 0.4 m^3 of air. The elemental analysis was performed in the fine mode fraction ($PM_{2.5}$), which is the focus of this study.

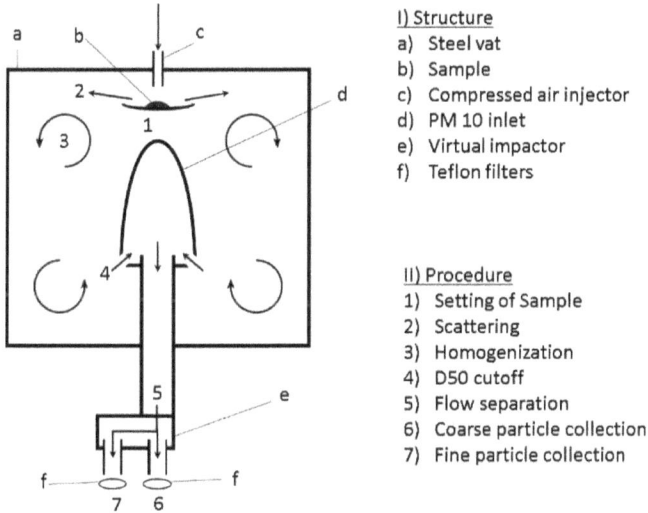

I) Structure
a) Steel vat
b) Sample
c) Compressed air injector
d) PM 10 inlet
e) Virtual impactor
f) Teflon filters

II) Procedure
1) Setting of Sample
2) Scattering
3) Homogenization
4) D50 cutoff
5) Flow separation
6) Coarse particle collection
7) Fine particle collection

Figure 1. Structure of the resuspension chamber, including the dichotomous sampler, and the procedure for collecting the material on the filters.

3. Analysis

The speciated emission inventory for road dust was constructed based on the material deposited on the filters. Although the particles may be shattered, the material deposited on the membrane filters retains the characteristics of the road dust that were generated by the abrasion of tires and wind on the road at the sampling site.

For the evaluation of the mass deposited on the filters, we submitted the samples to gravimetric analysis using a balance with a resolution of 1 µg (Mettler-Toledo MX-5). The filters were stored for 48 h in a temperature- and humidity-controlled environment (24 °C; 40% relative humidity). To quantify the accumulated mass, each filter was weighed before and after sampling. Measurements were corrected by subtracting the fluctuation observed on a set of blank filters. Ten blank filters were used for each set of samples. Concentrations of black carbon were evaluated through reflectance analysis of each filter with a smoke stain reflectometer (model 43; Diffusion Systems Ltd., London, UK). In brief, the filters were exposed to known amounts of visible light and a sensor registered the reflected fraction, thus inferring the superficial concentration of light-absorbing material, which can be classified as equivalent black carbon (BCe), as described by Bond et al. [17]. The material deposited on the filters was also submitted to analysis in an energy-dispersive X-ray fluorescence spectrometer (Epsilon 5; PANalytical, Almelo, The Netherlands). The spectrometer has three-dimensional, polarizing geometry with a supercritical water anode X-ray tube, up to 15 secondary targets as the polarizing scatterers. Details of the methodology for analyzing the trace-elements using this setup were described previously [15]. The typical detection limits for X-ray fluorescence analysis can be found elsewhere [18].

Data Processing Analysis

The profiles, composed of the mean concentrations for each element and for each sampling site, were defined as representing distinct sources. We then compared those sources with each other by calculating the coefficient of divergence (COD). The COD, as presented by [19], was defined as a measure of the data point spread between the road dust resuspension data sets and was applied to help determine whether any two profiles could be considered similar. The COD was calculated through the use of the following expression:

$$COD_{jk} = \sqrt{\frac{1}{p}\sum_{i=1}^{p}\left(\frac{x_{ij} - x_{ik}}{x_{ij} + x_{ik}}\right)^2} \tag{1}$$

where j and k represent two source profiles; p is the number of investigated components; and x_{ij} and x_{ik} represent the average mass concentrations of a compound i for the profiles j and k, respectively. The sources j and k can be considered similar if the COD_{jk} is close to zero and significantly different if the COD_{jk} is close to 1. In this analysis, we considered the following compounds: BCe, Na, Mg, Al, Si, P, S, Cl, K, Ca, Ti, V, Cr, Mn, Fe, Ni, Cu, Zn, Se, Br, Rb, Sr, Zr, Cd, Sb and Pb. Following the example set by [20], we excluded elements that accounted for less than 0.001% of the total mass, which resulted in the exclusion of As. According to those authors,

312

a COD above the threshold of 0.3 indicates that the profiles were different from each other, whereas COD values that were mostly lower than or near 0.3 indicate that the profiles are mostly similar to each other and can be substituted for each other.

4. Results and Discussion

Table 2 shows the composition of the road dust samples collected from each of the sampling sites according to their order of importance in explaining the mass concentration. The trace elements Al_2O_3, SiO_2, $CaCO_3$, K_2O, Fe_2O_3 and ZnO were considered to be oxides and collectively accounted for an average of 60% of the total mass, the main contributors being black carbon and the oxides Al, Si, Fe and Ca. The tunnel profiles present some remarkable differences: at the JQ site, Fe was a major contributor, accounting for 10% of the mass concentration; and at the RA site, the contribution of BCe was ten times higher inside the tunnel than outside of it.

Some elements, such as Na and P, were not considered to be in the oxidized state due to the uncertainties in the composition. The unexplained portion of the mass was due to components not measured in the samples, mainly water and organic compounds. In atmospheric samples, Ynoue and Andrade [10] demonstrated that organic material accounted for approximately 50% of the mass and black carbon (measured by optical reflectance) accounted for 30%, and they also showed that sulfate is present in the form of $(NH_4)_2SO_4$. Those authors showed that, within the explained fraction, crustal elements were the most abundant species, and that oxides of aluminum and silicon explained more than 10% and 20% of the mass, respectively. The compounds BCe, Na, $(NH_4)_2SO_4$, Cl, CuO, ZnO, As, Se, Br, Sb and Pb were more abundant in samples from inside the tunnels, whereas MgO, Al_2O_3, SiO_2, TiO and Rb were more abundant in samples from outside of the tunnels, as well as in those from the IAG site.

Compounds such as BCe and sulfur are characteristic of vehicle emissions in São Paulo, mainly from heavy-duty vehicles. This is similar to the situation in other parts of the world where policies to control diesel exhaust (mandatory use of diesel particle filters, for instance) have not been implemented. Strict regulations on diesel emissions, involving the mandatory use of retrofit systems, have been put in place only in recent years. One important program to control pollutant emissions from the light-duty fleet was the implementation of policies mandating the use of three-way catalysts, although the catalysts may also produce ammonia under reducing conditions. Typically, Cu and Zn are used as tracers of gasoline and ethanol emissions, respectively. Silva et al., [21], analyzing PM_{10} collected from ethanol and gasohol exhaust emissions in a dynamometer study performed with the fuels consumed in São Paulo, found different groups of metal elements in gasohol exhaust: Mn, Pt, Ni, Cu, Pb, Cr and Zn. Copper is added as an antioxidant, and Zn is

313

associated with the use of additives and lubricants, as found also by Morawska and Zhang [22].

Within the road tunnels, vehicle emissions are the dominant sources. Some elements, such as Al, Si and Ti, are mostly associated with road dust composition. Outside-tunnel samples were rich in those elements, because the outdoor pavement is more likely to have a dust component, due to soil transport mechanisms, than is the pavement within the tunnels. It is of note that diesel fuel contains Si and emissions of Si are therefore also associated with the burning of diesel [23,24].

Table 2. Composition of the road dust as a percentage of the total mass of the resuspended material collected on the Teflon filters in the resuspension chamber.

Compound	IAG	RA_i	RA_o	JQ_i	JQ_o
BCe	1.96 ± 0.39	10.75 ± 4.3	1.52 ± 0.17	3.99 ± 1.99	1.04 ± 0.19
Al_2O_3	32.22 ± 6.44	19.27 ± 7.71	32.57 ± 3.58	13.99 ± 7	32.76 ± 5.9
SiO_2	33.41 ± 6.68	12.53 ± 5.01	30.68 ± 3.37	19.52 ± 9.76	29.36 ± 5.29
Fe_2O_3	4.54 ± 0.91	3.76 ± 1.5	6.18 ± 0.68	9.74 ± 4.87	4.75 ± 0.85
K_2O	2.15 ± 0.43	1.41 ± 0.56	1.78 ± 0.2	1.55 ± 0.78	1.32 ± 0.24
$CaCO_3$	2.85 ± 0.57	6.27 ± 2.51	3.98 ± 0.44	8.78 ± 4.39	7.77 ± 1.4
MgO	2.34 ± 0.47	1.11 ± 0.44	2.23 ± 0.24	1.97 ± 0.98	2.29 ± 0.41
$(NH_4)_2SO_4$	1.68 ± 0.34	10.99 ± 4.4	1.51 ± 0.17	6.77 ± 3.39	1.14 ± 0.21
Na	0.34 ± 0.07	0.73 ± 0.29	0.27 ± 0.03	0.7 ± 0.35	0.2 ± 0.04
TiO	0.61 ± 0.12	0.23 ± 0.09	0.57 ± 0.06	0.43 ± 0.22	0.65 ± 0.12
ZnO	0.11 ± 0.02	0.44 ± 0.18	0.2 ± 0.02	0.73 ± 0.36	0.13 ± 0.02
MnO_2	0.07 ± 0.01	0.11 ± 0.05	0.12 ± 0.01	0.16 ± 0.08	0.06 ± 0.01
P	0.06 ± 0.01	0.26 ± 0.11	0.44 ± 0.05	0.14 ± 0.07	0.04 ± 0.01
Cl	0.06 ± 0.01	0.74 ± 0.3	0.19 ± 0.02	0.53 ± 0.27	0.02 ± 0.004
CuO	0.04 ± 0.01	0.23 ± 0.09	0.06 ± 0.01	0.28 ± 0.14	0.05 ± 0.01
V_2O_5	0.01 ± 0.002	--	0.02 ± 0.002	0.02 ± 0.011	0.01 ± 0.002
Cr	--	--	0.05 ± 0.005	0.04 ± 0.018	0.03 ± 0.005
NiO	0.01 ± 0.002	0.04 ± 0.016	0.03 ± 0.004	0.01 ± 0.006	0.02 ± 0.003
Rb	0.01 ± 0.002	--	0.01 ± 0.001	0.004 ± 0.002	0.01 ± 0.001
Sr	0.02 ± 0.004	0.02 ± 0.007	0.03 ± 0.004	0.06 ± 0.032	0.02 ± 0.004
Cd	0.01 ± 0.001	--	0.01 ± 0.001	0.02 ± 0.011	0.01 ± 0.001
Sb	0.01 ± 0.003	0.08 ± 0.033	0.01 ± 0.001	0.1 ± 0.048	0.02 ± 0.004
Pb	0.03 ± 0.005	0.07 ± 0.028	0.01 ± 0.001	0.05 ± 0.026	0.01 ± 0.002
As	0.001 ± 0.0002	0.01 ± 0.0041	0.002 ± 0.0003	0.009 ± 0.0047	0.0004 ± 0.0001
Se	0.002 ± 0.0005	0.03 ± 0.01	--	0.01 ± 0.01	0.003 ± 0.001
Br	0.001 ± 0.0003	0.005 ± 0.0021	0.001 ± 0.0001	0.005 ± 0.0027	0.002 ± 0.0003

Values are expressed as average $\pm$ standard deviation. JQ_i and JQ_o, respectively, inside and outside the Jânio Quadros Tunnel (traveled primarily by light-duty and small, diesel-powered utility vehicles); RA_i and RA_o, respectively, inside and outside the Rodoanel Tunnel (traveled by a significant number of heavy-duty vehicles); and IAG, Institute of Astronomy, Geophysics, and Atmospheric Sciences (local street traveled by a mix of light- and heavy-duty vehicles).

The COD results, shown in Table 3, present another form of grouping to these profiles. To define which groups could be considered similar and could be substitutes for each other, being considered representative of the resuspended road dust, we used the classification system devised by Kong *et al.* [20]. According to the COD values, the pairings between the IAG, JQ_o, and RA_o sites show that they are similar, even though they were collected from roadways that were quite different (hierarchy-wise) and in distinct areas of the MASP. This result is relevant, given that different sampling sites show the same profile for pavement dust emission. Previous studies applying principal component analysis to fine particle concentrations in various capital cities in Brazil, including São Paulo [3], have shown that it is very difficult to distinguish among road dust resuspension, vehicle exhaust, and the different urban sources. This was the motivation to improve the knowledge of resuspended road dust in order to subtract this source from the database before performing the receptor model analysis.

Table 3. Coefficient of divergence values for the different profiles analyzed, considering a similarity threshold of 0.3, as proposed by Kong *et al.* [20].

Site	RA_i	RA_o	JQ_i	JQ_o
IAG	0.557	**0.340**	0.533	**0.305**
RA_i	X	0.450	**0.279**	0.555
RA_o	X	X	0.414	**0.321**
JQ_i	X	X	X	0.501

Profile pairings with a COD below (or near) 0.3 are similar, whereas pairings with a COD above 0.3 are different. RA_i and RA_o, respectively, inside and outside the Rodoanel Tunnel (traveled by a significant number of heavy-duty vehicles); JQ_i and JQ_o, respectively, inside and outside the Jânio Quadros Tunnel (traveled primarily by light-duty and small, diesel-powered utility vehicles); and IAG, Institute of Astronomy, Geophysics, and Atmospheric Sciences (local street traveled by a mix of light- and heavy-duty vehicles).

The pairing of the profiles from inside the tunnels (RA_i and JQ_i) also resulted in COD values < 0.3, and those profiles are therefore also similar, despite the differences shown in Table 2, and could represent the dust profile inside traffic tunnels in São Paulo [19,20,25,26]. The pairings of the respective in- and outside-tunnel profiles for both tunnels resulted in COD values > 0.3. The concentrations of fine particles were higher in the in-tunnel samples than in the outside-tunnel samples, which also differed in composition, despite being collected from roadways traveled by the same types of vehicles. The profiles derived from in-tunnel samples are different from those derived from outside-tunnel samples, including those collected at the IAG site. The compounds presenting the highest contribution to the mass concentration were black carbon, Ca and Pb. Pairings of the mismatched profiles RA_o-JQ_i and RA_i-JQ_o resulted in COD values > 0.3, as did pairings of the IAG profile with both in-tunnel profiles, emphasizing how distinct they are. Considering the results obtained with the COD analysis, it was possible to establish a representative profile for road dust resuspension by using the average values for the JQ_o, RA_o and IAG samples, as presented in Table 4.

315

In the samples collected from the three outside-tunnel sites, we observed high contributions of the crustal elements Al_2O_3, SiO_2, Fe_2O_3, K_2O, $CaCO_3$ and MgO (Table 4). These compounds are also important tracers for tires, break wear and pavement abrasion which are the main sources of road dust [8,9].

In future multivariate analyses, that profile will likely be used as a standard road dust resuspension profile for the MASP. The contribution of BCe to that profile was one order of magnitude less than was its contribution to the in-tunnel profile. The profile encountered here was compared with data found for other urban sites in the United States and Spain. In its emissions profile database (SPECIATE 4.0), the United States Environmental Protection Agency (EPA) presented a composite of five paved road dust profiles from the cities of San Antonio and Laredo, Texas (BVPVRD01, BVPVRD02, BVPVRD03, BVPVRD04 and BVPVRD05). Amato *et al.* [27] analyzed road dust at different sites in Spain and compiled a profile for an urban site in the city of Seville. Table 4 compares the urban profile for the MASP (derived from the present study), the EPA composite urban profile, and the urban profile for Seville.

Table 4. Profile of resuspended road dust from the three outside-tunnel sites evaluated in Brazil, together with comparable profiles for cities in Texas and for the city of Seville, Spain.

Compound/Element	Urban Resuspended Road Dust Profiles		
	Brazil	United States	Spain
	% of Total $PM_{2.5}$	% of Total $PM_{2.5}$	% of Total $PM_{2.5}$
BCe	1.50 ± 0.16	2.54 ± 0.28	5.3 ± 2.1
Si	10.37 ± 1.72	19.13 ± 2.10	--
Al	8.61 ± 1.42	6.81 ± 0.75	2.1 ± 0.6
Fe	5.15 ± 0.82	2.63 ± 0.29	2.5 ± 0.5
Ca	2.81 ± 0.47	30.11 ± 3.30	11.2 ± 3.1
Mg	1.38 ± 0.23	0.54 ± 0.06	1.9 ± 1.3
K	1.32 ± 0.22	1.42 ± 0.16	0.7 ± 0.2
Ti	0.61 ± 0.10	0.31 ± 0.03	1.6 ± 0.3
S	0.35 ± 0.06	3.04 ± 0.33	0.4 ± 0.1
Na	0.27 ± 0.04	0.02 ± 0.002	0.3 ± 0.1
P	0.18 ± 0.02	0.28 ± 0.03	0.1 ± 0.0
Zn	0.14 ± 0.02	0.22 ± 0.02	1.3 ± 0.3
Mn	0.07 ± 0.01	0.05 ± 0.01	0.37 ± 0.13
Cu	0.04 ± 0.01	0.03 ± 0.003	0.77 ± 0.27
Cl	0.04 ± 0.005	0.15 ± 0.02	0.9 ± 1.5
Cr	0.03 ± 0.002	0.02 ± 0.002	0.145 ± 0.061
Sr	0.0254 ± 0.0040	0.1537 ± 0.0169	--
Pb	0.0158 ± 0.0027	--	--
Sb	0.0118 ± 0.0020	--	--
Ni	0.0113 ± 0.0016	0.0090 ± 0.0010	--
V	0.01 ± 0.001	0.02 ± 0.002	0.0057 ± 0.001

Values are expressed as average $\pm$ standard deviation. Data for the United States (five sites within the cities of San Antonio and Laredo, Texas) were obtained from the SPECIATE 4.0 emissions profile database of the United States Environmental Protection Agency. Data for Spain (an urban site in the city of Seville) were obtained from Amato *et al.* [27].

Comparing the profiles obtained for the fine fraction of resuspended road dust in urban areas, we can see that the MASP profile presents more aluminum and iron and less calcium than do the profiles obtained in the United States and Spain. The contributions of S, Na, P and Mg in the MASP were similar to those reported for Spain, whereas the contributions of K, Mn, Zn and Cu were similar to those reported for the United States. Although comparable for some species, the MASP sites had their own characteristics, demonstrating the need to expand the current road dust emission profiles. That would lead to better identification of this source and the evaluation of its contribution to the concentration of fine particulate matter.

5. Conclusions

Here, we have presented the inorganic composition of the fine (PM$_{2.5}$) fraction of resuspended road dust. The road dust samples were collected from (inside and outside) two tunnels and from one suburban street. The profiles defined here are representative of the inorganic fraction of resuspended urban road dust in the MASP. Our study was part of an initiative to provide more information regarding emission profiles in cities in South America. These profiles are important to the estimation of the sources responsible for particulate air pollution. We demonstrated that the profiles inside the tunnels were different from those outside the tunnels, as well as from that defined for the local street. We also found that the outside-tunnel profiles were similar and can be substitutes for each other. Therefore, the average profile for resuspended urban dust might be useful in quantifying the contribution of other vehicle emissions, including exhaust emissions. The in-tunnel profiles will improve the identification of this source of fine particulate matter. The profile found here provides important information for the identification of compounds in the fine fraction of particulate matter originating from the resuspension of road dust and from vehicle exhaust. This profile can be used in receptor model studies to estimate the impact that resuspended dust has on the fine particle mass concentration. Our results can be valuable for studies analyzing the toxicity of the PM content of road dust in urban areas that are especially affected by vehicle emissions.

Acknowledgments: The authors thank the Research Program on Global Climate Change of the Fundação de Amparo à Pesquisa do Estado de São Paulo (FAPESP, São Paulo Research Foundation) for the financial support provided (Grant No. 2008/58104-8, NUANCE project). This study was also supported by the Brazilian agencies Conselho Nacional de Desenvolvimento Científico e Tecnológico (CNPq, National Council for Scientific and Technological Development) and Coordenação de Aperfeiçoamento de Pessoal de Nível Superior (CAPES, Office for the Advancement of Higher Education). We are also grateful to the São Paulo State Companhia de Tecnologia de Saneamento Ambiental (CETESB, Environmental Protection Agency) for providing logistical support.

Author Contributions: Ivan Hetem conceived the experimental setup and Maria de Fatima Andrade with Ivan Hetem analysed the data.

Conflicts of Interest: The authors declare no conflict of interest.

References

1. Saiki, M.; Santos, J.O.; Alves, E.R.; Genezini, F.A.; Marcelli, M.P.; Saldiva, P.H.N. Correlation study of air pollution and cardio-respiratory diseases through NAA of an atmospheric pollutant biomonitor. *J. Radioanal. Nucl. Chem.* **2014**, *299*, 773–779.
2. Loomis, D.; Pereira, L.A.; Conceição, G.M.; Arcas, R.M.; Kishi, H.S.; Singer, J.M.; Böhm, G.M.; Saldiva, P.H.N. Association Between Air Pollution and Intrauterine Mortality in São Paulo, Brazil. *Environ. Health Perspect.* **1998**, *106*, 325–329.

3.	Andrade, M.F.; Miranda, R.M.; Fornaro, A.; Kerr, A.; Oyama, B.; André, P.A.; Saldiva, P.H. Vehicle emissions and PM$_{2.5}$ mass concentrations in six Brazilian cities. *Air Qual. Atmos. Health* **2012**, *5*, 79–88.

4.	Pant, P.; Harrison, R. Estimation of the contribution of road traffic emissions to particulate matter concentration form field measurements: A review. *Atmos. Environ.* **2013**, *77*, 78–97.

5.	Abbasi, S. Technical note: Experiences of studying airborne wear particles from road and rail transport. *Aerosol Air Qual. Res.* **2013**, *13*, 1161–1169.

6.	Grieshop, A.P.; Lipsky, E.M.; Pekney, N.J.; Takahama, S.; Robinson, A.L. Fine particle emission factors from vehicles in a highway tunnel: Effects of fleet composition and season. *Atmos. Environ.* **2006**, *40*, 287–298.

7.	Apeagyei, E.; Bank, M.S.; Spengler, J.D. Distribution of heavy metals in road dust along an urban-rural gradient in Massachusetts. *Atmos. Environ.* **2011**, *45*, 2310–2323.

8.	Wåhlin, P.; Berkowicz, R.; Palmgren, F. Characterization of traffic-generated particulate matter in Copenhagen. *Atmos. Environ.* **2006**, *40*, 2151–2159.

9.	Kupiainen, K.J.; Tervahattu, H.; Räisänen, M.; Mäkelä, T.; Aurela, M.; Hillamo, R. Size and composition of airborne particles from pavement wear, tires, and traction sanding. *Environ. Sci. Technol.* **2005**, *39*, 699–706.

10.	Ynoue, R.Y.; Andrade, M.F. Size resolved mass balance of aerosol particles over São Paulo Metropolitan Area, Brazil. *J. Aerosol Sci. Technol.* **2004**, *38*, 52–62.

11.	Sánchez-Ccoyllo, O.R.; Ynoue, R.Y.; Martins, L.D.; Astolfo, R.; Miranda, R.M.; Freitas, E.D.; Borges, A.S.; Fornaro, A.; Freitas, H.; Moreira, A.; *et al.* Vehicular particulate matter emissions in road tunnels in Sao Paulo, Brazil. *Environ. Monit. Assess.* **2009**, *149*, 241–249.

12.	Pérez-Martínez, P.J.; Miranda, R.M.; Nogueira, T.; Guardani, M.L.; Fornaro, A.; Ynoue, R.; Andrade, M.F. Emission factors of air pollutants from vehicles measured inside road tunnels in São Paulo: Case study comparison. *Int. J. Environ. Sci. Tech.* **2014**, *11*, 2155–2168.

13.	Martins, L.D.; Andrade, M.F.; Freitas, E.D.; Pretto, A.; Gatti, L.V.; Albuquerque, E.L.; Tomaz, E.; Guardani, M.L.; Martins, M.; Junior, O.M.A. Emission factors for gas-powered vehicles traveling through road tunnels in Sao Paulo, Brazil. *Environ. Sci. Technol.* **2006**, *40*, 6722–6729.

14.	São Paulo State Companhia de Tecnologia de Saneamento Ambiental. *Qualidade Do Ar No Estado De São Paulo 2013*; Série Relatórios; CETESB Report: Sao Paulo, Brazil, 2014.

15.	Brito, J.; Rizzo, L.V.; Herckes, P.; Vasconcellos, P.C.; Caumo, S.E.S.; Fornaro, A.; Ynoue, R.Y.; Artaxo, P.; Andrade, M.F. Physical-chemical characterisation of the particulate matter inside two road tunnels in the São Paulo Metropolitan Area. *Atmos. Chem. Phys.* **2013**, *13*, 12199–12213.

16.	Amato, F.; Viana, M.; Richard, A.; Furger, M.; Prevot, A.S.H.; Nava, S.; Lucarelli, F.; Bukoviwecki, N.; Alastuey, A.; Reche, C.; *et al.* Size and time-resolved roadside enrichment of atmospheric particulate pollutants. *Atmos. Chem. Phys.* **2011**, *11*, 2917–2931.

319

17. Bond, T.C.; Doherty, S.J.; Fahey, D.W.; Forster, P.M.; Berntsen, T.; DeAngelo, B.J.; Flanner, M.G.; Ghan, S.; Krächer, B.; Koch, D.; *et al.* Bounding the role of black carbon in the climate system: A scientific assessment. *J. Geophys. Res. Atmos.* **2013**, *118*, 5380–5552.

18. Spolnik, Z.; Belikov, K.; van Meel, K.; Adriaenssens, E.; de Roeck, F.; van Grieken, R. Optimization of measurement conditions of an energy dispersive X-ray fluorescence spectrometer with high-energy polarized beam excitation for analysis of aerosol filters. *Appl. Spectrosc.* **2005**, *59*, 1465–1469.

19. Wongphatarakul, V.; Friedlander, S.K.; Pinto, J.P. A comparative study of $PM_{2.5}$ ambient aerosol chemical databases. *Environ. Sci. Technol.* **1998**, *32*, 3926–3934.

20. Kong, S. Similarities and Differences in $PM_{2.5}$, PM_{10} and TSP Chemical Profiles of Fugitive Dust Sources in a Coastal Oilfield City in China. *Aerosol Air Qual. Res.* **2014**, *14*.

21. Silva, M.F.; Assunção, J.V.; Andrade, M.F.; Pesquero, C.R. Characterization of Metal and Trace Element Contents of Particulate Matter (PM_{10}) Emitted by Vehicles Running on Brazilian Fuels—Hydrated Ethanol and Gasoline with 22% of Anhydrous Ethanol. *J. Toxicol. Environ. Health A* **2010**, *73*, 901–909.

22. Morawska, L.; Zhang, J.J. Combustion sources of particles. 1. Health relevance and source signatures. *Chemosphere* **2002**, *49*, 1045–1058.

23. Taylor, D. Fuel Oils, Lubricant Oils and Their Treatment. In *Introduction to Marine Engineering*; Butterworth-Heinemann: Oxford, UK, 1996; pp. 150–162. Available online: http://books.google.com/ (accessed on 20 October 2015).

24. Giordano, S.; Adamo, P.; Spagnuolo, V.; Vaglieco, B.M. Instrumental and bio-monitoring of heavy metal and nanoparticle emissions from diesel engine exhaust in controlled environment. *J. Environ. Sci.* **2010**, *22*, 1357–1363.

25. Feng, Y.C.; Xue, Y.H.; Chen, X.H.; Wu, J.H.; Zhu, T.; Bai, Z.P. Source apportionment of Ambient Total Suspended Particulates and Coarse Particulate Matter in Urban Areas of Jiaozuo, China. *J. Air Waste Manag. Assoc.* **2007**, *57*, 561–575.

26. Kong, S.F.; Shi, J.W.; Lu, B.; Qiu, W.G.; Zhang, B.S.; Peng, Y.; Zhang, B.W.; Bai, Z.P. Characteristic of PAHs with PM_{10} Fraction for Ashes from Coke Production, Iron Smelt, Heating Station and Power Plant Stacks in Liaoning Province, China. *Atmos. Environ.* **2011**, *45*, 3777–3785.

27. Amato, F.; Alastuey, A.; de la Rosa, J.; Gonzalez Castanedo, Y.; Sánchez de la Campa, A.M.; Pandolfi, M.; Lozano, A.; Contreras González, J.; Querol, X. Trends of road dust emissions contributions on ambient air particulate levels at rural, urban and industrial sites in southern Spain. *Atmos. Chem. Phys.* **2014**, *14*, 3533–3544.

MDPI AG

St. Alban-Anlage 66

4052 Basel, Switzerland

Tel. +41 61 683 77 34

Fax +41 61 302 89 18

http://www.mdpi.com

Atmosphere Editorial Office

E-mail: atmosphere@mdpi.com

http://www.mdpi.com/journal/atmosphere

www.ingramcontent.com/pod-product-compliance
Lightning Source LLC
Chambersburg PA
CBHW051924190326

41458CB00026B/6399